AF371536

# Statistics for Industry and Technology


*Series Editor*

*N. Balakrishnan*
McMaster University
Department of Mathematics and Statistics
1280 Main Street West
Hamilton, Ontario L8S 4K1
Canada

*Editorial Advisory Board*

*Max Engelhardt*
EG&G Idaho, Inc.
Idaho Falls, ID 83415

*Harry F. Martz*
Group A-1 MS F600
Los Alamos National Laboratory
Los Alamos, NM 87545

*Gary C. McDonald*
NAO Research & Development Center
30500 Mound Road
Box 9055
Warren, MI 48090-9055

*Peter R. Nelson*
Department of Mathematical Sciences
Clemson University
Martin Hall
Box 341907
Clemson, SC 29634-1907

*Kazuyuki Suzuki*
Communication & Systems Engineering Department
University of Electro Communications
1-5-1 Chofugaoka
Chofu-shi
Tokyo 182
Japan

# Scan Statistics and Applications

Joseph Glaz
N. Balakrishnan

*Editors*

**Birkhäuser**
Boston • Basel • Berlin

Joseph Glaz
Department of Statistics
University of Connecticut at Storrs
Storrs, CT 06269-3120
USA

N. Balakrishnan
Department of Mathematics
and Statistics
McMaster University
Hamilton, Ontario L8S 4K1
Canada

**Library of Congress Cataloging-in-Publication Data**
Scan statistics and applications / [edited by] Joseph Glaz, N.
  Balakrishnan.
      p.    cm.—(Statistics for industry and technology)
    Includes bibliographical references and index.
    ISBN 0-8176-4041-X (hardcover : alk. paper)
    1. Order statistics.  I. Glaz, Joseph.  II. Balakrishnan, N.,
  1956–  .  III. Series.
  QA278.7.S25  1999
  519.5—dc21
                                                      99-20200
                                                      CIP

AMS Subject Classifications: 62N, 62P

Printed on acid-free paper.
© 1999 Birkhäuser Boston

*Birkhäuser*  ®

All rights reserved. This work may not be translated or copied in whole or in part without the written
permission of the publisher (Birkhäuser Boston, c/o Springer-Verlag New York, Inc., 175 Fifth Avenue,
New York, NY 10010, USA), except for brief excerpts in connection with reviews or scholarly analysis.
Use in connection with any form of information storage and retrieval, electronic adaptation, computer
software, or by similar or dissimilar methodology now known or hereafter developed is forbidden.
The use of general descriptive names, trade names, trademarks, etc., in this publication, even if the
former are not especially identified, is not to be taken as a sign that such names, as understood by the
Trade Marks and Merchandise Marks Act, may accordingly be used freely by anyone.

ISBN 0-8176-4041-X
ISBN 3-7643-4041-X

Typeset by the editors in LaTeX.
Cover design by Vernon Press, Boston, MA.
Printed and bound by Edwards Brothers, Inc., Ann Arbor, MI.
Printed in the United States of America.

9 8 7 6 5 4 3 2 1

# Contents

**3 Ratchet Scan and Disjoint Statistics**                                        **67**
*Joachim Krauth*

PART IV: APPLICATIONS

**11 A Start-Up Demonstration Test Using a Simple Scan-Based Statistic**  **251**

*Markos V. Koutras and N. Balakrishnan*



**12 Applications of the Scan Statistic in DNA Sequence Analysis**  **269**

*Ming-Ying Leung and Traci E. Yamashita*

# Preface

The study of scan statistics and their applications to many different scientific and engineering problems have received considerable attention in the literature recently. In addition to challenging theoretical problems, the area of scan statistics has also found exciting applications in diverse disciplines such as archaeology, astronomy, epidemiology, geography, material science, molecular biology, reconnaissance, reliability and quality control, sociology, and telecommunication. This will be clearly evident when one goes through this volume.

In this volume, we have brought together a collection of experts working in this area of research in order to review some of the developments that have taken place over the years and also to present their new works and point out some open problems. With this in mind, we selected authors for this volume with some having theoretical interests and others being primarily concerned with applications of scan statistics. Our sincere hope is that this volume will thus provide a comprehensive survey of all the developments in this area of research and hence will serve as a valuable source as well as reference for theoreticians and applied researchers. Graduate students interested in this area will find this volume to be particularly useful as it points out many open challenging problems that they could pursue. This volume will also be appropriate for teaching a graduate-level special course on this topic.

Our thanks go to all the authors who showed great enthusiasm for this project and shared some of their recent research on scan statistics. We fully appreciate their cooperation in sending their manuscripts on time and in helping us with the review of some of the materials. Next we express our sincere thanks to Mrs. Debbie Iscoe for the excellent typesetting of the entire volume. Final thanks go to Mr. Wayne Yuhasz (Editor, Birkhäuser, Boston) for taking a keen interest in this project.

Storrs, Connecticut *Joseph Glaz*
Hamilton, Ontario *N. Balakrishnan*
October 1998

# Contributors

Alm, Sven Erick
> Department of Mathematics, Uppsala University, P.O. Box 480, S-751 06 Uppsala, Sweden
> e-mail: *sea@math.uu.se*

Balakrishnan, N.
> Department of Mathematics and Statistics, McMaster University, Hamilton, Ontario L8S 4K1, Canada
> e-mail: *bala@mcmail.cis.mcmaster.ca*

Chen, Jie
> Computing Service Department, University of Massachusetts-Boston, 100 Morrissey Boulevard, Boston MA 02125
> e-mail: *chen_j@umbsky.cc.umb.edu*

Chen, S.C.
> Department of Applied Mathematics, National Donghwa University, Hualian, Taiwan, R.O.C.

Fu, James C.
> Department of Statistics, University of Manitoba, Winnipeg, Manitoba R3T 2N2, Canada
> e-mail: *fu@ccm.umanitoba.ca*

Glaz, Joseph
> Department of Statistics, U-120, University of Connecticut, 196 Auditorium Rd., Storrs, CT 06269-3120
> e-mail: *glaz@uconnvm.uconn.edu*

Huffer, Fred W.
> Department of Statistics, Florida State University, Tallahassee, FL 32306
> e-mail: *huffer@stat.fsu.edu*

*Koutras, Markos V.*
Department of Mathematics, University of Athens, Panepistemiopolis, Athens, 15784 Greece
e-mail: *mkoutras@atlas.uoa.gr*

*Krauth, Joachim*
Director of the Institute of Psychology, Heinrich Heine University, Universitatsstrasse 1, 40225 Düsseldorf, Germany
e-mail: *hebben@clio.rz.uni-duesseldorf.de*

*Kulldorff, Martin*
National Cancer Institute, Biometry Branch, DCPC EPN 344, 6130 Executive Blvd., Bethesda, MD 20892-7354
e-mail: *martink@helix.nih.gov*

*Leung, Ming-Ying*
Division of Mathematics and Statistics, The University of Texas at San Antonio, San Antonio, TX 78249
e-mail: *leung@sphere.math.utsa.edu*

*Lin, Chien-Tai*
Department of Mathematics, Tamkang University, Tamsui 25137, Taiwan, R.O.C.
e-mail: *chien@math.tku.edu.tw*

*Lou, Wendy, W.Y.*
Department of Biomathematical Sciences, Box 1023, Mount Sinai School of Medicine, 1 Gustave Levy Place, New York, NY 10029
e-mail: *lou@msvax.mssm.edu*

*Månsson, Marianne*
Department of Mathematics, Chalmers University of Technology, S-412 96 Göteborg, Sweden
e-mail: *marianne@math.chalmers.se*

*Naus, Joseph I.*
Department of Statistics, Rutgers University, Hill Center for Mathematical Sciences, Busch Campus, New Brunswick, NJ 08903
e-mail: *naus@stat.rutgers.edu*

*Wallenstein, Sylvan*
Department of Biomathematical Sciences, Box 1023, Mount Sinai School of Medicine, 1 Gustave Levy Place, New York, NY 10029
e-mail: *wallenst@msvax.mssm.edu*

*Yamashita, Traci E.*
Department of Epidemiology, Johns Hopkins School of Hygiene and Public Health, Baltimore, MD 21205-1999

# List of Tables

# List of Figures

# PART I

## INTRODUCTION AND PRELIMINARIES

# 1

# Introduction to Scan Statistics

**Joseph Glaz and N. Balakrishnan**

*University of Connecticut, Storrs, CT*
*McMaster University, Hamilton, Ontario, Canada*

**Abstract:** In this chapter, we define discrete and continuous scan statistics in one-dimensional as well as multidimensional cases. We then mention some related applications and open problems. We also present a brief account of order statistics which naturally arise in the study of scan statistics.

**Keywords and phrases:** Order statistics, discrete scan statistics, continuous scan statistics, Bonferroni inequalities, circular scan statistics, product-type inequalities, conditional scan statistic

## 1.1  Introduction

In this chapter, we introduce scan statistics and present some basic definitions and results concerning them. We will present the "bare-bones" details on scan statistics with which the reader of this volume should first of all become familiar. With this as background, we feel that the reader will be able to understand and appreciate all the other developments on scan statistics that are presented in subsequent chapters of this volume.

Since order statistics come in naturally in the study of scan statistics, we first present a brief and basic account of order statistics in Section 3. In Section 4, we define discrete scan statistics in the one-dimensional case. In Section 5, discrete scan statistics are defined for the multidimensional case. In these sections, we also describe briefly all the results presented in subsequent chapters of this volume and mention some related applications as well as open problems in this area of research. In Section 6, we define continuous scan statistics in the one-dimensional case. Finally, in Section 7, continuous scan statistics are defined for the multidimensional case. Once again, we also briefly outline in these sections the results presented in other chapters of this volume and point out some interesting applications as well as open problems in this area.

## 1.2    A Quick Glimpse

The chapters in this volume, in addition to illustrating some applications of scan statistics, all deal with methods of evaluating distributions of scan statistics in the discrete or the continuous case. In the discrete case, distributions of scan statistics for the one-dimensional case have been examined by Chen and Glaz in Chapter 2, Krauth in Chapter 3, Naus in Chapter 4, Koutras and Balakrishnan in Chapter 11, and Fu, Lou, and Chen in Chapter 13; distributions of scan statistics for the multidimensional case have been discussed by Chen and Glaz in Chapter 2 and Naus in Chapter 4. Similarly, in the continuous case, distributions of scan statistics for the one-dimensional case have been investigated by Alm in Chapter 5, Huffer and Lin in Chapters 6 and 7, Wallenstein in Chapter 8, Lin in Chapter 9, and Leung and Yamashita in Chapter 12; distributions of scan statistics for the multidimensional case have been discussed by Alm in Chapter 5, Månsson in Chapter 10, and Kulldorff in Chapter 14.

As will be evident from this volume, scan statistics have found many important applications. For example, scan statistics are used to test the null hypothesis of uniformity against a clustering alternative. Scan statistics have also been used in several different scientific and engineering problems. The diverse areas of applications highlighted in this volume include archaeology (Chapter 14 by Kulldorff), astronomy (Chapter 14 by Kulldorff), epidemiology (Chapter 3 by Krauth, Chapter 8 by Wallenstein, and Chapter 14 by Kulldorff), linguistics (Chapter 4 by Naus), geography (Chapter 14 by Kulldorff), material science (Chapter 5 by Alm), molecular biology and genetics (Chapter 4 by Naus, Chapter 12 by Leung and Yamashita, and Chapter 13 by Fu, Lou, and Chen), reconnaissance (Chapter 2 by Chen and Glaz, and Chapter 14 by Kulldorff), reliability and quality control (Chapter 2 by Chen and Glaz, Chapter 4 by Naus, and Chapter 11 by Koutras and Balakrishnan), telecommunication (Chapter 5 by Alm), and sociology (Chapter 2 by Chen and Glaz, and Chapter 4 by Naus).

In addition to such diverse and fascinating applications, the study of scan statistics also involves many sophisticated and intricate theoretical methods and techniques. Among these are included Bonferroni-type inequalities (Chapter 2 by Chen and Glaz, Chapter 3 by Krauth, Chapter 7 by Huffer and Lin, and Chapter 9 by Lin), compound Poisson approximations (Chapter 2 by Chen and Glaz, Chapter 7 by Huffer and Lin, Chapter 9 by Lin, and Chapter 12 by Leung and Yamashita), finite Markov chain embedding (Chapter 13 by Fu, Lou, and Chen), Karlin-McGregor approach to multiple candidate ballot problems (Chapter 8 by Wallenstein), large deviation results (Chapter 13 by Fu, Lou, and Chen), linear programming (Chapter 7 by Huffer and Lin), Markov chain modeling (Chapter 7 by Huffer and Lin, and Chapter 9 by Lin), Monte Carlo approach to testing of hypotheses (Chapter 14 by Kulldorff), order statistics and spacings (Chapter 6 by Huffer and Lin), Poisson approximations (Chapter 2 by Chen and Glaz, Chapter 5 by Alm, Chapter 10 by Månsson, and Chapter 12 by Leung and Yamashita), probability generating functions (Chapter 11 by Koutras and Balakrishnan), product-type approximations and inequalities (Chapter 2 by Chen and Glaz, and Chapter 4 by Naus), Chen–Stein method (Chapter 10 by Månsson), and symbolic computing (Chapter 6 by Huffer and Lin, and Chapter 9 by Lin).

With such a plethora of theoretical methods and techniques and fascinating and diverse applications, the field of scan statistics will provide a rich and exciting area of research. We sincerely hope that this volume will provide a comprehensive survey of all the developments in this area and thus serve as a valuable source as well as reference for theoreticians and applied researchers involved (or even interested) in this topic of research.

---

## 1.3  Order Statistics

Let $X_1, X_2, \ldots, X_n$ be $n$ arbitrarily distributed random variables. Let us denote $X_{1:n} \leq X_{2:n} \leq \cdots \leq X_{n:n}$ for the variables obtained by arranging the $n$ $X_i$'s in nondecreasing order of magnitude. Then, $X_{i:n}$ $(i = 1, 2, \ldots, n)$ are called *order statistics*. Note that, in this definition, we need the $X_i$'s to be neither independent nor identically distributed. However, most of the developments in the area of order statistics have been based on the assumption that the variables $X_i$'s are independent and identically distributed; see, for example, Sarhan and Greenberg (1962), David (1981), Arnold and Balakrishnan (1989), Balakrishnan and Cohen (1991), Arnold, Balakrishnan, and Nagaraja (1992), and Balakrishnan and Rao (1998a,b).

Let us now denote the population cumulative distribution function by $F(x)$ and the probability density (mass) function by $f(x)$. Then, it can be readily shown that the cumulative distribution function of $X_{i:n}$ $(1 \leq i \leq n)$ is

$$F_{i:n}(x) = P(X_{i:n} \leq x) = \sum_{r=i}^{n} \binom{n}{r} \{F(x)\}^r \{1 - F(x)\}^{n-r} \qquad (1.1)$$

and, in particular,

$$F_{1:n}(x) = 1 - \{1 - F(x)\}^n \qquad (1.2)$$

and

$$F_{n:n}(x) = \{F(x)\}^n. \qquad (1.3)$$

Instead of the familiar binomial form in (1.1), the cumulative distribution function of $X_{i:n}$ $(1 \le i \le n)$ can also be written in a negative binomial form as

$$F_{i:n}(x) = \sum_{r=0}^{n-i} \binom{n-1-r}{i-1} \{F(x)\}^i \{1-F(x)\}^{n-i-r}. \tag{1.4}$$

If the population is absolutely continuous, then we can differentiate the expression of the cumulative distribution function in (1.1) to readily obtain the probability density function of $X_{i:n}$ $(1 \le i \le n)$ as

$$f_{i:n}(x) = \frac{n!}{(i-1)!(n-i)!} \{F(x)\}^{i-1} \{1-F(x)\}^{n-i} f(x). \tag{1.5}$$

In this case, we can also similarly write the joint density function of $X_{i_1:n}, X_{i_2:n}, \ldots, X_{i_k:n}$ $(1 \le i_1 < i_2 < \cdots < i_k \le n)$ as

$$\begin{aligned}
f_{i_1,i_2,\ldots,i_k:n}&(x_1, x_2, \ldots, x_k) \\
&= C\{F(x_1)\}^{i_1-1}\{F(x_2)-F(x_1)\}^{i_2-i_1-1} \\
&\quad \times \cdots \times \{F(x_k)-F(x_{k-1})\}^{i_k-i_{k-1}-1} \\
&\quad \times \{1-F(x_k)\}^{n-i_k} f(x_1)f(x_2)\cdots f(x_k), \quad x_1 < x_2 < \cdots < x_k, \tag{1.6}
\end{aligned}$$

where

$$C = \frac{n!}{(i_1-1)!(i_2-i_1-1)!\cdots(i_k-i_{k-1}-1)!(n-i_k)!}. \tag{1.7}$$

If the population is discrete, then from (1.1) we can immediately write the probability mass function of $X_{i:n}$ $(1 \le i \le n)$ as

$$\begin{aligned}
f_{i:n}(x) &= F_{i:n}(x) - F_{i:n}(x-) \\
&= \sum_{r=i}^{n} \binom{n}{r} [\{F(x)\}^r \{1-F(x)\}^{n-r} \\
&\qquad\qquad -\{F(x-)\}^r \{1-F(x-)\}^{n-r}] \tag{1.8}
\end{aligned}$$

or, equivalently, as

$$f_{i:n}(x) = \frac{n!}{(i-1)!(n-i)!} \int_{F(x-)}^{F(x)} u^{i-1}(1-u)^{n-i} du. \tag{1.9}$$

Similarly, we can express the joint probability mass function of $X_{i_1:n}, X_{i_2:n}, \ldots X_{i_k:n}$ $(1 \le i_1 < i_2 < \cdots < i_k \le n)$ as

$$\begin{aligned}
f_{i_1,i_2,\ldots,i_k:n}&(x_1, x_2, \ldots, x_k) \\
&= C \int_B \left\{ \prod_{r=1}^{k} (u_r - u_{r-1})^{i_r-i_{r-1}-1} \right\} (1-u_k)^{n-i_k} du_1 \ldots du_k, \tag{1.10}
\end{aligned}$$

where $i_0 = 0$, $u_0 = 0$, $C$ is as defined in (1.7), and $B$ is the $k$-dimensional space given by

$$B = \{(u_1,\ldots,u_k) : u_1 \le u_2 \le \cdots \le u_k,\ F(x_r-) \le u_r \le F(x_r)$$
$$\text{for } r = 1, 2, \ldots, k\}.$$

From the probability density (mass) functions given above, we can readily write down the formulas for moments and joint moments of order statistics. Explicit expressions are available for a number of parent distributions and for others numerical computations are necessary; see, for example, Harter and Balakrishnan (1996). Several recurrence relations and identities satisfied by distributions as well as moments of order statistics are available in the literature; see David (1981) and Arnold and Balakrishnan (1989). These sources will also provide extensive reviews of various bounds and approximations connected with distributions and moments of order statistics.

Bounds are also available on probabilities associated with order statistics with many of the well-known ones being of the Bonferroni type. Specifically, let $E_1, E_2, \ldots, E_n$ be $n$ events. Then, the Boole's formula gives

$$P\left(\bigcup_{i=1}^{n} E_i\right) = \sum_i P(E_i) - \sum\sum_{i<j} P(E_i \cap E_j) + \cdots + (-1)^{n-1} P\left(\bigcap_{i=1}^{n} E_i\right)$$

$$(1.11)$$

with the sum of an odd number of terms providing an upper bound and the sum of an even number of terms providing a lower bound for the probability of occurrence of at least one of the $E_i$'s. If we take the event $E_i$ to be the event $\{X_i > x\}$ and assume that the joint distribution of $X_i$'s is symmetrical in the $X_i$ (which is more general than the assumption that the $X_i$'s are i.i.d.), (1.11) becomes

$$P(X_{n:n} > x) = nP(X_1 > x) - \binom{n}{2} P(X_1 > x, X_2 > x) + \cdots$$

$$+ (-1)^{n-1} P(X_1 > x, \ldots, X_n > x). \qquad (1.12)$$

These give rise to a series of inequalities, termed as *Bonferroni inequalities*; for example, the first of these inequalities are

$$\sum_i P(E_i) - \sum\sum_{i<j} P(E_i \cap E_j) \le P\left(\bigcup_{i=1}^{n} E_i\right) \le \sum_i P(E_i).$$

$$(1.13)$$

Strictly speaking, since $\sum_i P(E_i)$ can exceed 1, we may modify the upper bound in (1.13) to be $\min(\sum_i P(E_i), 1)$. This upper bound can be still further sharpened as shown by Kounias (1968). Since

$$P\left(\bigcup_{i=1}^{n} E_i\right) \le P(E_i) + \sum_{\substack{j=1\\j\neq i}}^{n} P(E_i^c \cap E_j),$$

we have

$$P\left(\bigcup_{i=1}^{n} E_i\right) \le P(E_i) + \sum_{\substack{j=1\\j\neq i}}^{n} \{P(E_j) - P(E_i \cap E_j)\}$$

$$= \sum_{i=1}^{n} P(E_i) - \sum_{\substack{j=1\\j\neq i}}^{n} P(E_i \cap E_j)$$

and consequently,

$$P\left(\bigcup_{i=1}^{n} E_i\right) \le \min_{1 \le i \le n} \left(\sum_{j=1}^{n} P(E_j) - \sum_{\substack{j=1\\j\neq i}}^{n} P(E_i \cap E_j)\right). \tag{1.14}$$

Notice that in the special case when $P(E_i) = P(E_1)$ for all $i$ and $P(E_i \cap E_j) = P(E_1 \cap E_2)$ for all $i \neq j$, (1.14) simplifies to

$$P\left(\bigcup_{i=1}^{n} E_i\right) \le nP(E_1) - (n-1)P(E_1 \cap E_2). \tag{1.15}$$

Various further refinements, improvements and generalizations are available in the literature. We refer the interested readers to Galambos and Simonelli (1996).

Order statistics also possess some very interesting general properties. For example, the sequence of order statistics arising from an absolutely continuous distribution forms a Markov chain whereas the sequence of order statistics arising from a discrete distribution does not form a Markov chain. Furthermore, in the case of absolutely continuous distribution, the conditional distribution of $X_{j:n}$ given $X_{i:n} = x$ (for $j > i$) is exactly the same as the distribution of the $(j-i)$th order statistic in a sample of size $n-i$ from the population distribution truncated on the left at $x$; similarly, the conditional distribution of $X_{i:n}$ given $X_{j:n} = x$ (for $i < j$) is exactly the same as the distribution of the $i$th order statistic in a sample of size $j-1$ from the population distribution truncated on the right at $x$.

In addition to these nice general properties, order statistics from a few specific distributions also possess some very interesting and useful distributional properties. For example, in the case of the standard exponential distribution, the *normalized spacings* defined by

$$Z_i = (n - i + 1)(X_{i:n} - X_{i-1:n}), \quad i = 1, 2, \ldots, n, \tag{1.16}$$

with $X_{0:n} \equiv 0$, are all independent and identically distributed once again as standard exponential. Also, in this exponential case, the sequence of order statistics form an additive Markov chain.

In various parts of this volume, some of these properties and results of order statistics will be utilized in the study of scan statistics.

---

## 1.4 Discrete Scan Statistics in the One-Dimensional Case

Let $X_1, \ldots, X_N$ be a sequence of integer valued random variables. For $2 \le m \le N - 1$ consider the moving sums of $m$ consecutive observations. The *linear unconditional discrete scan statistic* is defined as

$$S_m = \max_{1 \le t \le N-m+1} Y_t,$$

where $Y_t = \sum_{i=t}^{t+m-1} X_i$. The *circular* unconditional discrete scan statistics is defined as

$$S_m^* = \max_{1 \le t \le N} Y_t.$$

$S_m$ and $S_m^*$ are used in testing the null hypothesis that the observations are identically distributed as $F_0$, while under the alternative hypothesis, for some $1 \le i \le N - m + 1$ and $I_m = \{i, \ldots, i + m - 1\}$, $X_j$, $j \in I_m$, are distributed as $F_1$, and $X_j$, $j \in \{1, \ldots, N\}\backslash I_m$ are distributed as $F_0$. For independent and identically distributed (i.i.d.) integer valued observations, the generalized likelihood ratio test rejects the null hypothesis in favor of the alternative hypothesis if $S_m \ge k$ or $S_m^* \ge k$, where $k$ is determined from a specified probability of type I error [Glaz and Naus (1991)].

Most of the research has focused on the case when $X_1, \ldots, X_N$ are i.i.d. non-negative integer valued random variables and most of the results have been developed in particular for 0-1 i.i.d. Bernoulli trials. The case of an arbitrary sequence of i.i.d. integer valued random variables has been discussed by Glaz and Naus (1991); the nonidentical case for independent 0-1 Bernoulli trials has been treated by Wallenstein, Naus and Glaz (1994), and Koutras and Alexandrou (1995); the dependent case for 0-1 Markov trials has been investigated by Glaz (1983) and Koutras and Alexandrou (1995); and the case of the multinomial random vector has been discussed by Krauth (1992a,b) and Wallenstein, Weinberg and Gould (1989). In this volume, the one-dimensional discrete scan statistics are discussed by Chen and Glaz in Chapter 2, Krauth in Chapter 3, Naus in Chapter 4, Koutras and Balakrishnan in Chapter 11, and Fu, Lou, and Chen in Chapter 13. Many interesting applications, related references and open problems are presented in these chapters.

To implement a testing procedure based on the scan statistics for the hypotheses specified above, accurate approximations for $P(S_m \geq k)$ or $P(S_m^* \geq k)$ are needed. In what follows, we will outline the difficulties and the methods that are currently available for evaluating $P(k; m, N) = P(S_m \geq k)$ only. These methods can be extended to handle the circular scan statistics [Chen and Glaz (1998)] and will not be discussed here. Since

$$P(S_m \geq k) = P\left\{ \bigcup_{t=1}^{N-m+1} (Y_t \geq k) \right\},$$

one has to evaluate the probability of a union or an intersection of the sequence of dependent events $\{(Y_t \geq k)\}_{t=1}^{N-m+1}$. Exact formula for $P(S_m \geq k)$ is available only for 0-1 i.i.d. Bernoulli trials [Naus (1974)], and even for this special case it is computable only for a restricted range of parameters. Koutras and Alexandrou (1995), using the Markov Chain embedding technique from Fu and Koutras (1994), have presented recursive equations that can be used to evaluate the tail probabilities for discrete scan statistics for a sequence of 0-1 trials. This computational approach is also feasible for a restricted range of parameters. Therefore, accurate approximations and inequalities are especially needed.

Two types of approximations have been developed for $P(S_m \geq k)$. The first is the Poisson-type approximations. This method is using the fact that if we define

$$I_t = \begin{cases} 1 & if \quad Y_t \geq k \\ 0 & if \quad Y_t < k, \end{cases}$$

then, under certain conditions, the distribution of $\sum_{t=1}^{N-m+1} Y_t$ can be approximated by a Poisson distribution with mean $\lambda = \sum_{t=1}^{N-m+1} E(I_t)$ [Darling and Waterman (1986) and Goldstein and Waterman (1992)]. Refinement of this method to account for clumping and compound-Poisson approximations are presented by Chen and Glaz in Chapter 2 of this volume.

The second method used in approximating $P(S_m \geq k)$ is the product-type approximation [Naus (1982) and Glaz and Naus (1991)]. For $N > 3m$, a third-order product-type approximation is given by

$$P(S_m \geq k) \approx 1 - q_{3m} \left( \frac{q_{3m}}{q_{3m-1}} \right)^{N-3m},$$

where for $j \geq m$, $q_j = 1 - P(k; m, j)$ [Glaz and Naus (1991)]. This method exploits the positive dependence structure between the successive elements in the sequence of random variables $\{Y_t\}_{t=1}^{N-m+1}$. In most cases, the product-type approximations are the most accurate ones [Chen and Glaz in Chapter 2, and Chen (1998)].

Two methods that provide accurate inequalities for $P(S_m \geq k)$ have been discussed. The first is the method of product-type inequalities introduced in Glaz and Naus (1991). For $N > 3m$, a third-order product-type inequality is given by

$$1 - q_{3m}[1 - (q_{2m-1} - q_{2m})]^{N-3m} \leq P(S_m \geq k) \leq 1 - \frac{q_{3m}}{\left[1 + \frac{q_{2m-1} - q_{2m}}{q_{3m}}\right]^{N-3m}}.$$

The second method is the Bonferroni-type approach investigated for the discrete scan statistics in Chen (1998). A related upper Bonferroni-type inequality has been presented in Glaz and Naus (1991). From the numerical results, it is evident that in most cases the product-type inequalities produce more accurate results.

Let $X_1, \ldots, X_N$ be a sequence of integer valued random variables. Suppose we are given that $\sum_{i=1}^{N} X_i = a$. Then, conditional on $\sum_{i=1}^{N} X_i = a$, $S_m$ $(S_m^*)$ is referred to as a linear (circular) *conditional* discrete scan statistic. These statistics have applications in many areas of science including meteorology [Moye *et al.* (1988)], minefield detection [Glaz (1996)], molecular biology [Altschul and Erickson (1988), Arratia, Goldstein, and Gordon (1989), Fu and Curnow (1990), Karlin *et al.* (1983), and Naus and Sheng (1997)], quality control and reliability theory [Balakrishnan, Balasubramanian, and Viveros (1993), Chao, Fu, and Koutras (1995), Fu and Koutras (1994), Glaz (1983), Greenberg (1970), and Saperstein (1972, 1973)], and radar detection [Bogush (1972) and Nelson (1978)]. For the special case of 0-1 i.i.d. Bernoulli trials, Naus (1974, Theorems 1 and 2) has derived exact formulae for

$$P\left(S_m \geq k \,\Big|\, \sum_{i=1}^{N} X_i = a\right),$$

in terms of a sum of determinants of $L \times L$ matrices where $L = \left[\frac{N}{m}\right]$, and $[x]$ is the integer part of $x$. That formula can be evaluated only for a restricted range of parameters. Exact formula is not available for the circular case. Product-type and Poisson-type approximations for the distribution of these linear and circular conditional scan statistics for 0-1 i.i.d. Bernoulli, binomial and Poisson models have been discussed by Chen (1998) and Chen and Glaz in Chapter 2 in this volume. Bonferroni-type inequalities have been investigated by Chen *et al.* (1998) just for the 0-1 i.i.d. Bernoulli trials. Inequalities for general i.i.d. nonnegative integer valued random variables are still not available for the conditional case.

## 1.5   Discrete Scan Statistics in the Multidimensional Case

Let $X_{ij}$, $1 \leq i \leq N_1, 1 \leq j \leq N_2$, be a sequence of integer valued random variables. The scanning window here is a discrete rectangle of size $m_1 \times m_2$. The *two-dimensional unconditional discrete scan statistic* is defined as

$$S_{m_1,m_2} = \max_{1 \leq i_1 \leq N_1 - m_1 + 1} \max_{1 \leq i_2 \leq N_2 - m_2 + 1} Y_{i_1,i_2},$$

where

$$Y_{i_1,i_2} = \sum_{j=i_2}^{i_2+m_2-1} \sum_{i=i_1}^{i_1+m_1-1} X_{ij}.$$

Areas of applications for this two-dimensional scan statistic include astronomy [Darling and Waterman (1986)], computer science [Pfaltz (1983)], ecology [Cressie (1991) and Koen (1991)], epidemiology [Cressie (1991)], image analysis [Rosenfeld (1978)], pattern recognition [Panayirci and Dubes (1983)], reliability theory [Barbour, Chryssaphinou, and Roos (1996), Koutras, Papadopoulos, and Papastavridis (1993), and Salvia and Lasher (1990)], and minefield detection via remote sensing [Muises and Smith (1992) and Smith (1991)]. We are interested in developing accurate approximations and inequalities for $P(S_{m_1,m_2} \geq k)$. In this case, exact results are not yet available. For the special case of 0-1 i.i.d. Bernoulli trials Darling and Waterman (1986) have discussed a Poisson approximation. For 0-1 i.i.d. Bernoulli trials, for the special case of $k = m^2$, some approximations have been discussed by Barbour, Chryssaphinou, and Roos (1996), Koutras, Papadopoulos, and Papastavridis (1993), Roos (1993), and Sheng and Naus (1996). Refined Poisson-type and product-type approximations and Bonferroni-type inequalities have been investigated by Chen and Glaz (1996) for the i.i.d. Bernoulli, binomial and Poisson models. In this volume, Chen and Glaz (in Section 5 of Chapter 2) have presented a survey of these results. Moreover, approximations are derived for the expected size and standard deviation of $S_{m_1,m_2}$. Numerical results indicate that these approximations perform well. In this volume, Naus (in Section 4 of Chapter 4) has mentioned the two-dimensional scan statistic and has presented interesting references for a different method of scanning two-dimensional sequences of symbols.

To conclude this brief introduction to the two-dimensional discrete scan statistic, we would like to mention that no approximations or inequalities are available for the two-dimensional conditional discrete scan statistic. Also, no results are available for higher dimensional discrete scan statistics. In particular, the three-dimensional scan statistics could be of interest in astronomy, biology, medicine and oceanography.

---

## 1.6 Continuous Scan Statistics in the One-Dimensional Case

Let $\{X_t, t \geq 0\}$ be a Poisson process with intensity $\lambda$ on $(0, \infty)$. $X_t$ is the number of points (events) that have occurred in the interval $(0, t]$. For $0 < w < T$, let the associated scanning process $Y_t(w) = X_{t+w} - X_t$ denote the number of points (events) that have occurred in the interval $(t, t + w)$. An *unconditional one-dimensional scan statistic* is defined as

$$S_w = S_w(\lambda, T) = \max_{0 < t \leq T - w} Y_t(w),$$

where $(0, T]$ is the total interval in which the Poisson process is observed. Applications of this scan statistic have been discussed in many problems in science and engineering, including epidemiology [Wallenstein and Neff (1987)], molecular biology [Karlin and Brendel (1992)], material science [Newell (1963)], queueing theory [Glaz (1981)], visual perception [Glaz (1979), Leslie (1969), Ikeda (1965) and Van de Grind (1971)], and telecommunication [Alm (1983)]. For convenience, in the above definition of the scan statistic, three parameters have been used. In fact, $\lambda w$ and $w/T$ are sufficient. Moreover, without loss of generality, one can assume that $T = 1$, by redefining $w$ to be $w/T$ and $\lambda$ to be $\lambda T$.

The exact distribution of $S_w$ has been derived by Wallenstein and Naus (1974) and Huntington and Naus (1975). The expression they derived can be evaluated for a restricted range of parameters. Therefore, accurate approximations and inequalities have been investigated by several authors [Alm (1983), Naus (1982), Samuel-Cahn (1983), and Wallenstein and Neff (1987)]. For $L \geq 2$, $\lambda' = \lambda w$ and $T = Lw$, Naus (1982) has derived accurate product-type approximations for $P^*(k; \lambda T, w/T) = P(S_w \geq k)$ in terms of exact results for $P^*(k; i\lambda', 1/i)$, $i = 2, 3$ :

$$P^*(k; L\lambda', 1/L) \approx 1 - Q_3^*(Q_3^*/Q_2^*)^{L-3},$$

where $Q_i^* = 1 - P^*(k; i\lambda', 1/i)$, $i = 2, 3$. The above approximation is quite accurate even if $T/w$ is not an integer and we use $L = [T/w]$. Moreover, Naus (1982) has developed a product-type approximation for the circular unconditional scan statistic.

Alm (1983), by analyzing the primary and secondary upcrossings of level $n$ by $a$ in the scanning process associated with the Poisson process, arrived at the following approximation:

$$P^*(k; \lambda T, w/T) = 1 - F_{\lambda w}(n-1) \exp\left[-\left(1 - \frac{\lambda w}{n}\right)\lambda(T-w)\, p_{\lambda w}(n-1)\right],$$

where $F_{\lambda w}(n-1)$ and $p_{\lambda w}(n-1)$ are the cumulative distribution function and the probability mass function, respectively, of a Poisson random variable with mean $\lambda w$, evaluated at $n-1$. This approximation performs well and is easy to evaluate. Other accurate approximations for $P^*(k; \lambda T, w/T)$ have been derived by Samuel-Cahn (1983) and Wallenstein and Neff (1987). Accurate product-type inequalities for $P^*(k; \lambda T, w/T)$ have been derived by Janson (1984).

Large deviation approximations for the distribution of scan statistics for a one-dimensional Poisson process on the unit interval have been derived by Loader (1991). His approach also yields approximations for a certain class of nonhomogeneous Poisson processes and in the case when the scanning window is not fixed. Approximations for the distribution of the scan statistic for a certain class of nonhomogeneous Poisson process and their applications to evaluate the power of testing procedures based on scan statistics have been discussed by Wallenstein, Naus, and Glaz (1995).

Let $\{X_t, t \geq 0\}$ be a Poisson process with intensity $\lambda$ on $(0, \infty)$. If one conditions on the number of observations in the interval $(0, T]$ or the time interval in which the $(n+1)$th observation is recorded, after a proper rescaling, the null hypothesis of testing uniformity reduces to testing that the times (or locations) of the $n$ observations follow a uniform distribution on the interval $(0, 1]$.

Let $X_1, \ldots, X_N$ be a sequence of independent uniformly distributed observations in the interval $(0, 1]$. For $0 < w < 1$ and $0 < t \leq 1 - w$, let $Y_t(w)$ be the number of observations in the interval $(t, t + w]$. Define

$$S_w = \max_{0 < t \leq 1-w} Y_t(w).$$

$S_w$ is known as the *linear conditional scan statistic* and is used to test the null hypothesis that the observations are uniformly distributed in $(0, 1]$ against a clustering alternative. For a certain class of alternatives, $S_w$ has several optimal properties, including being a generalized likelihood ratio test [Cressie (1977, 1978, 1979, 1984) and Naus (1966b)]. The null hypothesis of uniformity is rejected if $S_w$ exceeds the value $k$, which corresponds to a specified level of significance. The importance of this scan statistic arises from the applications in different disciplines, including geology [Conover, Bement, and Iman (1979) and Shepard, Creasey, and Fisher (1981)], medicine [Ederer, Myers, and Mantel (1964), Wallenstein (1980), and Wallenstein and Neff (1987)], nuclear physics [Orear and Cassel (1971)], photography [Hamilton, Lawton, and Trabka (1972)] and radio-optics [Trusov (1970)].

To implement the testing procedure based on $S_w$, one naturally has to evaluate the tail probabilities

$$P(k|N, w) = P(S_w \geq k).$$

Exact formulae for $P(k|N, w)$ that are computable for a restricted range of the parameters have been derived by Naus (1965, 1966a), Wallenstein and Naus (1974), and Huntington and Naus (1975). Approximations for $P(k|N, w)$ have been discussed extensively in the statistical literature; see, for example, Berman and Eagleson (1985), Gates and Wescott (1984), Glaz (1989, 1992), Glaz, Naus, Roos, and Wallenstein (1994), Naus (1982), Wallenstein and Neff (1987), and Huffer and Lin (1997). Huffer and Lin (in Chapters 6 and 7) and Lin (in Chapter 9) have reviewed several approximations mentioned in these references and have presented numerical results comparing their performance.

Berman and Eagleson (1985) have derived a second-order upper Bonferroni-type inequality for $P(k|n, w)$ based on the order statistics representation of $S_w$. That inequality has been extended to order $k$ in Glaz (1989, 1992). These high-order upper inequalities perform well. Lower Bonferroni-type inequalities investigated by Glaz (1989, 1992) perform poorly. Recently, Huffer and Lin (1997) utilized the moments of the number of intervals of length at most $w$ containing $k$ observations to derive accurate inequalities for $P(k|N, w)$ that are valid for $k \leq 10$. Huffer and Lin (in Chapters 6 and 7) and Lin (in Chapter 9) have reviewed the inequalities investigated in the references listed above and have presented numerical results comparing their performance. Chen *et al.* (1998) have derived tight second-order Bonferroni-type inequalities. These inequalities are based on the scanning window representation of the scan statistic and are valid for all values of the parameters. In turn, Chen *et al.* (1998) have derived accurate inequalities for $E(S_w)$. Moreover, these inequalities can be easily extended to the *circular* conditional scan statistic,

$$S_w^* = \max_{0 < t < 1} Y_t(w),$$

in which case $n$ observations are uniformly distributed on the unit circle. The scan statistic on the circle, its use as a test for randomness, and other related problems have been discussed by many researchers including Ajne (1968), Cressie (1977, 1978, 1980, 1984), Hüsler (1982), Kokic (1987), Naus (1982), and Takacs (1996). Exact results for the distribution of $S_w^*$ are available only for $w = 1/2$ [Ajne (1968)] and $w = 1/3$ [Takacs (1996)]. Approximations for $P(S_w^* \geq k)$ have been investigated in Naus (1982). Chen *et al.* (1998) have presented Bonferroni-type inequalities for the distribution as well as the expected value of this circular scan statistic.

## 1.7   Continuous Scan Statistics in the Multidimensional Case

For a two-dimensional Poisson process $X$ with intensity $\lambda$, the *two-dimensional continuous unconditional scan statistic* in a rectangular set $A$ is defined as

$$S_W = S_W(\lambda, A) = \max_{x \in R^2} X(W(x) \cap A),$$

where $A = [0, T_1] \times \{0, T_2\}$, and $W$ is a general scanning set. Applications in astronomy, biology, and logistics are mentioned in Naus (1965). Alm (1997) has discussed applications to structural mechanics and risk analysis. Alm (1997, 1998) has derived accurate Poisson approximations for the distribution of $S_W$ for rectangular scanning sets. Moreover, these approximations are extended to the three-dimensional case and extensions of these approximations to the $n$ dimensional case are also indicated. These results are extended to general scanning sets. In the two-dimensional case, examples are given for the circular and triangular scanning sets. A review of these results and their applications along with a discussion on simulations, that are quite complex for multidimensional scan statistics, has been presented by Alm in Chapter 5.

If one conditions on $X(A) = N$, then the $N$ points are uniformly distributed over the region $A$. In this setting, the two-dimensional *conditional* scan statistic has been discussed by Eggleton and Kermack (1944) and Mack (1949). Naus (1965) has discussed upper and lower bounds for the distribution of this scan statistic for rectangular scanning sets. For a fixed rectangular set $W$, Loader (1991) has derived large deviation approximations for the distribution of the conditional two-dimensional scan statistic. These approximations are based on a result that characterizes the local behavior of a two-dimensional stationary Poisson process in a rectangular region. Moreover, a modified scan statistic based on a likelihood ratio principle has been investigated for varying window widths.

Let $N$ points be independently and uniformly distributed in a $d$-dimensional rectangle $A \subset R^d$, $d \geq 2$. Let $W \subset A$ be a convex set, which is small relatively to $A$. Månsson (in Chapter 10) has investigated the distribution of a multiple scan statistic defined as

$$\xi(d, N, m, W) = \sum_{i=1}^{\binom{N}{m}} I_i,$$

where

$$I_i = \begin{cases} 1 & \text{if there exists } x \in A \text{ for which } W(x) \text{ covers the } i\text{th} \\ & \text{subset of } m \text{ points} \\ 0 & \text{otherwise,} \end{cases}$$

where $W(x)$ denotes the translate of $W$ by $x \in R^d$. Poisson approximations using the Stein–Chen method have been derived for the distribution of this multiple scan statistic. In Chapter 10, multiple scan statistics in other settings have also been reviewed. Lin (in Chapter 9) has discussed approximations for a related multiple scan statistic in the one-dimensional case.

---

# References


1. Ajne, B. (1968). A simple test for uniformity of a circular distribution, *Biometrika*, **55**, 343–354.

2. Alm, S. E. (1983). On the distribution of the scan statistic of a Poisson process, *Probability and Mathematical Statistics, Essays in Honour of Carl-Gustav Esseen*, 1–10.

3. Alm, S. E. (1997). On the distribution of scan statistics of a two-dimensional Poisson process, *Advances in Applied Probability*, **29**, 1–18.

4. Alm, S. E. (1998). Approximation and simulation of the distributions of scan statistics for Poisson process in higher dimensions, *Extremes*, **1**, 111–126.

5. Altschul, S. F. and Erickson, B. W. (1988). Significance levels for biological sequence comparisons using non-linear similarity functions, *Bulletin of Mathematical Biology*, **50**, 77–92.

6. Arnold, B. C. and Balakrishnan, N. (1989). *Relations, Bounds and Approximations for Order Statistics*, Lecture Notes in Statistics, **53**, Springer-Verlag, New York.

7. Arnold, B. C., Balakrishnan, N. and Nagaraja, H. N. (1992). *A First Course in Order Statistics*, John Wiley & Sons, New York.

8. Arratia, R., Goldstein, L. and Gordon, L. (1989). Two moments suffice for Poisson approximations: The Chen-Stein method, *Annals of Probability*, **17**, 9–25.

9. Balakrishnan, N., Balasubramanian, K. and Viveros, R. (1993). On sampling inspection plans based on the theory of runs, *The Mathematical Scientist*, **18**, 113–126.

10. Balakrishnan, N. and Cohen, A. C. (1991). *Order Statistics and Inference: Estimation Methods*, Academic Press, San Diego, California.

11. Balakrishnan, N. and Rao, C. R. (Eds.) (1998a). *Handbook of Statistics – 16: Order Statistics: Theory and Methods*, North-Holland, Amsterdam, The Netherlands.

12. Balakrishnan, N. and Rao, C. R. (Eds.) (1998b). *Handbook of Statistics – 17: .Order Statistics: Applications*, North-Holland, Amsterdam, The Netherlands.

13. Barbour, A. D., Chryssaphinou, O. and Roos, M. (1996). Compound Poisson approximation in system reliability, *Naval Research Logistics*, **43**, 251–264.

14. Berman, M. and Eagleson, G. K. (1985). A useful upper bound for the tail probabilities of the scan statistic when the sample size is large, *Journal of the American Statistical Association*, **80**, 886–889.

15. Bogush, Jr., A. J. (1972). Correlated clutter and resultant properties of binary signals, *IEEE Transactions on Aerospace and Electronic Systems*, **9**, 208–213.

16. Chao, M. T., Fu, J. C. and Koutras, M. V. (1995). Survey of reliability studies of consecutive-k-out-of-n: F and related systems, *IEEE Transactions on Reliability*, **44**, 120–127.

17. Chen, J. (1998). Approximations and Inequalities for Discrete Scan Statistics, *Ph.D. Dissertation*, University of Connecticut, Storrs, CT.

18. Chen, J. and Glaz, J. (1996). Two-dimensional discrete scan statistics, *Statistics & Probability Letters*, **31**, 59–68.

19. Chen, J. and Glaz, J. (1998). Approximations for discrete scan statistics on the circle, *Submitted for publication*.

20. Chen, J., Glaz, J., Naus, J. and Wallenstein, S. (1998). Bonferroni-type inequalities for conditional scan statistics, *Under preparation*.

21. Conover, W. J., Bement, T. R. and Iman, R. L. (1979). On a method for detecting clusters of possible uranium deposits, *Technometrics*, **21**, 277-282.

22. Cressie, N. (1977). On some properties of the scan statistic on the circle and the line, *Annals of Probability*, **14**, 272–283.

23. Cressie, N. (1978). Power results for tests based on higher order gaps, *Biometrika*, **65**, 214–218.

24. Cressie, N. (1979). An optimal statistic based on higher order gaps, *Biometrika*, **66**, 619–627.

25. Cressie, N. (1980). The asymptotic distribution of the scan statistic under uniformity, *Annals of Probability*, **8**, 828–840.

26. Cressie, N. (1984). Using the scan statistic to test uniformity, *Colloquia Mathematica Societatis János Bolyai*, **45**, pp. 87–100, Debrecen, Hungary.

27. Cressie, N. (1991). *Statistics for Spatial Data*, John Wiley & Sons, New York.

28. Darling, R. W. R. and Waterman, M. S. (1986). Extreme value distribution for the largest cube in random lattice, *SIAM Journal on Applied Mathematics*, **46**, 118–132.

29. David, H. A. (1981). *Order Statistics*, Second edition, John Wiley & Sons, New York.

30. Ederer, F., Myers, M. H. and Mantel, N. (1964). A statistical problem in space and time: Do leukemia cases come in clusters?, *Biometrics*, **20**, 626–638.

31. Eggleton, P. and Kermack, W. O. (1944). A problem in the random distribution of particles, *Proceedings of the Royal Society of Edinburgh, Section A*, **62**, 103–115.

32. Fu, J. and Koutras, M. (1994). Distribution theory of runs: A Markov chain approach, *Journal of the American Statistical Association*, **89**, 1050–1058.

33. Fu, Y. X. and Curnow, R. N. (1990). Locating a changed segment in a sequence of Bernoulli variables, *Biometrika*, **77**, 295–304.

34. Galambos, J. and Simonelli, I. (1996). *Bonferroni-type Inequalities with Applications*, Springer-Verlag, New York.

35. Gates, D. J. and Westcott, M. (1984). On the distributions of scan statistics, *Journal of the American Statistical Association*, **79**, 423–429.

36. Glaz, J. (1979). Expected waiting time for a visual response, *Biological Cybernetics*, **35**, 39–41.

37. Glaz, J. (1981). Clustering of events in a stochastic process, *Journal of Applied Probability*, **18**, 268–275.

38. Glaz, J. (1983). Moving window detection for discrete data, *IEEE Transactions on Information Theory*, **IT-29**, 457–462.

39. Glaz, J. (1989). Approximations and bounds for the distribution of the scan statistic, *Journal of the American Statistical Association*, **84**, 560–566.

40. Glaz, J. (1992). Approximations for tail probabilities and moments of the scan statistic, *Computational Statistics and Data Analysis*, **14**, 213–227.

41. Glaz, J. (1996). Discrete scan statistics with applications to minefield detection, *Proceedings of the Conference of SPIE*, Orlando, FL, **2765**, 420–429.

42. Glaz, J. and Naus, J. (1991). Tight bounds and approximations for scan statistic probabilities for discrete data, *Annals of Applied Probability*, **1**, 306–318.

43. Glaz, J., Naus, J., Roos, M. and Wallenstein, S. (1984). Poisson approximations for the distribution and moments of ordered m-spacings, *Journal of Applied Probability*, **31**, 271–281.

44. Goldstein, L. and Waterman, M.S. (1992). Poisson, compound Poisson and process approximations for testing statistical significance in sequence comparisons, *Bulletin of Mathematical Biology*, **54**, 785–812.

45. Greenberg, I. (1970). The first occurrence of $n$ successes in $N$ trials, *Technometrics*, **12**, 627–634.

46. Hamilton, J. F., Lawton, W. H. and Trabka, E. A. (1972). Some spatial and temporal point processes in photographic science, *Stochastic Processes: Statistical Analysis, Theory and Applications* (Ed., P. A. W. Lewis), pp. 817–867, New York: Wiley Interscience.

47. Harter, H. L. and Balakrishnan, N. (1996). *CRC Handbook of Tables for the Use of Order Statistics in Estimation*, CRC Press, Boca Raton, FL.

48. Huffer, F. and Lin, C. T. (1997). Approximating the distribution of the scan statistic using moments of the number of clumps, *Journal of the American Statistical Association*, **92**, 1466–1475.

49. Huntington, R. and Naus, J. (1975). A simpler expression for Kth nearest neighbor coincidence probabilities, *Annals of Probability*, **3**, 894–896.

50. Hüsler, J. (1982). Random coverage of the circle and asymptotic distributions, *Journal of Applied Probability*, **19**, 578–587.

51. Ikeda, S. (1965). On Bouman-Velden-Yamamoto's asymptotic evaluation formula for the probability of visual response in a certain experimental research in quantum biophysics of vision, *Annals of the Institute of Statistical Mathematics*, **17**, 295–310.

52. Janson, S. (1984). Bounds on the distributions of extremal values of a scanning process, *Stochastic Processes and Their Applications*, **18**, 313–328.

53. Karlin, S. and Brendel, V. (1992). Chance and statistical significance in Protein and DNA sequence analysis, *Science*, **257**, 39–49.

54. Karlin, S., Ghandour, G., Ost, F., Tavare, S. and Korn, L. J. (1983). New approaches for computer analysis of nucleic acid sequences, *Proceedings of the National Academy of Sciences*, **80**, 5660–5664.

55. Koen, C. (1991). A computer program package for the statistical analysis of spatial point processes in a square, *Biometrical Journal*, **33**, 493–503.

56. Kokic, P. N. (1987). On tests of uniformity for randomly distributed arcs on the circle, *The Australian Journal of Statistics*, **29**, 179–187.

57. Koutras, M. V. and Alexandrou, V. A. (1995). Runs, scans and urn model distributions: A unified Markov chain approach, *Annals of the Institute of Statistical Mathematics*, **47**, 743–766.

58. Koutras, M. V., Papadopoulos, G. K. and Papastavridis, S. G. (1993). Reliability of 2-dimensional consecutive-k-out-of-n: F systems, *IEEE Transactions on Reliability*, **R-42**, 658–661.

59. Kounias, E. G. (1968). Bounds for the probability of a union of events, with applications, *Annals of Mathematical Statistics*, **39**, 2154–2158.

60. Krauth, J. (1992a). Bounds for the upper-tail probabilities of the circular ratchet scan statistic, *Biometrics*, **48**, 1177–1185.

61. Krauth, J. (1992b). Bounds for the upper-tail probabilities of the linear ratchet scan statistic, In *Analyzing and Modeling Data and Knowledge* (Ed., M. Schader), pp. 51–61, Springer-Verlag, Berlin.

62. Leslie, R. T. (1969). Recurrence times of clusters of Poisson points, *Journal of Applied Probability*, **6**, 372–388.

63. Loader, C. R. (1991). Large-deviation approximations to the distribution of scan statistics, *Advances in Applied Probability*, **23**, 751–771.

64. Mack, C. (1949). The expected number of aggregates in a random distribution of $n$ points, *Proceedings of the Cambridge Philosophical Society*, **46**, 285–292.

65. Moye, L. A., Kapadia, A. S., Cech, I. M. and Hardy, R. J. (1988). The theory of runs with applications to drought prediction, *Journal of Hydrology*, **103**, 127–137.

66. Muises, R. R. and Smith, C. M. (1992). Nonparametric minefield detection and localization, *Technical Report CSS-TM*, **591-91**, Coastal Systems Station, Naval Surface Warfare Center.

67. Naus, J. (1965). The distributions of the size of the maximum cluster of points on a line, *Journal of the American Statistical Association*, **60**, 532–538.

68. Naus, J. (1966a). Some probabilities, expectations, and variances for the size of the largest clusters and smallest intervals, *Journal of the American Statistical Association*, **61**, 1191–1199.

69. Naus, J. (1966b). A power comparison of two tests of non-random clusters, *Technometrics*, **8**, 493–517.

70. Naus, J. (1974). Probabilities for a generalized birthday problem, *Journal of the American Statistical Association*, **69**, 810–815.

71. Naus, J. (1982). Approximations for distributions of scan statistics, *Journal of the American Statistical Association*, **77**, 377–385.

72. Naus, J. and Sheng, K. N. (1997). Matching among multiple random sequences, *Bulletin of Mathematical Biology*, **59**, 483–496.

73. Nelson, J. B. (1978). Minimal order models for false alarm calculations on sliding windows, *IEEE Transactions on Aerospace and Electronic Systems*, **15**, 352–363.

74. Newell, G. F. (1963). Distribution for the smallest distance between any pair of Kth nearest-neighbor random points on a line, *Time Series Analysis*, Proceedings of the Conference, Brown University (Ed., M. Rosenblatt), New York: Academic Press.

75. Orear, J. and Cassel, D. (1971). Applications of statistical inference to physics, In *Foundation of Statistical Inference* (Eds., V. Godambe and D. Sprott), pp. 280–289, Toronto: Holt, Rinehart and Winston.

76. Panayirci, E. and Dubes, R. C. (1983). A test for multidimensional clustering tendency, *Pattern Recognition*, **16**, 433–444.

77. Pfaltz, J. L. (1983). Convex clusters in discrete $m$-dimensional space, *Journal of Computation*, **12**, 746–750.

78. Roos, M. (1993). Stein–Chen method for compound Poisson approximation, *Ph.D. Dissertation*, University of Zurich, Zurich, Switzerland.

79. Rosenfeld, (1978). Clusters in digital pictures, *Information Control*, **39**, 19–34.

80. Salvia, A. A. and Lasher, W. C. (1990). 2-dimensional consecutive-k-out-of-n: F models, *IEEE Transactions on Reliability*, **R-39**, 382–385.

81. Samuel-Cahn, E. (1983). Simple approximations to the expected waiting time for a cluster of any given size for point processes, *Advances in Applied Probability*, **15**, 21–38.

82. Saperstein, B. (1972). The generalized birthday problem, *Journal of the American Statistical Association*, **67**, 425–428.

83. Saperstein, B. (1973). On the occurrence of $n$ successes within $N$ Bernoulli trials, *Technometrics*, **15**, 809–818.

84. Sarhan, A. E. and Greenberg, B. G. (Eds.) (1962). *Contributions to Order Statistics*, John Wiley & Sons, New York.

85. Sheng, K. N. and Naus, J. I. (1996). Matching rectangles in 2-dimensions, *Statistics & Probability Letters*, **26**, 83–90.

86. Shepard, J., Creasey, J. W. and Fisher, N. I. (1981). Statistical analysis of spacings between geological discontinuities in coal mines, with applications to short-range forecasting of mining conditions, *Australian Coal Geol.*, **3**, 71–80.

87. Smith, C. M. (1991). Two-dimensional minefield simulation, *Technical Report* NCSM-TM-558-91, Coastal Systems Center, Naval Surface Warfare Center.

88. Takács, L. (1996). On a test for uniformity of a circular distribution, *Mathematical Methods of Statistics*, **5**, 77–98.

89. Trusov, A. G. (1970). Estimation of the optimal signal arrival time under conditions of photon counting in free space, *Proceedings of the IEEE Radio-Optics*, **19**, 137–139.

90. Van de Grind, W. A., Koenderink, J. J., Van der Heyde, G. L., Landman, H. and Bowman, M. A. (1971). Adapting coincidence scalars and neural modeling studies of vision, *Kybernetik*, **8**, 85–105.

91. Wallenstein, S. (1980). A test for detection of clustering over time, *American Journal of Epidemiology*, **11**, 367–372.

92. Wallenstein, S. and Naus, J. (1974). Probabilities for the size of the largest clusters and smallest intervals, *Journal of the American Statistical Association*, **69**, 690–697.

93. Wallenstein, S., Naus, J. and Glaz, J. (1994). Power of the scan statistic in detecting a changed segment in a Bernoulli sequence, *Biometrika*, **81**, 595–601.

94. Wallenstein, S., Naus, J. and Glaz, J. (1995). Power of the scan statistics, *Proceedings Section Epidemiology*, Annual ASA Meeting Toronto, Canada, pp. 70–75.

95. Wallenstein, S. and Neff, N. (1987). An approximation for the distribution of the scan statistic, *Statistics in Medicine*, **12**, 1–15.

96. Wallenstein, S., Weinberg, C. R. and Gould, M. (1989). Testing for a pulse in seasonal event data, *Biometrics*, **45**, 817–830.

# PART II
## DISCRETE SCAN STATISTICS

# 2

# Approximations for the Distribution and the Moments of Discrete Scan Statistics

**Jie Chen and Joseph Glaz**

*University of Massachusetts-Boston, Boston, MA*
*University of Connecticut, Storrs, CT*

**Abstract:** Let $X_1, \ldots, X_N$ be a sequence of independent and identically distributed nonnegative integer valued random variables. For $2 \leq m \leq N$, consider the moving sums of $m$ consecutive observations. The discrete scan statistic is defined as the maximum value of these moving sums. Conditional on the sum of all the observations, we refer to this scan statistic as the conditional scan statistic.

In this chapter, we review the approximations for the distributions of the conditional and unconditional discrete scan statistics. Moreover, new approximations are derived for the distribution, the expected size and the standard deviation of scan statistics. Numerical results and simulation studies are presented to evaluate the performance of these approximations.

Let $X_{i,j}$, $1 \leq i \leq N_1$, $1 \leq j \leq N_2$, be a sequence of independent and identically distributed nonnegative integer valued random variables. The observation $X_{i,j}$ denotes the number of events that have occurred in the $(i, j)$th location. For $2 \leq m_i \leq N_i$, $i = 1, 2$, the two-dimensional discrete scan statistic is defined as the maximum number of events in any of the $m_1 \times m_2$ consecutive rectangular windows.

In this chapter, we review the approximations for the distributions of the two-dimensional discrete scan statistics. Based on these approximations, we derive accurate approximations for the expected size and standard deviation of the scan statistic. Numerical results and a simulation study are presented to evaluate the performance of these approximations.

**Keywords and phrases:** Compound Poisson approximations, expected size of a scan statistic, generalized run, multiple scan statistics, Poisson approximations, product-type approximations, standard deviation of a scan statistic, testing for randomness, two-dimensional scan statistic

## 2.1   Introduction

Let $X_1, \ldots, X_N$ be a sequence of independent nonnegative integer valued random variables. In this chapter, we present accurate approximations for the distribution and moments of discrete scan statistics defined in terms of moving sums of a fixed length. For the special case of 0-1 Bernoulli trials, the discrete scan statistic generalizes the notion of the longest run of $1's$ that has been studied extensively in statistical literature; see, for example, Balasubramanian, Viveros, and Balakrishnan (1993), Banjevic (1990), Chryssaphinou, and Papastavridis (1990), Fu (1986), Fu and Hu (1987), Fu and Koutras (1994a), Godbole (1990, 1991, 1993), Gordon, Schilling, and Waterman (1986), Hirano and Aki (1993), Karlin and Ost (1987), Koutras and Alexandrou (1996, 1997), Koutras and Papastavridis (1993), Mott, Kirkwood, and Curnow (1990), Philippou and Makri (1986), and Schwager (1983).

The discrete scan statistics are used for testing the null hypothesis of uniformity against an alternative hypothesis of clustering, that specifies an increased occurrence of events in a connected subsequence of the observations [Glaz and Naus (1991)]. Under the null hypothesis, the random variables $X_1, \ldots, X_N$ are identically distributed as $F_0$, while under the alternative hypothesis, for some $0 \le i_0 \le N - m + 1$ and $I_m = \{i_0, \ldots, i_0 + m - 1\}$, $X_i$ $(i \in I_m)$ are distributed as $F_1$, and $X_i$ $(i \in \{1, \ldots, N\} \backslash I_m)$ are distributed as $F_0$. In this chapter, we derive accurate approximations for the distribution of scan statistic for $F_0$ being any discrete distribution. In this case, one can easily show [Glaz and Naus (1991, Section 1.3)] that the discrete scan statistic defined in (2.1) is a generalized likelihood ratio test for testing the above hypothesis. In this chapter, numerical examples are presented only for $F_0$ being a binomial or a Poisson distribution. Approximations for the power function of the scan statistic for the case of 0-1 Bernoulli trials have been studied by Wallenstein, Naus and Glaz (1994).

The discrete scan statistics have applications to many areas of science including analysis of DNA and protein sequences [Altschul and Erickson (1988), Arratia, Goldstein, and Gordon (1989), Fousler and Karlin (1987), Fu and Curnow (1990), Gotoh (1990), Karlin $et$ $al.$ (1989), Naus and Sheng (1997), Sheng and Naus (1994, 1996), and Waterman (1995)], epidemiology [Krauth (1992), and Wallenstein, Weinberg, and Gould (1989)], minefield detection [Glaz (1995)], quality control and reliability theory [Balakrishnan, Balasubramanian, and Viveros (1993), Chao, Fu, and Koutras (1995), Fu and Koutras (1994), Glaz (1983), Greenberg (1970), Koutras and Papastavridis (1993), Mosteller (1941), Saperstein (1972), and Viveros and Balakrishnan (1993)], radar detection [Bogush (1972) and Nelson (1978)], and sociology [Schwager (1983)].

In Section 2.2 of this chapter, we review the product-type approximations for the tail probabilities of the discrete scan statistic. A new higher-order approximation is derived. In this section, we also investigate several Poisson and compound-Poisson approximations for the distribution of the discrete scan statistic. Based on these approximations, we derive approximations for the expected size and the standard deviation of the scan statistic. Approximations for a multiple scan statistic are also mentioned here. The new approximations for the distributions of discrete scan statistics presented in Sections 2.2 and 2.3 are based on the approach used in Chen and Glaz (1997) to derive approximations for discrete scan statistics on the circle.

In Section 2.3, for the special case of 0-1 Bernoulli trials, we extend the approximations discussed in Section 2.2 to the conditional case, when the total number of events that have occurred is known. In Section 2.4, we discuss how one can employ a Monte Carlo approach to derive approximations for the conditional scan statistic for the binomial and Poisson models.

In Section 2.5, we review the approximations studied in Chen and Glaz (1995) for the two-dimensional scan statistics. Based on these approximations, we derive approximations for the expected size and the standard deviation of the two-dimensional scan statistic.

To evaluate the performance of the approximations discussed in this chapter, numerical results along with a simulation study are presented in Section 2.6 for selected values of the parameters for the binomial and Poisson models. Concluding remarks are presented in Section 2.7.

---

## 2.2 Scan Statistic for i.i.d. Discrete Random Variables

### 2.2.1 Product-type approximations

Let $X_1, \ldots, X_N$ be i.i.d. nonnegative integer valued random variables. For integers $2 \leq m < N$ and $k \geq 2$, define the discrete scan statistic as

$$S_m = \max\{X_i + \ldots + X_{i+m-1}; 1 \leq i \leq N - m + 1\}. \tag{2.1}$$

We are interested in approximating $P(k; m, N) = P(S_m \geq k)$.
For $m \leq j \leq N$, let

$$q_j = 1 - P(k; m, j). \tag{2.2}$$

For $N > 3m$, the product-type approximation

$$P(S_m \geq k) \approx 1 - q_{3m} \left( \frac{q_{3m}}{q_{3m-1}} \right)^{N-3m} \tag{2.3}$$

that has been derived by Glaz and Naus (1991), has been evaluated recently by Chen and Glaz (1996) using an algorithm from Karwe and Naus (1997).

We now present a product-type approximation recommended by Naus (1982) and Glaz and Naus (1991) based on a different representation of the event $(S_m \leq k - 1)$.

For $N = lm$, $l \geq 2$ and $1 \leq i \leq l - 1$, let

$$B_i = \bigcap_{j=1}^{m+1} \left( X_{(i-1)m+j} + \ldots + X_{im+j-1} \leq k - 1 \right). \tag{2.4}$$

Then $(S_m \leq k - 1) = \bigcap_{i=1}^{l-1} B_i$. The following product-type approximation for $P(S_m \leq k - 1)$ [Naus (1982)] can be obtained:

$$
\begin{aligned}
P(S_m \leq k - 1) &= P\left( \bigcap_{i=1}^{l-1} B_i \right) = P(B_1) \prod_{i=2}^{l-1} P\left( B_i \Big| \bigcap_{j=1}^{i-1} B_j \right) \\
&\approx P(B_1) \prod_{i=2}^{l-1} P(B_i | B_{i-1}) = P(B_1 \cap B_2) \prod_{i=2}^{l-2} \frac{P(B_i \cap B_{i+1})}{P(B_i)}.
\end{aligned}
\tag{2.5}
$$

Since for $1 \leq i \leq l - 2$, $P(B_i) = q_{2m}$ and $P(B_i \cap B_{i+1}) = q_{3m}$, we get

$$P(S_m \geq k) \approx 1 - q_{3m} \left( \frac{q_{3m}}{q_{2m}} \right)^{l-3}. \tag{2.6}$$

For the special case of i.i.d. Bernoulli trails, employing the approach leading to approximation (2.6), we can obtain the following higher-order approximation:

$$P(S_m \geq k) \approx 1 - q_{4m} \left( \frac{q_{4m}}{q_{3m}} \right)^{l-4}, \tag{2.7}$$

where

$$q_{4m} = 1 - \sum_{a=k}^{4m} [1 - q(4m|a)] \binom{4m}{a} p^a (1-p)^{4m-a}, \tag{2.8}$$

and $q(4m|a)$ is the probability that any consecutive $m$ trials contain at most $k - 1$ 1's when the total number of 1's in $4m$ trials is equal to $a$. The term $q(4m|a)$ is obtained from Naus (1974, Theorem 1).

Numerical results for product-type approximations (2.3), (2.6) and (2.7) for selected values of $N, m$ and $k$ and parameters of binomial and Poisson distributions are given in Section 2.6.

If $N$ is not a multiple of $m$, an adjustment is needed for approximations (2.6) and (2.7). Suppose $N = lm - \nu$ where $1 \leq \nu \leq m - 1$. Following an approach similar to Eq. (2.5), we recommend the following modification for the approximation (2.6):

$$
\begin{aligned}
P(S_m \geq k) &\approx 1 - q_{3m} \left( \frac{q_{3m}}{q_{2m}} \right)^{l-4} \left( \frac{q_{3m-\nu}}{q_{2m}} \right) \\
&= 1 - q_{3m-\nu} \left( \frac{q_{3m}}{q_{2m}} \right)^{l-3}.
\end{aligned}
\tag{2.9}
$$

If $N$ is not a multiple of $m$, we recommend to modify approximation (2.7) as follows:

$$P(S_m \geq k) \approx 1 - q^*_{4m-\nu} \left(\frac{q_{4m}}{q_{3m}}\right)^{l-4}, \qquad (2.10)$$

where $q^*_{4m-\nu} = (\nu/m)q_{3m} + (1 - (\nu/m))\, q_{4m}, 1 \leq \nu \leq m - 1$. In approximation (2.10) we have approximated $q_{4m-\nu}$ by $q^*_{4m-\nu}$, since there are no algorithms to evaluate $q_{4m-\nu}$. The performance of approximations (2.9) and (2.10) are similar to those of (2.6) and (2.7), respectively, and so in our numerical examples we consider only $n = lm, l \geq 4$.

### 2.2.2 Poisson approximations

Let $X_1, \ldots, X_N$ be i.i.d. nonnegative integer valued random variables. For $1 \leq j \leq N - m + 1$, define

$$I_j = \begin{cases} 1 & \text{if } X_j + \ldots + X_{j+m-1} \geq k \\ 0 & \text{otherwise.} \end{cases} \qquad (2.11)$$

Then,

$$P(S_m \leq k - 1) = P\left(\sum_{j=1}^{N-m+1} I_j = 0\right). \qquad (2.12)$$

Under quite general conditions, the distribution of $\sum_{j=1}^{N-m+1} I_j$ converges to the Poisson distribution with mean

$$\lambda = E\left(\sum_{j=1}^{N-m+1} I_j\right) = (N - m + 1)(1 - q_m); \qquad (2.13)$$

see Darling and Waterman (1986) and Goldstein and Waterman (1992). A simple Poisson approximation is given by

$$P(S_m \geq k) \approx 1 - \exp(-\lambda). \qquad (2.14)$$

For $k < m$, approximation (2.14) can be inaccurate. The reason for this is that the events $(X_j + \ldots + X_{j+m-1} \geq k), 1 \leq j \leq N - m + 1$, tend to clump [Goldstein and Waterman (1992)]. Employing a local declumping approach [Glaz *et al.* (1994)], we derive below a more accurate Poisson approximation.

For $1 \leq j \leq N - m + 1$, let

$$I_j^* = \begin{cases} 1 & \text{if } (I_j = 1) \cap \left\{\cap_{t=j-m+1}^{j-1} (I_t = 0)\right\} \\ 0 & \text{otherwise,} \end{cases} \qquad (2.15)$$

where for $t \leq 0$ we define the event $(I_t = 0)$ to be the entire space. By defining this new set of indicators $I_j^*, 1 \leq j \leq N-m+1$, we are not allowing the starting points of two generalized runs to be too close to each other. For the problem at hand, we have

$$\sum_{j=1}^{N-m+1} I_j = 0 \iff \sum_{j=1}^{N-m+1} I_j^* = 0. \tag{2.16}$$

We approximate

$$P\left(\sum_{j=1}^{N-m+1} I_j^* = 0\right) \approx \exp(-\lambda^*), \tag{2.17}$$

where $A_j = (I_j = 0)$ and

$$
\begin{aligned}
\lambda^* &= E\left(\sum_{j=1}^{N-m+1} I_j^*\right) \\
&= P(A_1^c) + \sum_{j=2}^{m-1} P\left\{A_j^c \cap \left(\bigcap_{i=1}^{j-1} A_i\right)\right\} + (N - m + 2)P\left\{A_m^c \cap \left(\bigcap_{j=1}^{m-1} A_j\right)\right\} \\
&= 1 - q_{2m-2} + (N - 2m + 2)(q_{2m-2} - q_{2m-1}). \tag{2.18}
\end{aligned}
$$

It follows from Eqs. (2.14)–(2.18) that the improved Poisson approximation for $P(S_m \geq k)$ based on the local declumping approach described above is given by

$$P(S_m \geq k) \approx 1 - \exp(-\lambda^*), \tag{2.19}$$

where $\lambda^*$ is given in (2.18).

We now proceed to examine new Poisson-type approximations based on the events $B_i$ defined in (2.4). For $1 \leq i \leq l - 1$, let

$$J_i = \begin{cases} 1 & \text{if } B_i^c \text{ occur} \\ 0 & \text{otherwise.} \end{cases} \tag{2.20}$$

Then,

$$P(S_m \leq k - 1) = P\left(\sum_{i=1}^{l-1} J_i = 0\right). \tag{2.21}$$

A Poisson approximation can be obtained as

$$P(S_m \geq k) \approx 1 - \exp(-\lambda_B), \tag{2.22}$$

where

$$\lambda_B = E\left(\sum_{i=1}^{l-1} J_i\right) = (l - 1)(1 - q_{2m}). \tag{2.23}$$

Employing a local declumping approach that is similar to the one described above, we get the following Poisson approximation. For $1 \leq i \leq l - 1$, let

$$J_i^* = \begin{cases} 1 & \text{if } B_i^c \cap B_{i-1} \text{ occur} \\ 0 & \text{otherwise.} \end{cases}$$

Approximate

$$P\left(\sum_{i=1}^{l-1} J_i^* = 0\right) \approx \exp(-\lambda_B^*), \qquad (2.24)$$

where

$$\lambda_B^* = E\left(\sum_{i=1}^{l-1} J_i^*\right) = 1 - q_{2m} + (l-2)(q_{2m} - q_{3m}). \qquad (2.25)$$

It follows from (2.25) that a Poisson approximation for $P(S_m \geq k)$ is given by

$$P(S_m \geq k) \approx 1 - \exp(-\lambda_B^*), \qquad (2.26)$$

where $\lambda_B^*$ is as defined in (2.25). The performance of Poisson-type approximations given in Eqs. (2.14), (2.19), (2.22) and (2.26) for selected values of $n, m, k$ and parameters of binomial and Poisson distributions is evaluated in Section 2.6.

### 2.2.3 Compound Poisson approximations

We now examine six compound Poisson approximations for $P(S_m \geq k)$. Following the approach in Roos (1993a,b) and Glaz *et al.* (1994), we approximate the distribution of $\sum_{j=1}^{N-m+1} I_j$ by $\sum_{j=1}^{N-m+1} jM_j$, where $M_j$ are independent Poisson random variables with mean $\lambda_j$. In practice, only a small number of $\lambda_j$'s is used and even then they have to be approximated. The first approximation is based on the clump heuristic of Aldous (1989) as it was applied by Glaz *et al.* (1994). A set of $m$ $\lambda_j$'s is used with the following approximations:

$$\lambda_1 \approx \lambda_1^* = (N - m + 1)\pi[1 - 2p + p^2 - mp^{m+1} + (m+1)p^m], \qquad (2.27)$$

$$\lambda_j \approx \lambda_j^* = (N - m + 1)\pi(1 - p)^2 p^{j-1}, \quad j = 2, \ldots, m, \qquad (2.28)$$

where

$$\pi = P(I_1 = 1) = 1 - q_m,$$

$$p = P(I_1 = 1, I_2 = 2)/P(I_1 = 1) = (1 - 2q_m + q_{m+1})/(1 - q_m)$$

and the $q_j$'s are as defined in (2.2) and $I_j$'s are as defined in (2.11). This yields the following approximation:

$$P(S_m \geq k) \approx 1 - \exp\left(-\sum_{i=1}^{m} \lambda_i^*\right), \qquad (2.29)$$

where $\lambda_i^*$'s are as given in (2.27) and (2.28).

The second compound Poisson approximation is based on Barbour, Holst, and Janson (1992, Theorem 10.N and Corollary 10.N.1). It follows from their work that

$$P\left(\sum_{j=1}^{N-m+1} I_j = 0\right) \approx \exp[-(N - m + 1)P(I_1 = 1, I_2 = 0, \ldots, I_m = 0)]. \qquad (2.30)$$

Therefore,

$$P(S_m \geq k) \approx 1 - \exp[-(N - m + 1)(q_{2m-2} - q_{2m-1})], \qquad (2.31)$$

The third compound Poisson approximation is using

$$\lambda_j \approx \lambda_j^* = (N - m + 1)\pi(1 - p)^2 p^{j-1}, \qquad j = 1, \ldots, 2m - 1. \qquad (2.32)$$

This yields the following approximation:

$$P(S_m \geq k) \approx 1 - \exp\left(-\sum_{i=1}^{2m-1} \lambda_i\right) = 1 - \exp\left\{(N - m + 1)\pi(1 - p)(1 - p^{2m-1})\right\}.$$
$$(2.33)$$

In the fourth compound Poisson approximation we have

$$\begin{aligned}
\lambda_1 \approx \lambda_1^* &= \sum_{i=2}^{\infty} \lambda_i - \sum_{i=2}^{2m-1} \lambda_i \\
&= (N - m + 1)\pi\left\{(1 - p)^2 + p^{2m-1}\left[2m - (2m + 1)p\right]\right\}
\end{aligned}$$

and

$$\lambda_j \approx \lambda_j^* = (N - m + 1)\pi(1 - p)^2 p^{j-1}, \qquad j = 2, \ldots, 2m - 1. \qquad (2.34)$$

This yields the following approximation:

$$P(S_m \geq k) \approx 1 - \exp\left(-\sum_{i=1}^{2m-1} \lambda_i\right) \qquad (2.35)$$

The fifth compound Poisson approximation is based on the Roos (1993b, Lemma 3.3.4), where

$$\lambda_i = (N - m + 1)\pi(1 - p)^2 p^{i-1}, \qquad i = 1, \ldots, m - 1, \qquad (2.36)$$

$$\lambda_i = \frac{(N - m + 1)\pi}{i}\left[2(1 - p)p^{i-1} + (2m - i - 2)(1 - p)^2 p^{i-1}\right],$$
$$i = m, \ldots, 2m - 2 \qquad (2.37)$$

and

$$\lambda_{2m-1} = \frac{(N - m + 1)(1 - q_m)p^{2m-2}}{2m - 1}. \qquad (2.38)$$

This yields the following approximation:

$$P(S_m \geq k) \approx 1 - \exp\left(-\sum_{i=1}^{2m-1} \lambda_i\right). \qquad (2.39)$$

The sixth compound Poisson approximation is based on the representation of the event $(S_m \leq k - 1)$ via the events $B_i$ given in (2.4) and Roos (1993b, 1994). It follows that

$$P(S_m \geq k) \approx 1 - \exp\left(-\sum_{i=1}^{3} \lambda_i\right), \tag{2.40}$$

where for $1 \leq i \leq 3$ the $\lambda_i$'s are given by [Roos (1993b)]

$$\lambda_i = \frac{1}{i} \, P(B_1^c) \left\{2\pi_{1,i} + (l - 3)\pi_{2,i}\right\}, \tag{2.41}$$

$$\pi_{1,i} = P\left(J_2 = i - 1 | J_1 = 1\right), \tag{2.42}$$

$$\pi_{2,i} = P\left(J_1 + J_3 = i - 1 | J_2 = 1\right), \tag{2.43}$$

and $J_i$ is as defined in (2.20). To evaluate approximation (2.40), we have to evaluate

$$\pi_{1,1} = P\left(J_2 = 0 | J_1 = 1\right) = \frac{q_{2m} - q_{3m}}{1 - q_{2m}}, \tag{2.44}$$

$$\pi_{1,2} = P\left(J_2 = 1 | J_1 = 1\right) = \frac{1 - 2q_{2m} + q_{3m}}{1 - q_{2m}}, \tag{2.45}$$

$$\pi_{2,1} = P\left(J_1 + J_3 = 0 | J_2 = 1\right) = \frac{q_{2m}^2 - q_{4m}}{1 - q_{2m}}, \tag{2.46}$$

$$\pi_{2,2} = P\left(J_1 + J_3 = 1 | J_2 = 1\right) = \frac{2(q_{2m} - q_{3m} + q_{4m} - q_{2m}^2)}{1 - q_{2m}} \tag{2.47}$$

and

$$\pi_{2,3} = P\left(J_1 + J_3 = 2 | J_2 = 1\right) = \frac{1 - 3q_{2m}^2 + 2q_{3m} - q_{4m} + q_{2m}^2}{1 - q_{2m}}. \tag{2.48}$$

In Section 2.6, for selected values of $N, m, k$ and parameters of binomial and Poisson distributions we evaluate the accuracy of these compound Poisson approximations and compare their performance with those of other approximations derived in this section.

### 2.2.4 Approximations for the expected size and standard deviation of the scan statistic

Since $S_m$ is a discrete random variable, we have

$$E(S_m) = \sum_{k=1}^{m} P(S_m \geq k) \tag{2.49}$$

and

$$Var(S_m) = 2 \sum_{k=1}^{m} k P(S_m \geq k) - E(S_m)\{1 + E(S_m)\}. \tag{2.50}$$

Therefore, approximations for $P(S_m \geq k)$ will yield approximations for $E(S_m)$ and $Var(S_m)$. In Section 2.6, we present approximations for $E(S_m)$ and $SD = [Var(S_m)]^{1/2}$, denoted by $\hat{E}(S_m)$ and $\hat{SD}(S_m)$, respectively, based on the product-type approximation (2.6), improved Poisson approximation (2.26) and the best compound Poisson approximation (2.39), for selected values of the parameters for the binomial and Poisson models. To evaluate the performance of these approximations, we present the simulated values for $E(S_m)$ and $SD(S_m)$, denoted by $E^*(S_m)$ and $SD^*(S_m)$, respectively, based on $10,000$ trials.

### 2.2.5  A multiple scan statistic

We now discuss approximations for a multiple scan statistic defined as

$$\xi = \sum_{j=1}^{N-m+1} I_j, \tag{2.51}$$

where $I_j$ is as defined in (2.11). A product-type approximation for this statistic is extremely complex and of limited value. For a product-type approximation for a multiple scan statistic for continuous observations, see Glaz and Naus (1983). Since Poisson approximations usually give poor results for $P(\xi \geq l)$, $l \geq 2$ [Glaz *et al.* (1994)], we investigate the performance of several compound Poisson approximations for $P(\xi \geq l)$. The compound Poisson approximations for the multiple scan statistics are given by

$$P(\xi \geq l) \approx 1 - \sum_{j=0}^{l-1} \left( \sum_{\beta_1+2\beta_2+3\beta_3+4\beta_4+5\beta_5=j} \prod_{i=1}^{5} \frac{\lambda_i^{\beta_i}}{\beta_i!} \right) \exp\left( -\sum_{i=1}^{5} \lambda_i \right), \tag{2.52}$$

where $\beta_i$ are nonnegative integers and several choices of $\lambda_i$ are given in Eqs. (2.38)–(2.40). In Section 2.6, for selected values of $N, m, k$ and $p$ we evaluate the accuracy of these compound Poisson approximations for the Bernoulli model.

---

## 2.3  Scan Statistic for i.i.d. Bernoulli Trials When the Number of Successes Is Known

### 2.3.1  Product-type approximations

Let $X_1, \ldots, X_N$ be a sequence of $N$ i.i.d. 0-1 Bernoulli trials. Suppose we know that $a$ successes (1's) and $N - a$ failures (0's) have been observed. In this case, the joint distribution of the 0-1 trials assigns equal probabilities to all the $\binom{N}{a}$ arrangements of $a$ 1's and $N - a$ 0's:

$$P\left( X_1 = x_1, \ldots, X_N = x_N \middle| \sum_{i=1}^{N} X_i = a \right) = \frac{1}{\binom{N}{a}}. \tag{2.53}$$

We are interested in approximating

$$P(k; m, N, a) = P\left(S_m \geq k \,\middle|\, \sum_{i=1}^{N} X_i = a\right),\qquad(2.54)$$

where $S_m$ is as defined in (2.1).

For integers $k, l, m \geq 2$ and $N = ml$, an exact formula for $P(k; m, N, a)$ has been derived by Naus (1974, Theorem 1) as

$$P(k; m, N, a) = 1 - \frac{(m!)^l}{\binom{N}{a}} \, \Sigma_{\sigma \in S_k} \det |d_{ij}|,\qquad(2.55)$$

where

$$d_{ij} = \frac{1}{c_{ij}!(m - c_{ij})!} = \begin{cases} 0 & \text{if } c_{ij} < 0 \text{ or } c_{ij} > m \\ & \text{otherwise}, \end{cases}$$

$$
\begin{aligned}
c_{ij} &= (j - i)k - \sum_{r=1}^{j-1} N_r + N_i & \text{for } i < j \\
&= (j - i)k + \sum_{r=j}^{i} N_r & \text{for } i \geq j,
\end{aligned}
$$

where $N_i$ is the number of 1's in trials $(i - 1)m + 1, \ldots, im$, $1 \leq i \leq \ell$, and any negative factorial is defined to be 0. For the special case when $k > a/2$, a simple formula is given by Naus (1974, Corollary 2) as

$$P(k; m, N, a) = \frac{2 \sum_{s=k}^{a} \binom{m}{s}\binom{N-m}{a-s}}{\binom{N}{a}} + (lk - a - 1)\,\frac{\binom{m}{k}\binom{N-m}{a-k}}{\binom{N}{a}}.\qquad(2.56)$$

If $N, m, l$ are large and $k < a/2$, the computation of $P(k; m, N, a)$ using Eq. (2.55) becomes impractical. In this section, we derive product-type approximations for $P(k; m, N, a)$ that are valid for any values of the parameters.

Following Naus (1974, Section 1), consider the total number of 1's in $m$ consecutive trials along the entire sequence. For $N = lm$, $l \geq 2$ and $1 \leq i \leq l-1$, define the events

$$E_i = \bigcap_{j=1}^{m+1} \left(Y_{(i-1)m+j} + \ldots + Y_{im+j-1} \leq k - 1\right),\qquad(2.57)$$

where $Y_1, \ldots, Y_N$ is the sequence of 0-1 trials that contains $a$ 1's and $N - a$ 0's. Following an approach similar to the one in Eq. (2.5), we get

$$1 - P(k; m, N, a) \approx P\left(E_1 \cap E_2\right) \prod_{i=2}^{l-2} \frac{P\left(E_i \cap E_{i+1}\right)}{P\left(E_i\right)}\qquad(2.58)$$

and

$$1 - P(k; m, N, a) \approx P\left(E_1 \bigcap E_2 \bigcap E_3\right) \prod_{i=2}^{l-3} \frac{P\left(E_i \bigcap E_{i+1} \bigcap E_{i+2}\right)}{P\left(E_i \bigcap E_{i+1}\right)}. \tag{2.59}$$

Since for $1 \le i \le l-3$, $P(E_i) = q_{2m}(a)$, $P\left(E_i \bigcap E_{i+1}\right) = q_{3m}(a)$ and $P\left(E_i \bigcap E_{i+1} \bigcap E_{i+2}\right) = q_{4m}(a)$, for $r = 2, 3, 4$

$$q_{rm}(a) = \sum_{j=0}^{min(rk-r,a)} q(rm|j) \frac{\binom{rm}{j}\binom{N-rm}{a-j}}{\binom{N}{a}} \tag{2.60}$$

and

$$q(rm|j) = P(k; m, rm, j) \tag{2.61}$$

can be evaluated using (2.55). Substitution of (2.60) into Eqs. (2.58) and (2.59) yields the following product-type approximations:

$$P(k; m, N, a) \approx 1 - q_{3m}(a) \left(\frac{q_{3m}(a)}{q_{2m}(a)}\right)^{l-3}, \quad l \ge 4 \tag{2.62}$$

and

$$P(k; m, N, a) \approx 1 - q_{4m}(a) \left(\frac{q_{4m}(a)}{q_{3m}(a)}\right)^{l-4}, \quad l \ge 5. \tag{2.63}$$

In Section 2.6, for selected values of $l, k, m, N$ and $a$ we evaluate the performance of these product-type approximations.

### 2.3.2　Poisson approximations

Let $E_1, \ldots, E_{l-1}$ be the events defined in (2.57). For $1 \le j \le l-1$, set

$$H_j = \begin{cases} 1 & \text{if } E_j^c \text{ occur} \\ 0 & \text{otherwise.} \end{cases} \tag{2.64}$$

Then, a Poisson approximation for $P(k; m, N, a)$ is given by

$$P(k; m, N, a) \approx 1 - \exp(-\lambda_E), \tag{2.65}$$

where

$$\lambda_E = E\left(\sum_{j}^{l-1} H_j\right) = (l-1)\{1 - q_{2m}(a)\}. \tag{2.66}$$

Since the events $E_j$ might occur in clumps, we could employ the following declumping approach. For $1 \le j \le l-1$, let

$$H_j^* = \begin{cases} 1 & \text{if } E_j^c \bigcap E_{j-1} \text{ occur} \\ 0 & \text{otherwise,} \end{cases}$$

where $E_0$ is defined to be the entire space. Since

$$\sum_{j=1}^{l-1} H_j = 0 \iff \sum_{j=1}^{l-1} H_j^* = 0, \tag{2.67}$$

the following Poisson approximation for $P(k; m, N, a)$ will be studied:

$$P(k; m, N, a) \approx 1 - \exp(-\lambda_E^*), \tag{2.68}$$

where

$$\lambda_E^* = E\left(\sum_{j=1}^{l-1} H_j^*\right) = 1 - q_{2m}(a) + (l-2)\{q_{2m}(a) - q_{3m}(a)\}. \tag{2.69}$$

Numerical results for Poisson-type approximations will be presented in Section 2.6.

### 2.3.3 Compound Poisson approximations

In this section, we examine five compound Poisson approximations for $P(k; m, N, a)$. The first approximation is based on the clump heuristic of Aldous (1989) as it was applied by Glaz *et al.* (1994). A set of $m$ $\gamma_j$'s is used with the following approximations:

$$\gamma_1 \approx \gamma_1^* = (N-m+1)\pi(a)[1-2p(a)+p^2(a)-mp^{m+1}(a)+(m+1)p^m(a)], \tag{2.70}$$

$$\gamma_j \approx \gamma_j^* = (N-m+1)\pi(a)\{1-p(a)\}^2 p^{j-1}(a), \quad j = 2, \ldots, m, \tag{2.71}$$

where

$$\pi(a) = P(H_1 = 1) = 1 - q_m(a),$$

$$p(a) = P(H_1 = 1, H_2 = 2)/P(H_1 = 1) = \{1 - 2q_m(a) + q_{m+1}(a)\}/\{1 - q_m(a)\},$$

$q_m(a)$ is as defined in (2.60), $H_j$'s are as defined in (2.64), and

$$q_{m+1}(a) = \sum_{k_1=0}^{1}\sum_{k_3=0}^{1}\sum_{k_2=0}^{\min(k-1-k_1, k-1-k_3)} \frac{\binom{1}{k_1}\binom{m-1}{k_2}\binom{1}{k_3}\binom{N-m-1}{a-(k_1+k_2+k_3)}}{\binom{N}{a}}.$$

This yields the following approximation:

$$P(k; m, N, a) \approx 1 - \exp\left(-\sum_{i=1}^{m} \gamma_i^*\right), \tag{2.72}$$

where $\gamma_i^*$'s are given in Eqs. (2.70) and (2.71).

The second compound Poisson approximation uses

$$\gamma_j \approx \gamma_j^* = (N-m+1)\pi(a)\{1-p(a)\}^2 p^{j-1}(a), \quad j = 1, \ldots, 2m-1. \tag{2.73}$$

This yields the following approximation:

$$
\begin{aligned}
P(k; m, N, a) &\approx 1 - \exp\left(-\sum_{i=1}^{2m-1} \gamma_i^8\right)\\
&= 1 - \exp\left\{(N - m + 1)\pi(a)(1 - p(a))(1 - p^{2m-1}(a))\right\}.
\end{aligned}
$$

$$(2.74)$$

In the third compound Poisson approximation, we have

$$
\begin{aligned}
\gamma_1 \approx \gamma_1^* &= \sum_{i=2}^{\infty} \gamma_i - \sum_{i=2}^{2m-1} \gamma_i\\
&= (N - m + 1)\pi(a)\left\{(1 - p(a))^2 + p^{2m-1}(a)\left[2m - (2m+1)p(a)\right]\right\}
\end{aligned}
$$

and

$$
\gamma_j \approx \gamma_j^* = (N - m + 1)\pi(a)(1 - p(a))^2 p^{j-1}(a), \quad j = 2, \ldots, 2m - 1. \quad (2.75)
$$

This yields the following approximation:

$$
P(k; m, N, a) \approx 1 - \exp\left(-\sum_{i=1}^{2m-1} \gamma_i^*\right). \quad (2.76)
$$

The fourth compound Poisson approximation is based on Roos (1993b, Lemma 3.3.4), where

$$
\gamma_i = (N - m + 1)\pi(a)(1 - p(a))^2 p^{i-1}(a), \quad i = 1, \ldots, m - 1, \quad (2.77)
$$

$$
\begin{aligned}
\gamma_i &= \frac{(N - m + 1)\pi(a)}{i}\\
&\quad \times \left[2(1 - p(a))p^{i-1}(a) + (2m - i - 2)(1 - p(a))^2 p^{i-1}(a)\right],\\
&\qquad\qquad i = m, \ldots, 2m - 2, \quad (2.78)
\end{aligned}
$$

and

$$
\gamma_{2m-1} = \frac{(N - m + 1)(1 - q_m(a))p^{2m-2}(a)}{2m - 1}. \quad (2.79)
$$

This yields the following approximation:

$$
P(k; m, N, a) \approx 1 - \exp\left(-\sum_{i=1}^{2m-1} \gamma_i\right). \quad (2.80)
$$

The fifth compound Poisson approximation based on Roos (1993b, 1994) is given by

$$
P(k; m, N, a) \approx 1 - \exp\left(-\sum_{i=1}^{3} \gamma_i\right), \quad (2.81)
$$

where, for $i = 1, 2, 3$,

$$\gamma_i = \frac{1}{i} P(E_1^c) \left\{ 2\pi_{1,i}^* + (l - 3)\pi_{2,i}^* \right\} \tag{2.82}$$

$$P(E_1^c) = 1 - q_{2m}(a), \tag{2.83}$$

$$\pi_{1,i}^* = P\left(H_2 = i - 1 | H_1 = 1\right), \tag{2.84}$$

and

$$\pi_{2,i}^* = P\left(H_1 + H_3 = i - 1 | H_2 = 1\right). \tag{2.85}$$

The explicit formulae for $\pi_{1,i}^*$ and $\pi_{2,i}^*$ have the same general form as the one for $\pi_{1,i}$ and $\pi_{2,i}$ in Eqs. (2.44)–(2.48). The only difference is that we have to replace $q_{rm}$ by $q_{rm}(a)$, $2 \le r \le 4$, and $q_{2m}^2$ by $q_{2m,2m}(a)$. Note that the second compound Poisson approximation studied in Section 2.2.4, is not valid. In Section 2.6, we evaluate the performance of these compound Poisson approximations for selected values of the parameters $k, m, N$ and $a$.

### 2.3.4 Approximations for the expected size and standard deviation of the scan statistic

Approximations for $E(S_m | \sum_{i=1}^N X_i = a)$ and $SD(S_m | \sum_{i=1}^N X_i = a)$ are obtained from Eqs. (2.49) and (2.50), respectively, by replacing the approximations for $P(S_m \ge k)$ with approximations for $P(S_m \ge k | \sum_{i=1}^N = a)$.

In Section 2.6, we present approximations for $E(S_m | \sum_{i=1}^N = a)$ and $SD(S_m | \sum_{i=1}^N = a)$, denoted by $\hat{E}(S_m)$ and $\hat{SD}(S_m)$, respectively, based on the product-type approximation (2.63), improved Poisson approximation (2.68), and the best compound Poisson approximation (2.80), for selected values of the parameters of 0-1 i.i.d. Bernoulli model. To evaluate the performance of these approximations, we present the simulated values of $E(S_m)$ and $SD(S_m)$, denoted by $E^*(S_m)$ and $SD^*(S_m)$, respectively, based on $10,000$ trials.

---

## 2.4 Scan Statistics for Binomial and Poisson Distributions Conditional on the Total Number of Events

### 2.4.1 Poisson model

Let $X_1, \ldots, X_N$ be independent and identically distributed Poisson random variables. Suppose we know that the total number of events $\sum_{i=1}^N X_i = a$. In this case, the sequence of Poisson random variable has a multinomial distribution given by

$$P\left(X_1 = x_1, \ldots, X_N = x_N \middle| \sum_{i=1}^{N} X_i = a\right) = \binom{a}{x_1, x_2, \ldots, x_N}\left(\frac{1}{N}\right)^a. \quad (2.86)$$

We are interested in approximating the tail probability of the conditional scan statistic

$$P_p(k; m, N, a) = P\left(S_m \geq k \middle| \sum_{i=1}^{N} X_i = a\right).$$

Since there are no exact results available for $q_{rm}^p(a) = P_p(k; m, rm, a)$, their simulated values denoted by $\hat{q}_{rm}^p(a), r = m, m+1, \ldots, 4m$, based on $100,000$ trials, will be used to approximate $q_{rm}^p(a)$ in product-type approximation (2.63), Poisson approximations (2.65) and (2.68), and the compound Poisson approximation (2.81). The performance of these approximations will be examined in Section 2.6.

### 2.4.2   Binomial model

Let $X_1, \ldots, X_N$ be a sequence of $n$ independent and identically distributed binomial $(n, p)$ random variables. Suppose we know that the total number of successes $\sum_{i=1}^{N} X_i = a$ has been observed. In this case, the joint distribution of $X_1, \ldots, X_N$ conditional on $\sum_{i=1}^{N} X_i = a$ has a multivariate hypergeometric distribution given by

$$P\left(X_1 = x_1, \ldots, X_N = x_N \middle| \sum_{i=1}^{N} X_i = a\right) = \frac{\binom{n}{x_1}\binom{n}{x_2}\cdots\binom{n}{a-\sum_{i=1}^{N-1} x_i}}{\binom{nN}{a}}. \quad (2.87)$$

We are interested in approximating the tail probability of the conditional scan statistic

$$P_b(k; m, N, a) = P\left(S_m \geq k \middle| \sum_{i=1}^{N} X_i = a\right).$$

Since there are no exact results available for $q_{rm}^b(a) = P_b(k; m, rm, a)$, we use Patefield (1981) algorithm to simulate the values for $q_{rm}^b(a)$ denoted by $\hat{q}_{rm}^b(a)$, $r = m, m+1, \ldots, 4m$ based on $100,000$ trials. The performance of product-type approximation (2.63), Poisson approximations (2.65) and (2.68), and the compound Poisson approximation (2.81) will be studied in Section 2.6 using $\hat{q}_{rm}^b(a), r = m, m+1, \ldots, 4m$, instead of $q_{rm}^b(a)$.

## 2.5 Two-Dimensional Scan Statistics

### 2.5.1 Product-type approximations

Let $Y_{i,j}$, $i = 1, \ldots, N_1$ and $j = 1, \ldots, N_2$, be i.i.d. nonnegative integer valued random variables. Let

$$S(i_1, i_2) = \sum_{j=i_2}^{i_2+m_2-1} \sum_{i=i_1}^{i_1+m_1-1} Y_{i,j}, \tag{2.88}$$

where $1 \le i_1 \le N_1 - m_1 + 1$ and $1 \le i_2 \le N_2 - m_2 + 1$. The two-dimensional scan statistic is defined as

$$S_{m_1,m_2} = max \left\{ S(i_1, i_2); 1 \le i_1 \le N_1 - m_1 + 1, 1 \le i_2 \le N_2 - m_2 + 1 \right\}. \tag{2.89}$$

For simplicity, we assume that $N_1 = N_2 = N$ and $m_1 = m_2 = m$. For $1 \le i_1, i_2 \le N - m + 1$, let us define the events

$$A_{i_1,i_2} = \left\{ \sum_{i=i_1}^{i_1+m-1} \sum_{j=i_2}^{i_2+m-1} Y_{i,j} \ge k \right\}. \tag{2.90}$$

Then,

$$P(S_{m,m} \ge k) = P\left( \bigcup_{i_1=1}^{N-m+1} \bigcup_{i_2=1}^{N-m+1} A_{i_1,i_2} \right). \tag{2.91}$$

To derive a product-type approximation for $P(S_{m,m} \ge k)$ the following approach was used by Chen and Glaz (1996). Let

$$P(S_{m,m} \le k - 1) = P\left( \bigcap_{i_2=1}^{N-m+1} \bigcap_{i_1=1}^{N-m+1} A_{i_1,i_2}^c \right). \tag{2.92}$$

Then, for a fixed value of $1 \le i_1 \le N - m + 1$, one can approximate accurately [Glaz and Naus (1991)]

$$P\left( \bigcap_{i_2=1}^{N-m+1} A_{i_1,i_2}^c \right) \approx q_{2m} \left( \frac{q_{2m}}{q_{2m-1}} \right)^{N-2m}, \tag{2.93}$$

where for $1 \le l \le m + 1$,

$$q_{m+l-1} = P(A_{1,1}^c \cap A_{1,2}^c \cdots \cap A_{1,l}^c). \tag{2.94}$$

Since we have to scan $N - m + 1$ rectangular $m \times n$ adjacent regions, the following product-type approximation for (2.92) is used:

$$\left[ q_{2m} \left( \frac{q_{2m}}{q_{2m-1}} \right)^{N-2m} \right] \left[ \left( \frac{q_{2m}}{q_{2m-1}} \right) \left( \frac{q_{2m}}{q_{2m-1}} \right)^{N-2m} \right]^{N-m}$$
$$= q_{2m-1} \left( \frac{q_{2m}}{q_{2m-1}} \right)^{(N-2m+1)(N-m+1)} . \tag{2.95}$$

Eq. (2.95) uses approximation (2.93) for $i_1 = 1$, and for $2 \le i_1 \le N - m + 1$, it uses an adjusted approximation (2.93) with $q_{2m}$ replaced by $q_{2m}/q_{2m-1}$ to account for the dependence of the events $\{A^c_{i_1,i_2}; i_2 = 1, \ldots, N - m + 1\}$, for different values of $i_1$. Eq. (2.95) then yields

$$P(S_{m,m} \ge k) \approx 1 - q_{2m-1} \left( \frac{q_{2m}}{q_{2m-1}} \right)^{(N-2m+1)(N-m+1)} . \tag{2.96}$$

In Section 2.6, we present numerical results for the product-type approximation (2.96) for selected values of $N, m, k$ and the parameters of binomial and Poisson distributions.

### 2.5.2   Poisson approximations

Let $\Gamma = \{(i_1, i_2); 1 \le i_1 \le N - m + 1, 1 \le i_2 \le N - m + 1\}$ denote the index set of a collection of the integer valued random variables $\{I_\alpha; \alpha \in \Gamma\}$, where

$$I_\alpha = \begin{cases} 1 & \text{if } S(\alpha) \ge k \\ 0 & \text{otherwise.} \end{cases} \tag{2.97}$$

Then,

$$P(S_{m,m} \ge k) = 1 - P\left( \sum_{\alpha \in \Gamma} I_\alpha = 0 \right) .$$

Under quite general conditions, the distribution of $\sum_{\alpha \in \Gamma} I_\alpha$ converges to a Poisson distribution with mean $\lambda$, where

$$\lambda = E\left( \sum_{\alpha \in \Gamma} I_\alpha \right) = (N - m + 1)^2 (1 - q_m) \tag{2.98}$$

and $q_m = P(A^c_{1,1})$; see Darling and Waterman (1986). The Poisson approximation for this problem for the special case of $k = m^2$ has also been discussed by Barbour, Chryssaphinou, and Roos (1995), Koutras, Papadopoulos, and Papastavridis (1993), and Roos (1994). In Section 2.6, we evaluate the performance of the Poisson approximation given by

$$P(S_{m,m} \ge k) \approx 1 - \exp(-\lambda) \tag{2.99}$$

for selected values of $N, m$ and $k$ and parameters of binomial and Poisson distributions.

The Poisson approximation (2.99) is not expected to perform well when $k < m^2$, since the events $\{(S(\alpha) \geq k); \alpha \in \Gamma\}$ tend to clump. Employing a local declumping approach, Chen and Glaz (1996) derived a more accurate Poisson approximation as

$$P(S_{m,m} \geq k) \approx 1 - \exp(-\lambda^*), \qquad (2.100)$$

where

$$\lambda^* = 1 - q_{2m-2} + (N - 2m + 2)(N - m + 1)(q_{2m-2} - q_{2m-1}). \qquad (2.101)$$

In Section 2.6, numerical results are presented for this Poisson approximation for selected values of $N, m$ and $k$ and parameters of binomial and Poisson distributions.

### 2.5.3  A compound Poisson approximation

For the Bernoulli model, the compound Poisson approximations for $P(S_{m,m} \geq k)$ presented below are from Roos (1993b, 1994). Roos (1993b) recommended approximating the distribution of $\sum_{\alpha \in \Gamma} I_\alpha$ by the compound Poisson distribution of $M^* = \sum_{i=1}^{\infty} i M_i$, where $M_i$ are independent Poisson random variables $M^* = \sum_{i=1}^{\infty} M_i$ where $M_i$ are with mean $\lambda_i$. The constants $\lambda_i$ are given by [Roos (1993b)]

$$\lambda_i = \frac{1}{i} \, P(S(1,1) \geq k) \left\{ 4\pi_{1,i} + 4(N - m - 1)\pi_{2,i} + (N - m + 1)^2 \pi_{3,i} \right\}, \qquad (2.102)$$

where, for $1 \leq i \leq 5$,

$$\pi_{1,i} = P\left\{ I_{1,2} + I_{2,1} = i - 1 | I_{1,1} = 1 \right\}, \qquad (2.103)$$

$$\pi_{2,i} = P\left\{ I_{1,1} + I_{2,2} + I_{3,1} = i - 1 | I_{2,1} = 1 \right\}, \qquad (2.104)$$

and

$$\pi_{3,i} = P\left\{ I_{1,2} + I_{2,1} + I_{2,3} + I_{3,2} = i - 1 | I_{2,2} = 1 \right\}. \qquad (2.105)$$

It is tedious but routine to evaluate $\pi_{1,i}, \pi_{2,i}$ and $\pi_{3,i}$ and, therefore, we omit the derivations of their formulae. For details, see Chen and Glaz (1995). For the simpler special case of $k = m^2$, Roos (1993b) has evaluated the compound Poisson approximation. In Section 2.6, numerical results are presented for the compound Poisson approximation given by

$$P(S_{m,m} \geq k) \approx 1 - \exp\left( -\sum_{i=1}^{5} \lambda_i \right) \qquad (2.106)$$

for selected values of $N, m$ and $k$ and for selected values of $p$ of the Bernoulli distribution.

### 2.5.4 Approximations for the expected size and standard deviation of the scan statistic

Since $S_{m,m}$ is a discrete random variable, we can write

$$E(S_{m,m}) = \sum_{k=1}^{m^2} P(S_{m,m} \geq k) \qquad (2.107)$$

and

$$Var(S_{m,m}) = 2\sum_{k=1}^{m^2} kP(S_{m,m} \geq k) - E(S_{m,m})\left[1 + E(S_{m,m})\right]. \qquad (2.108)$$

Hence, approximations for $P(S_{m,m} \geq k)$ will yield approximations for $E(S_{m,m})$ and $Var(S_{m,m})$. In Section 2.6, we present approximations for $E(S_{m,m})$ and $SD = [Var(S_{m,m})]^{1/2}$, denoted by $\hat{E}(S_{m,m})$ and $\hat{SD}(S_{m,m})$, respectively, based on the product-type approximation (2.96), improved Poisson approximation (2.100), and the compound Poisson approximation (2.106), for selected values of the parameters for the binomial and Poisson models. To evaluate the performance of these approximations, we present simulated values of $E(S_{m,m})$ and $SD(S_{m,m})$, denoted by $E^*(S_{m,m})$ and $SD^*(S_{m,m})$, respectively, based on $10,000$ trials.

### 2.5.5 A multiple scan statistic

We now discuss approximations for a two-dimensional multiple scan statistic defined as

$$\xi^* = \sum_{i_2=1}^{N_2-m_2+1} \sum_{i_1=1}^{N_1-m_1+1} I(i_1, i_2), \qquad (2.109)$$

where $I(i_1, i_2)$ is as defined in (2.97).

A product-type approximation for this statistic is extremely complex and of limited value. Poisson approximations give poor results for $P(\xi^* \geq l)$, $l \geq 2$. Based on the compound Poisson approximation for $P(S_m^* \geq k)$ discussed above, the following compound Poisson approximation for the multiple scan statistic has been studied by Chen and Glaz (1996):

$$P(\xi^* \geq l) \approx 1 - \sum_{j=0}^{l-1}\left(\sum_{\beta_1+2\beta_2+3\beta_3+4\beta_4+5\beta_5=j} \prod_{i=1}^{5}\frac{\lambda_i^{\beta_i}}{\beta_i!}\right)\exp\left(-\sum_{i=1}^{5}\lambda_i\right), \qquad (2.110)$$

where $\beta_i$ are nonnegative integers and $\lambda_i$ are as given in (2.102). Numerical results for selected values of $p$ of the Bernoulli model are given in Section 2.6.

## 2.6 Numerical Examples

In this section, we present numerical examples for the distribution, expected size and the standard deviation of the discrete scan statistics discussed in this chapter. In Tables 2.1 and 2.2, numerical examples are presented for i.i.d. 0-1 Bernoulli trials. From these examples, it is evident that the product-type approximation (2.7) is the most accurate one. The usual Poisson approximation (2.14) performs poorly. The Poisson approximation (2.22) based on the events $B_i$ given in (2.4) performs better. The reason for this is that the events $B_i$ incorporate the dependence structure of the distribution of the scan statistic. The local declumping approach used in Poisson-type approximations (2.19) and (2.26) improve these approximations significantly. The best compound Poisson approximation given by (2.40) performs better in most cases than any other Poisson-type approximation.

In Tables 2.3 and 2.4, numerical examples are presented for the binomial and Poisson models. From these examples, it is evident that the product-type approximation (2.6) is the most accurate one. In these tables, we report only the most accurate Poisson, improved Poisson, and compound Poisson approximations. The compound Poisson approximation (2.40) is the most accurate among these approximations.

In Tables 2.5 and 2.6, numerical results are presented for the conditional scan statistics for the 0-1 Bernoulli trials. Here too, the product-type approximation (2.63) and the compound Poisson approximation (2.81) are the best performers. The product-type approximation is the most accurate approximation. In Tables 2.7 and 2.8, a Monte Carlo approach was used to generate approximations for the conditional scan statistic for the binomial and Poisson models. The product-type approximation (2.63) performs quite well.

In Table 2.9, several compound Poisson approximations for a multiple scan statistic for the 0-1 Bernoulli trials have been evaluated for selected values of the parameters. It is clear that the approximation (2.39) is the most accurate one. Still, there is room for improvement.

In Tables 2.10–2.12, numerical results are presented for the two-dimensional discrete scan statistic. Again, it is evident that the product-type approximation (2.96) is the most accurate one. The improved Poisson approximation (2.100) and compound Poisson approximation (2.106) are not as consistent in their performance here as in the one-dimensional case. In Table 2.13, numerical results are presented for a two-dimensional multiple scan statistic for 0-1 Bernoulli trials. The compound Poisson approximation performs reasonably well in some cases, but in others it performs rather poorly.

In Tables 2.14 and 2.15, numerical results are presented for the approximations for expected size and standard deviation of scan statistics discussed in this chapter. The approximations based on product-type approximations for tail probabilities of scan statistics are the best. These approximations perform remarkably well, especially for the expected size of the scan statistic. For one-dimensional scan statistics, the approximations for the expected size have relative error less than or equal to .01, and for the standard deviation the relative error is less than or equal to .05. For the two-dimensional scan statistic, the relative error for the expected size is less than or equal to .06, and for the standard deviation it is less than or equal to .12. The second best approximation is based on the best compound Poisson approximation for tail probabilities of scan statistics.

---

## 2.7  Concluding Remarks

In this chapter, two methods of approximation for the distribution of scan statistics have been discussed. The method of product-type approximations, when it is applicable, produces the most accurate results. The second method, of Poisson-type approximations, produced varying results. The most frequently studied Poisson approximations, that do not take into account the dependence structure of the sequence of moving sums, perform poorly. The use of local declumping in Poisson approximations or the use of compound Poisson approximations result in a significant improvement in the accuracy of the approximations.

For the one- or two-dimensional multiple scan statistic studied in this chapter, the method of Poisson-type approximations is the only method currently available. For 0-1 Bernoulli trials, the compound Poisson approximations (2.39) and (2.110) are the most accurate ones. For the two-dimensional case, a more accurate approximation than (2.110) is needed. Also, it would be interesting to study compound Poisson approximation for other discrete distributions.

For one-dimensional conditional scan statistics, no analytical approximations are available except the case of 0-1 Bernoulli trials. Also, there are no approximations available for the conditional scan statistic for the two-dimensional case.

The multiple scan statistics mentioned in this chapter are based on moving sums of fixed length of observations. It would be interesting to investigate different approaches to account for multiple clusters. Also, power studies are needed to investigate the performance of scan statistics for testing the null hypothesis of uniformity against various alternatives.

**Acknowledgments.** This research work was supported in part by the Office of Naval Research Contract No. N00014-94-1-0061 and the University of Connecticut Research Foundation.

**Table 2.1:** Product-type and Poisson-type approximations to $P(S_m \geq k)$ for i.i.d. Bernoulli model

| $N$ | $m$ | $p$ | $k$ | $\hat{P}(S_m \geq k)$ | (2.7) | (2.14) | (2.19) | (2.22) | (2.26) |
|---|---|---|---|---|---|---|---|---|---|
| 100 | 10 | .05 | 2 | 0.7482 | 0.7476 | 0.9996 | 0.6740 | 0.8483 | 0.6504 |
| | | | 3 | 0.2283 | 0.2272 | 0.6489 | 0.2200 | 0.2942 | 0.2178 |
| | | | 4 | 0.0315 | 0.0311 | 0.0893 | 0.0310 | 0.0394 | 0.0310 |
| | | | 5 | 0.0026 | 0.0025 | 0.0058 | 0.0025 | 0.0031 | 0.0025 |
| | | .10 | 3 | 0.7473 | 0.7461 | 0.9983 | 0.6766 | 0.8313 | 0.6530 |
| | | | 4 | 0.2814 | 0.2816 | 0.6879 | 0.2711 | 0.3503 | 0.2676 |
| | | | 5 | 0.0533 | 0.0542 | 0.1382 | 0.0539 | 0.0672 | 0.0537 |
| | | | 6 | 0.0060 | 0.0063 | 0.0133 | 0.0063 | 0.0076 | 0.0063 |
| | 20 | .05 | 2 | 0.8715 | 0.8713 | 1.0000 | 0.6852 | 0.8781 | 0.6423 |
| | | | 3 | 0.5009 | 0.5009 | 0.9978 | 0.4360 | 0.5691 | 0.4210 |
| | | | 4 | 0.1718 | 0.1714 | 0.7242 | 0.1636 | 0.2085 | 0.1616 |
| | | | 5 | 0.0379 | 0.0385 | 0.1882 | 0.0381 | 0.0464 | 0.0380 |
| | | | 6 | 0.0062 | 0.0063 | 0.0263 | 0.0063 | 0.0073 | 0.0062 |
| | | .10 | 4 | 0.7264 | 0.7265 | 1.0000 | 0.5988 | 0.7584 | 0.5671 |
| | | | 5 | 0.3949 | 0.3953 | 0.9697 | 0.3556 | 0.4515 | 0.3454 |
| | | | 6 | 0.1491 | 0.1500 | 0.5981 | 0.1441 | 0.1783 | 0.1426 |
| | | | 7 | 0.0425 | 0.0418 | 0.1757 | 0.0413 | 0.0494 | 0.0412 |
| | | | 8 | 0.0091 | 0.0090 | 0.0331 | 0.0090 | 0.0104 | 0.0090 |
| 500 | 10 | .01 | 2 | 0.3263 | 0.3270 | 0.8769 | 0.3237 | 0.4518 | 0.3227 |
| | | | 3 | 0.0159 | 0.0159 | 0.0544 | 0.0159 | 0.0213 | 0.0159 |
| | | | 4 | 0.0003 | 0.0004 | 0.0010 | 0.0004 | 0.0005 | 0.0004 |
| | | .05 | 3 | 0.7399 | 0.7405 | 0.9965 | 0.7268 | 0.8500 | 0.7221 |
| | | | 4 | 0.1558 | 0.1541 | 0.3965 | 0.1534 | 0.1967 | 0.1532 |
| | | | 5 | 0.0136 | 0.0134 | 0.0308 | 0.0134 | 0.0166 | 0.0134 |
| | | | 6 | 0.0007 | 0.0007 | 0.0014 | 0.0007 | 0.0009 | 0.0007 |
| | 20 | .01 | 2 | 0.5214 | 0.5218 | 0.9997 | 0.5053 | 0.6716 | 0.5007 |
| | | | 3 | 0.0645 | 0.0638 | 0.3829 | 0.0635 | 0.0858 | 0.0635 |
| | | | 4 | 0.0043 | 0.0038 | 0.0203 | 0.0038 | 0.0048 | 0.0038 |
| | | .05 | 3 | 0.9761 | 0.9765 | 1.0000 | 0.9496 | 0.9936 | 0.9384 |
| | | | 4 | 0.6452 | 0.6466 | 0.9995 | 0.6246 | 0.7541 | 0.6178 |
| | | | 5 | 0.1961 | 0.1984 | 0.7101 | 0.1962 | 0.2481 | 0.1956 |
| | | | 6 | 0.0359 | 0.0352 | 0.1465 | 0.0352 | 0.0432 | 0.0352 |
| | | | 7 | 0.0044 | 0.0045 | 0.0162 | 0.0045 | 0.0053 | 0.0045 |

**Table 2.2:** Compound Poisson approximations to $P(S_m \geq k)$ for i.i.d. Bernoulli model

| $N$ | $m$ | $p$ | $k$ | $\hat{P}(S_m \geq k)$ | (2.29) | (2.31) | (2.33) | (2.35) | (2.39) | (2.40) |
|---|---|---|---|---|---|---|---|---|---|---|
| 100 | 10 | .05 | 2 | 0.7482 | 0.6452 | 0.9890 | 0.7130 | 0.8474 | 0.7390 | 0.6739 |
|  |  |  | 3 | 0.2283 | 0.2123 | 0.3590 | 0.2372 | 0.2489 | 0.2393 | 0.2193 |
|  |  |  | 4 | 0.0315 | 0.0303 | 0.0374 | 0.0328 | 0.0329 | 0.0328 | 0.0310 |
|  |  |  | 5 | 0.0026 | 0.0025 | 0.0027 | 0.0026 | 0.0026 | 0.0026 | 0.0025 |
|  |  | .10 | 3 | 0.7473 | 0.6533 | 0.9494 | 0.7529 | 0.7983 | 0.7607 | 0.6723 |
|  |  |  | 4 | 0.2814 | 0.2630 | 0.3740 | 0.3060 | 0.3095 | 0.3069 | 0.2696 |
|  |  |  | 5 | 0.0533 | 0.0527 | 0.0625 | 0.0591 | 0.0592 | 0.0591 | 0.0538 |
|  |  |  | 6 | 0.0060 | 0.0062 | 0.0068 | 0.0067 | 0.0067 | 0.0067 | 0.0063 |
|  | 20 | .05 | 2 | 0.8715 | 0.5725 | 1.0000 | 0.7429 | 0.9781 | 0.7893 | 0.6906 |
|  |  |  | 3 | 0.5009 | 0.3868 | 0.9138 | 0.4940 | 0.5900 | 0.5022 | 0.4318 |
|  |  |  | 4 | 0.1718 | 0.1497 | 0.3069 | 0.1853 | 0.1919 | 0.1859 | 0.1626 |
|  |  |  | 5 | 0.0379 | 0.0357 | 0.0517 | 0.0423 | 0.0424 | 0.0423 | 0.0380 |
|  |  |  | 6 | 0.0062 | 0.0060 | 0.0072 | 0.0068 | 0.0068 | 0.0068 | 0.0062 |
|  |  | .10 | 4 | 0.7264 | 0.5293 | 0.9813 | 0.7276 | 0.7929 | 0.7336 | 0.5911 |
|  |  |  | 5 | 0.3949 | 0.3233 | 0.6187 | 0.4408 | 0.4502 | 0.4419 | 0.3509 |
|  |  |  | 6 | 0.1491 | 0.1340 | 0.2073 | 0.1763 | 0.1768 | 0.1764 | 0.1432 |
|  |  |  | 7 | 0.0425 | 0.0391 | 0.0517 | 0.0490 | 0.0490 | 0.0490 | 0.0413 |
|  |  |  | 8 | 0.0091 | 0.0086 | 0.0105 | 0.0103 | 0.0103 | 0.0103 | 0.0090 |
| 500 | 10 | .01 | 2 | 0.3263 | 0.3213 | 0.6594 | 0.3275 | 0.3971 | 0.3386 | 0.3236 |
|  |  |  | 3 | 0.0159 | 0.0158 | 0.0225 | 0.0162 | 0.0166 | 0.0162 | 0.0159 |
|  |  |  | 4 | 0.0003 | 0.0004 | 0.0004 | 0.0004 | 0.0004 | 0.0004 | 0.0004 |
|  |  | .05 | 3 | 0.7399 | 0.7241 | 0.9092 | 0.7680 | 0.7865 | 0.7714 | 0.7256 |
|  |  |  | 4 | 0.1558 | 0.1528 | 0.1861 | 0.1647 | 0.1653 | 0.1649 | 0.1533 |
|  |  |  | 5 | 0.0136 | 0.0134 | 0.0145 | 0.0141 | 0.0141 | 0.0141 | 0.0134 |
|  |  |  | 6 | 0.0007 | 0.0007 | 0.0008 | 0.0007 | 0.0007 | 0.0007 | 0.0007 |
|  | 20 | .01 | 2 | 0.5214 | 0.4971 | 0.9805 | 0.5221 | 0.7117 | 0.5399 | 0.5056 |
|  |  |  | 3 | 0.0645 | 0.0627 | 0.1408 | 0.0662 | 0.0717 | 0.0667 | 0.0635 |
|  |  |  | 4 | 0.0043 | 0.0038 | 0.0053 | 0.0039 | 0.0040 | 0.0039 | 0.0038 |
|  |  | .05 | 3 | 0.9761 | 0.9452 | 1.0000 | 0.9825 | 0.9950 | 0.9841 | 0.9499 |
|  |  |  | 4 | 0.6452 | 0.6183 | 0.8866 | 0.7038 | 0.7179 | 0.7053 | 0.6228 |
|  |  |  | 5 | 0.1961 | 0.1942 | 0.2705 | 0.2262 | 0.2270 | 0.2263 | 0.1959 |
|  |  |  | 6 | 0.0359 | 0.0349 | 0.0421 | 0.0397 | 0.0397 | 0.0397 | 0.0352 |
|  |  |  | 7 | 0.0044 | 0.0045 | 0.0050 | 0.0050 | 0.0050 | 0.0050 | 0.0045 |

**Table 2.3:** Comparison of five approximations to $P(S_m \geq k)$ for i.i.d. Poisson model

| $N$ | $m$ | $\theta$ | $k$ | $\hat{P}(S_m \geq k)$ | (2.6) | (2.22) | (2.26) | (2.40) |
|---|---|---|---|---|---|---|---|---|
| 100 | 10 | .10 | 3 | .7645 | .7667 | .8529 | .6676 | .6898 |
| | | | 4 | .3389 | .3472 | .4354 | .3257 | .3291 |
| | | | 5 | .0977 | .0965 | .1237 | .0949 | .0951 |
| | | | 6 | .0215 | .0195 | .0245 | .0195 | .0195 |
| | | | 7 | .0033 | .0032 | .0039 | .0032 | .0032 |
| | | .25 | 5 | .8918 | .8900 | .9317 | .7681 | .7977 |
| | | | 6 | .9328 | .6231 | .7144 | .5574 | .5686 |
| | | | 7 | .3205 | .3187 | .3909 | .3009 | .3034 |
| | | | 8 | .1265 | .1248 | .1558 | .1221 | .1224 |
| | | | 9 | .0402 | .0401 | .0495 | .0398 | .0399 |
| | | | 10 | .0117 | .0111 | .0135 | .0111 | .0111 |
| | | | 11 | .0027 | .0027 | .0033 | .0027 | .0027 |
| | 20 | .25 | 8 | .7563 | .7623 | .7811 | .5893 | .6151 |
| | | | 9 | .5609 | .5539 | .6022 | .4591 | .4704 |
| | | | 10 | .3570 | .3458 | .3952 | .3075 | .3113 |
| | | | 11 | .1906 | .1872 | .2198 | .1757 | .1767 |
| | | | 12 | .0950 | .0894 | .1056 | .0868 | .0870 |
| | | | 13 | .0376 | .0384 | .0450 | .0379 | .0379 |
| | | | 14 | .0147 | .0150 | .0174 | .0149 | .0149 |
| | | | 15 | .0068 | .0054 | .0062 | .0054 | .0054 |
| 500 | 10 | .25 | 7 | .8713 | .8673 | .9327 | .8463 | .8499 |
| | | | 8 | .5032 | .5060 | .6023 | .4971 | .4983 |
| | | | 9 | .1896 | .1954 | .2417 | .1940 | .1942 |
| | | | 10 | .0594 | .0578 | .0713 | .0577 | .0577 |
| | | | 11 | .0146 | .0145 | .0176 | .0145 | .0145 |
| | | | 12 | .0030 | .0033 | .0039 | .0033 | .0033 |
| | 20 | .25 | 11 | .6886 | .6870 | .7745 | .6561 | .6606 |
| | | | 12 | .4115 | .4110 | .4882 | .3993 | .4008 |
| | | | 13 | .1923 | .1996 | .2415 | .1968 | .1971 |
| | | | 14 | .0795 | .0829 | .0999 | .0824 | .0825 |
| | | | 15 | .0322 | .0307 | .0306 | .0306 | .0307 |

**Table 2.4:** Comparison of five approximations to $P(S_m \geq k)$ for i.i.d. binomial model

| $N$ | $m$ | $N$ | $p$ | $k$ | $\hat{P}(S_m \geq k)$ | (2.6) | (2.22) | (2.26) | (2.40) |
|---|---|---|---|---|---|---|---|---|---|
| 100 | 10 | 5 | .05 | 5 | .8791 | .8859 | .9263 | .7635 | .7923 |
| | | | | 6 | .6015 | .5983 | .6871 | .5365 | .5469 |
| | | | | 7 | .2920 | .2847 | .3486 | .2702 | .2721 |
| | | | | 8 | .1039 | .1008 | .1249 | .0990 | .0992 |
| | | | | 9 | .0290 | .0286 | .0350 | .0285 | .0285 |
| | | | | 10 | .0059 | .0069 | .0082 | .0069 | .0069 |
| | | | .10 | 9 | .7589 | .7624 | .8260 | .6669 | .6841 |
| | | | | 10 | .4972 | .4970 | .5772 | .4540 | .4602 |
| | | | | 11 | .2635 | .2577 | .3117 | .2459 | .2474 |
| | | | | 12 | .1164 | .1098 | .1340 | .1077 | .1079 |
| | | | | 13 | .0447 | .0399 | .0483 | .0397 | .0397 |
| | | | | 14 | .0124 | .0127 | .0152 | .0127 | .0127 |
| | | | | 15 | .0032 | .0036 | .0043 | .0036 | .0036 |
| | | 10 | .05 | 9 | .7788 | .7836 | .8443 | .6832 | .7020 |
| | | | | 10 | .5357 | .5342 | .6167 | .4846 | .4921 |
| | | | | 11 | .3014 | .2958 | .3572 | .2802 | .2822 |
| | | | | 12 | .1434 | .1369 | .1676 | .1335 | .1339 |
| | | | | 13 | .0598 | .0548 | .0667 | .0542 | .0543 |
| | | | | 14 | .0213 | .0195 | .0234 | .0194 | .0194 |
| | | | | 15 | .0048 | .0063 | .0074 | .0063 | .0063 |
| | 20 | 5 | .05 | 7 | .9040 | .9126 | .8921 | .6687 | .7171 |
| | | | | 8 | .7504 | .7569 | .7734 | .5848 | .6099 |
| | | | | 9 | .5491 | .5371 | .5839 | .4468 | .4574 |
| | | | | 10 | .3271 | .3227 | .3690 | .2888 | .2922 |
| | | | | 11 | .1769 | .1659 | .1945 | .1567 | .1575 |
| | | | | 12 | .0746 | .0744 | .0875 | .0725 | .0726 |
| | | | | 13 | .0278 | .0296 | .0345 | .0293 | .0293 |
| | | | | 14 | .0129 | .0106 | .0122 | .0106 | .0106 |
| | | | | 15 | .0027 | .0035 | .0039 | .0035 | .0035 |
| 500 | 10 | 5 | .05 | 7 | .8286 | .8290 | .9031 | .8082 | .8117 |
| | | | | 8 | .4268 | .4305 | .5163 | .4238 | .4247 |
| | | | | 9 | .1414 | .1432 | .1761 | .1424 | .1425 |
| | | | | 10 | .0355 | .0361 | .0439 | .0360 | .0360 |
| | | | | 11 | .0080 | .0076 | .0091 | .0076 | .0076 |
| | | | | 12 | .0014 | .0014 | .0016 | .0014 | .0014 |

**Table 2.5:** Comparison of six approximations to $P(k; m, N, a)$ for Bernoulli model for conditional case for $N = 100$

| $m$ | $a$ | $k$ | $\hat{P}(k; n, m, a)$ | (2.63) | (2.65) | (2.68) | (2.80) | (2.81) |
|---|---|---|---|---|---|---|---|---|
| 10 | 5 | 2 | .8941 | .7885 | .8355 | .6703 | .7207 | .6945 |
| | | 3 | .1678 | .1598 | .1988 | .1534 | .1573 | .1546 |
| | | 4 | .0098 | .0090 | .0110 | .0090 | .0090 | .0090 |
| | 10 | 3 | .8563 | .7665 | .8080 | .6583 | .7321 | .6779 |
| | | 4 | .2287 | .2210 | .2641 | .2095 | .2266 | .2115 |
| | | 5 | .0247 | .0261 | .0314 | .0260 | .0271 | .0260 |
| | | 6 | .0016 | .0016 | .0018 | .0016 | .0016 | .0016 |
| | 15 | 3 | .9997 | .9835 | .9786 | .8570 | .9566 | .8948 |
| | | 4 | .7316 | .6753 | .7237 | .5890 | .6690 | .6031 |
| | | 5 | .2031 | .1946 | .2302 | .1859 | .2057 | .1873 |
| | | 6 | .0261 | .0268 | .0318 | .0266 | .0284 | .0267 |
| | | 7 | .0018 | .0021 | .0024 | .0021 | .0021 | .0021 |
| | 20 | 4 | .9864 | .9445 | .9448 | .8141 | .9209 | .8426 |
| | | 5 | .5964 | .5441 | .5977 | .4857 | .5584 | .4946 |
| | | 6 | .1431 | .1402 | .1652 | .1356 | .1507 | .1363 |
| | | 7 | .0196 | .0187 | .0219 | .0186 | .0199 | .0186 |
| | | 8 | .0021 | .0014 | .0016 | .0014 | .0014 | .0014 |
| 20 | 5 | 3 | .5350 | .5167 | .5097 | .4155 | .4351 | .4278 |
| | | 4 | .0835 | .0824 | .0947 | .0795 | .0791 | .0798 |
| | | 5 | .0042 | .0043 | .0049 | .0043 | .0042 | .0043 |
| | 10 | 4 | .8714 | .8254 | .7487 | .6107 | .7183 | .6342 |
| | | 5 | .3465 | .3433 | .3563 | .2967 | .3358 | .3022 |
| | | 6 | .0752 | .0748 | .0844 | .0724 | .0785 | .0727 |
| | | 7 | .0102 | .0096 | .0108 | .0096 | .0100 | .0096 |
| | 15 | 5 | .9585 | .9218 | .8313 | .6744 | .8335 | .7037 |
| | | 6 | .5654 | .5496 | .5321 | .4417 | .5350 | .4532 |
| | | 7 | .0192 | .1875 | .2033 | .1729 | .2026 | .1746 |
| | | 8 | .0423 | .0414 | .0464 | .0407 | .0456 | .0407 |
| | | 9 | .0062 | .0062 | .0069 | .0062 | .0067 | .0062 |
| | 20 | 6 | .9820 | .9581 | .8694 | .7014 | .8885 | .7351 |
| | | 7 | .7079 | .6809 | .6360 | .5265 | .6614 | .5421 |
| | | 8 | .2948 | .2973 | .3101 | .2625 | .3229 | .2664 |
| | | 9 | .0882 | .0862 | .0955 | .0830 | .0981 | .0834 |
| | | 10 | .0163 | .0177 | .0197 | .0176 | .0199 | .0176 |
| | | 11 | .0037 | .0026 | .0029 | .0026 | .0029 | .0026 |

**Table 2.6:** Comparison of six approximations to $P(k; m, N, a)$ for Bernoulli model for conditional case for $N = 500$

| $m$ | $a$ | $k$ | $\hat{P}(k; n, m, a)$ | (2.63) | (2.65) | (2.68) | (2.80) | (2.81) |
|---|---|---|---|---|---|---|---|---|
| 10 | 25 | 3 | .7829 | .7221 | .8293 | .7025 | .7463 | .7063 |
|  |  | 4 | .1358 | .1314 | .1668 | .1306 | .1390 | .1307 |
|  |  | 5 | .0092 | .0096 | .0118 | .0096 | .0100 | .0096 |
|  | 50 | 4 | .8523 | .8119 | .8911 | .7902 | .8449 | .7940 |
|  |  | 5 | .2369 | .2314 | .2840 | .2292 | .2511 | .2295 |
|  |  | 6 | .0265 | .0270 | .0327 | .0270 | .0288 | .0270 |
|  |  | 7 | .0022 | .0018 | .0022 | .0018 | .0019 | .0018 |
|  | 75 | 4 | .9986 | .9974 | .9995 | .9909 | .9986 | .9928 |
|  |  | 5 | .8077 | .7686 | .8494 | .7487 | .8115 | .7519 |
|  |  | 6 | .2111 | .2215 | .2677 | .2196 | .2426 | .2198 |
|  |  | 7 | .0259 | .0280 | .0333 | .0279 | .0299 | .0279 |
|  |  | 8 | .0023 | .0020 | .0023 | .0020 | .0021 | .0020 |
|  | 100 | 5 | .9964 | .9879 | .9965 | .9757 | .9936 | .9784 |
|  |  | 6 | .6601 | .6535 | .7388 | .6383 | .7029 | .6405 |
|  |  | 7 | .1460 | .1598 | .1913 | .1588 | .1747 | .1590 |
|  |  | 8 | .0159 | .0180 | .0212 | .0180 | .0191 | .0180 |
|  |  | 9 | .0011 | .0011 | .0012 | .0011 | .0011 | .0011 |
| 20 | 25 | 4 | .6769 | .6204 | .7163 | .5898 | .6629 | .5953 |
|  |  | 5 | .1612 | .1623 | .2002 | .1599 | .1801 | .1603 |
|  |  | 6 | .0249 | .0233 | .0282 | .0233 | .0256 | .0233 |
|  |  | 7 | .0016 | .0023 | .0027 | .0023 | .0025 | .0023 |
|  | 50 | 6 | .6169 | .5675 | .6513 | .5425 | .6394 | .5465 |
|  |  | 7 | .1899 | .1822 | .2189 | .1793 | .2135 | .1797 |
|  |  | 8 | .0377 | .0372 | .0442 | .0371 | .0429 | .0371 |
|  |  | 9 | .0055 | .0056 | .0065 | .0056 | .0063 | .0056 |
|  | 75 | 7 | .8686 | .8105 | .8718 | .7672 | .8798 | .7744 |
|  |  | 8 | .4283 | .4022 | .4690 | .3893 | .4775 | .3911 |
|  |  | 9 | .1254 | .1213 | .1440 | .1201 | .1461 | .1203 |
|  |  | 10 | .0254 | .0258 | .0301 | .0258 | .0303 | .0258 |
|  |  | 11 | .0042 | .0042 | .0048 | .0042 | .0048 | .0042 |

**Table 2.7:** Comparison of six approximations to $P_p(k; m, N, a)$ for Poisson model for conditional case for $N = 100$

| $m$ | $a$ | $k$ | $\hat{P}_p(k; n, m, a)$ | (2.63) | (2.65) | (2.68) | (2.80) | (2.81) |
|---|---|---|---|---|---|---|---|---|
| 10 | 5 | 2 | .9075 | .7956 | .8454 | .6716 | .8418 | .7204 |
|  |  | 3 | .1943 | .1819 | .2413 | .1731 | .1962 | .1728 |
|  |  | 4 | .0137 | .0132 | .0177 | .0134 | .0172 | .0172 |
|  | 10 | 4 | .3060 | .2879 | .3496 | .2706 | .2929 | .2805 |
|  |  | 5 | .0514 | .0496 | .0662 | .0498 | .0503 | .0498 |
|  |  | 6 | .0055 | .0049 | .0063 | .0048 | .0055 | .0055 |
| 20 | 5 | 3 | .5578 | .5352 | .5287 | .4273 | .5096 | .4346 |
|  |  | 4 | .0970 | .0967 | .1115 | .0919 | .0894 | .0862 |
|  |  | 5 | .0058 | .0058 | .0061 | .0063 | .0049 | .0018 |
|  | 10 | 4 | .8876 | .8459 | .7667 | .6217 | .7732 | .7096 |
|  |  | 5 | .4082 | .3980 | .4095 | .3363 | .3881 | .3820 |
|  |  | 6 | .1070 | .1061 | .1212 | .1019 | .0891 | .0862 |
|  |  | 7 | .0176 | .0177 | .0205 | .0179 | .0232 | .0232 |
|  |  | 8 | .0019 | .0019 | .0020 | .0017 | .0024 | .0024 |

**Table 2.8:** Comparison of six approximations to $P_b(k; m, N, a)$ for binomial model for conditional case

| $N$ | $m$ | $n$ | $a$ | $k$ | $\hat{P}_b(k; n, m, a)$ | (2.63) | (2.65) | (2.68) | (2.80) | (2.81) |
|---|---|---|---|---|---|---|---|---|---|---|
| 100 | 10 | 5 | 50 | 10 | .4882 | .4490 | .5122 | .4117 | .4991 | .4989 |
| | | | | 11 | .2096 | .1973 | .2319 | .1913 | .2043 | .2043 |
| | | | | 12 | .0727 | .0739 | .0819 | .0688 | .0778 | .0778 |
| | | | | 13 | .0216 | .0212 | .0278 | .0209 | .0313 | .0313 |
| | | | | 14 | .0059 | .0054 | .0060 | .0055 | .0027 | .0027 |
| | | | 25 | 5 | .9672 | .9096 | .9185 | .7777 | .8953 | .8809 |
| | | | | 6 | .6247 | .5701 | .6308 | .5004 | .5780 | .5736 |
| | | | | 7 | .2331 | .2192 | .2585 | .2094 | .2276 | .2272 |
| | | | | 8 | .0625 | .0617 | .0748 | .0582 | .0778 | .0778 |
| | | | | 9 | .0126 | .0131 | .0158 | .0125 | .0154 | .0154 |
| | 20 | 5 | 50 | 14 | .8751 | .8455 | .7629 | .6287 | .8175 | .8169 |
| | | | | 15 | .6426 | .6232 | .5890 | .4874 | .6302 | .6299 |
| | | | | 16 | .3928 | .3849 | .3898 | .3280 | .4240 | .4240 |
| | | | | 17 | .2017 | .1964 | .2130 | .1814 | .2302 | .2302 |
| | | | | 18 | .0918 | .0902 | .0995 | .0872 | .1187 | .1187 |
| | | | | 19 | .0379 | .0376 | .0417 | .0378 | .0436 | .0436 |
| | | | | 20 | .0137 | .0132 | .0140 | .0133 | .0161 | .0161 |
| | | 10 | 50 | 14 | .8921 | .8591 | .7769 | .6366 | .8342 | .8335 |
| | | | | 15 | .6750 | .6516 | .6139 | .5066 | .6654 | .6653 |
| | | | | 16 | .4230 | .4111 | .4080 | .3487 | .4433 | .4433 |
| | | | | 17 | .2334 | .2293 | .2379 | .2097 | .2738 | .2738 |
| | | | | 18 | .1117 | .1108 | .1205 | .1041 | .1237 | .1237 |
| | | | | 19 | .0474 | .0475 | .0529 | .0467 | .0567 | .0567 |
| | | | | 20 | .0187 | .0185 | .0206 | .0177 | .0240 | .0240 |
| | | | | 21 | .0069 | .0073 | .0071 | .0072 | .0073 | .0073 |
| 500 | 20 | 5 | 100 | 9 | .8159 | .7418 | .8153 | .7064 | .8370 | .8369 |
| | | | | 10 | .4583 | .4172 | .4854 | .4032 | .4598 | .4598 |
| | | | | 11 | .1829 | .1646 | .1921 | .1682 | .1790 | .1790 |
| | | | | 12 | .0608 | .0513 | .0594 | .0537 | .0829 | .0829 |
| | | | | 13 | .0170 | .0146 | .0209 | .0143 | .0107 | .0096 |
| | | | | 14 | .0041 | .0031 | .0034 | .0029 | .0048 | .0048 |
| | 10 | 10 | 100 | 10 | .4682 | .3929 | .5027 | .4084 | .4755 | .4752 |
| | | | | 11 | .1945 | .1618 | .2261 | .1641 | .1949 | .1947 |
| | | | | 12 | .0651 | .0498 | .0666 | .0607 | .0741 | .0741 |
| | | | | 13 | .0183 | .0122 | .0181 | .0176 | .0284 | .0284 |
| | | | | 14 | .0052 | .0041 | .0048 | .0036 | .0039 | .0039 |

**Table 2.9:** Comparison of compound Poisson approximations to $P(\xi \geq l)$ for the Bernoulli model

| $N$ | $m$ | $p$ | $k$ | $l$ | $\hat{P}(\xi \geq l)$ | (2.31) | (2.35) | (2.39) | (2.40) |
|---|---|---|---|---|---|---|---|---|---|
| 100 | 10 | .05 | 2 | 2 | .7139 | .6520 | .7185 | .6836 | .5353 |
|     |    |     |   | 3 | .6834 | .5946 | .6371 | .6314 | .3018 |
|     |    |     |   | 4 | .6572 | .5407 | .5763 | .5824 | .1881 |
|     |    |     |   | 5 | .6114 | .4904 | .5230 | .5367 | .0957 |
|     |    |     | 3 | 2 | .1953 | .1834 | .1843 | .1857 | .1033 |
|     |    |     |   | 3 | .1614 | .1417 | .1424 | .1441 | .0189 |
|     |    |     |   | 4 | .1312 | .1094 | .1099 | .1118 | .0057 |
|     |    |     |   | 5 | .0976 | .0843 | .0847 | .0868 | .0010 |
|     |    |     | 4 | 2 | .0207 | .0213 | .0213 | .0213 | .0090 |
|     |    |     |   | 3 | .0154 | .0138 | .0138 | .0139 | .0003 |
|     |    |     |   | 4 | .0106 | .0090 | .0090 | .0090 | .0000 |
|     |    |     |   | 5 | .0066 | .0058 | .0058 | .0059 | .0000 |
|     | 20 | .05 | 4 | 2 | .1560 | .1586 | .1589 | .1594 | .0602 |
|     |    |     |   | 3 | .1413 | .1358 | .1360 | .1366 | .0087 |
|     |    |     |   | 4 | .1235 | .1163 | .1164 | .1170 | .0020 |
|     |    |     |   | 5 | .1060 | .0995 | .0997 | .1003 | .0003 |
| 500 | 10 | .01 | 2 | 2 | .3061 | .2752 | .2843 | .2871 | .2004 |
|     |    |     |   | 3 | .2675 | .2309 | .2359 | .2436 | .0489 |
|     |    |     |   | 4 | .2432 | .1935 | .1976 | .2067 | .0217 |
|     |    |     |   | 5 | .2145 | .1619 | .1654 | .1756 | .0046 |
|     |    | .05 | 4 | 2 | .1145 | .1111 | .1111 | .1113 | .0568 |
|     |    |     |   | 3 | .0825 | .0749 | .0749 | .0751 | .0070 |
|     |    |     |   | 4 | .0560 | .0504 | .0505 | .0507 | .0016 |
|     |    |     |   | 5 | .0384 | .0340 | .0340 | .0342 | .0002 |
|     | 20 | .05 | 5 | 2 | .1777 | .1851 | .1851 | .1852 | .0742 |
|     |    |     |   | 3 | .1443 | .1514 | .1514 | .1516 | .0117 |
|     |    |     |   | 4 | .1193 | .1238 | .1238 | .1240 | .0028 |
|     |    |     |   | 5 | .1068 | .1012 | .1012 | .1014 | .0004 |
|     |    |     | 6 | 2 | .0267 | .0297 | .0297 | .0297 | .0087 |
|     |    |     |   | 3 | .0210 | .0223 | .0223 | .0223 | .0003 |
|     |    |     |   | 4 | .0158 | .0167 | .0167 | .0167 | .0000 |
|     |    |     |   | 5 | .0150 | .0125 | .0125 | .0125 | .0000 |

**Table 2.10:** Comparison of five approximations to $P(S_{m,m} \geq k)$ for Bernoulli model

| $N$ | $m$ | $p$ | $k$ | $\hat{P}(S_{m,m} \geq k)$ | (2.96) | (2.99) | (2.100) | (2.106) |
|---|---|---|---|---|---|---|---|---|
| 25 | 5 | .05 | 6 | .1711 | .2206 | .4143 | .2310 | .2365 |
| | | | 7 | .0358 | .0386 | .0717 | .0408 | .0419 |
| | | | 8 | .0038 | .0050 | .0086 | .0054 | .0055 |
| | | .10 | 9 | .0940 | .1016 | .1827 | .1071 | .1115 |
| | | | 10 | .0206 | .0200 | .0342 | .0212 | .0223 |
| | | | 11 | .0037 | .0032 | .0051 | .0034 | .0036 |
| | 10 | .05 | 12 | .1570 | .1569 | .6652 | .1737 | .3443 |
| | | | 13 | .0704 | .0633 | .3126 | .0713 | .1428 |
| | | | 14 | .0259 | .0226 | .1118 | .0257 | .0505 |
| | | | 15 | .0104 | .0073 | .0343 | .0083 | .0160 |
| | | .10 | 20 | .0995 | .0865 | .3974 | .0971 | .1936 |
| | | | 21 | .0470 | .0394 | .1862 | .0445 | .0883 |
| | | | 22 | .0205 | .0166 | .0767 | .0188 | .0368 |
| | | | 23 | .0080 | .0065 | .0288 | .0074 | .0143 |
| 100 | 5 | .05 | 8 | .0959 | .1208 | .1653 | .1220 | .1079 |
| | | | 9 | .0130 | .0135 | .0174 | .0136 | .0123 |
| | | | 10 | .0012 | .0012 | .0016 | .0012 | .0011 |
| | | .10 | 11 | .0646 | .0789 | .1021 | .0797 | .0724 |
| | | | 12 | .0095 | .0110 | .0137 | .0111 | .0103 |
| | | | 13 | .0003 | .0013 | .0016 | .0013 | .0012 |
| | 10 | .05 | 16 | .0841 | .1383 | .2644 | .1399 | .1352 |
| | | | 17 | .0268 | .0398 | .0750 | .0402 | .0383 |
| | | | 18 | .0075 | .0102 | .0186 | .0103 | .0098 |
| | | | 19 | .0020 | .0024 | .0039 | .0024 | .0023 |
| | | .10 | 24 | .1018 | .1537 | .2798 | .1554 | .1511 |
| | | | 25 | .0377 | .0565 | .1025 | .0572 | .0550 |
| | | | 26 | .0130 | .0189 | .0335 | .0192 | .0183 |
| | | | 27 | .0057 | .0060 | .0103 | .0060 | .0056 |

**Table 2.11:** Comparison of four approximations to $P(S_{m,m} \geq k)$ for Poisson model

| $N$ | $m$ | $\theta$ | $k$ | $\hat{P}(S_{m,m} \geq k)$ | (2.96) | (2.99) | (2.100) |
|---|---|---|---|---|---|---|---|
| 25 | 5 | .25 | 14 | .5350 | .6382 | .8960 | .6526 |
|  |  |  | 15 | .2970 | .3556 | .5966 | .3698 |
|  |  |  | 16 | .1404 | .1626 | .2905 | .1708 |
|  |  |  | 17 | .0598 | .0649 | .1154 | .0686 |
|  |  |  | 18 | .0217 | .0236 | .0407 | .0250 |
|  |  |  | 19 | .0070 | .0080 | .0133 | .0084 |
|  |  |  | 20 | .0023 | .0025 | .0041 | .0027 |
| 25 | 5 | .50 | 26 | .1122 | .1241 | .2176 | .1306 |
|  |  |  | 27 | .0579 | .0603 | .1046 | .0637 |
|  |  |  | 28 | .0255 | .0277 | .0470 | .0293 |
|  |  |  | 29 | .0170 | .0122 | .0200 | .0129 |
|  |  |  | 30 | .0066 | .0051 | .0082 | .0054 |
| 100 | 5 | .50 | 30 | .1039 | .1233 | .1589 | .1246 |
|  |  |  | 31 | .0464 | .0520 | .0661 | .0525 |
|  |  |  | 32 | .0184 | .0207 | .0259 | .0209 |
|  |  |  | 33 | .0077 | .0079 | .0097 | .0080 |

**Table 2.12:** Comparison of four approximations to $P(S_{m,m} \geq k)$ for binomial$(5, p)$ model

| $N$ | $m$ | $p$ | $k$ | $P(\hat{S}_{m,m} \geq k)$ | (2.96) | (2.99) | (2.100) |
|---|---|---|---|---|---|---|---|
| 25 | 05 | .05 | 15 | .2420 | .2830 | .4853 | .2954 |
|  |  |  | 16 | .1068 | .1170 | .2077 | .1232 |
|  |  |  | 17 | .0383 | .0423 | .0734 | .0446 |
|  |  |  | 18 | .0132 | .0138 | .0232 | .0146 |
|  |  |  | 19 | .0052 | .0042 | .0067 | .0044 |
|  | 10 | .05 | 40 | .1259 | .1155 | .4991 | .1289 |
|  |  |  | 41 | .0814 | .0715 | .3267 | .0805 |
|  |  |  | 42 | .0495 | .0428 | .1983 | .0484 |
|  |  |  | 43 | .0297 | .0247 | .1135 | .0281 |
|  |  |  | 44 | .0200 | .0139 | .0622 | .0158 |
|  |  |  | 45 | .0100 | .0076 | .0329 | .0086 |
|  |  |  | 46 | .0064 | .0040 | .0169 | .0046 |
| 100 | 05 | .05 | 19 | .0843 | .1022 | .1326 | .1032 |
|  |  |  | 20 | .0252 | .0300 | .0383 | .0304 |
|  |  |  | 21 | .0071 | .0080 | .0098 | .0082 |
|  |  |  | 22 | .0014 | .0020 | .0022 | .0021 |
|  | 10 | .05 | 47 | .0862 | .1337 | .2396 | .1351 |
|  |  |  | 48 | .0476 | .0698 | .1248 | .0706 |
|  |  |  | 49 | .0249 | .0349 | .0617 | .0354 |
|  |  |  | 50 | .0119 | .0170 | .0291 | .0172 |

60          *Jie Chen and Joseph Glaz*

**Table 2.13:** A compound Poisson approximation to $P(\xi^* \geq l)$ for a Bernoulli model

| $N$ | $m$ | $p$ | $k$ | $l$ | $\check{P}(\xi^* \geq l)$ | (2.110) | $N$ | $m$ | $p$ | $k$ | $l$ | $\check{P}(\xi^* \geq l)$ | (2.110) |
|---|---|---|---|---|---|---|---|---|---|---|---|---|---|
| 25 | 5 | .05 | 6 | 2 | .1180 | .1637 | 100 | 5 | .05 | 7 | 2 | .3436 | .4203 |
| | | | | 3 | .0720 | .0807 | | | | | 3 | .2227 | .2528 |
| | | | | 4 | .0486 | .0327 | | | | | 4 | .1559 | .1429 |
| | | | | 5 | .0317 | .0131 | | | | | 5 | .0103 | .0778 |
| | | | 7 | 2 | .0168 | .0228 | | | | 8 | 2 | .0420 | .0516 |
| | | | | 3 | .0071 | .0076 | | | | | 3 | .0178 | .0158 |
| | | | | 4 | .0061 | .0017 | | | | | 4 | .0116 | .0039 |
| | | | | 5 | .0027 | .0003 | | | | | 5 | .0053 | .0010 |
| | | .10 | 9 | 2 | .0459 | .0600 | | | .10 | 10 | 2 | .1743 | .2064 |
| | | | | 3 | .0250 | .0217 | | | | | 3 | .0971 | .0906 |
| | | | | 4 | .0152 | .0062 | | | | | 4 | .0588 | .0369 |
| | | | | 5 | .0104 | .0017 | | | | | 5 | .0320 | .0145 |
| | | | 10 | 2 | .0090 | .0095 | | | | 11 | 2 | .0210 | .0272 |
| | | | | 3 | .0032 | .0025 | | | | | 3 | .0100 | .0065 |
| 25 | 10 | .05 | 12 | 2 | .1224 | .2912 | 100 | 10 | .05 | 15 | 2 | .1632 | .3071 |
| | | | | 3 | .0978 | .2026 | | | | | 3 | .1187 | .1942 |
| | | | | 4 | .0780 | .1200 | | | | | 4 | .0988 | .1092 |
| | | | | 5 | .0671 | .0644 | | | | | 5 | .0738 | .0586 |
| | | | 13 | 2 | .0500 | .1122 | | | | 16 | 2 | .0520 | .0936 |
| | | | | 3 | .0398 | .0678 | | | | | 3 | .0346 | .0478 |
| | | | | 4 | .0323 | .0323 | | | | | 4 | .0282 | .0197 |
| | | | | 5 | .0229 | .0127 | | | | | 5 | .0198 | .0071 |

**Table 2.14:** Approximations for the expected value and standard deviation of one-dimensional discrete scan statistic for $N = 100$

| | Simulation | | Product-Type | | Poisson | | Compound Poisson | |
|---|---|---|---|---|---|---|---|---|
| | $E(S_m^*)$ | $SD^*$ | $E(\hat{S}_m)$ | $\hat{SD}$ | $E(\hat{S}_m)$ | $\hat{SD}$ | $E(\hat{S}_m)$ | $\hat{SD}$ |
| p | Bernoulli $(p)$ model for $m = 20$ | | | | | | | |
| .05 | 2.5735 | 1.0175 | 2.5833 | 1.0323 | 1.9433 | 1.5754 | 2.4088 | 1.3130 |
| .10 | 4.2605 | 1.2371 | 4.2648 | 1.2340 | 3.7873 | 1.5509 | 4.2203 | 1.5039 |
| p | Binomial $(5,p)$ model for $m = 10$ | | | | | | | |
| .05 | 5.8734 | 1.2507 | 5.8951 | 1.2565 | 5.5639 | 1.5711 | 5.9899 | 1.3078 |
| .10 | 9.6003 | 1.5463 | 9.6134 | 1.5355 | 9.2047 | 1.9285 | 9.7719 | 1.5575 |
| $\theta$ | Poisson $(\theta)$ model for $m = 10$ | | | | | | | |
| .10 | 3.1962 | .9925 | 3.2173 | .9985 | 2.9534 | 1.2403 | 3.2290 | 1.0710 |
| .25 | 6.0051 | 1.3139 | 6.0002 | 1.3299 | 5.6594 | 1.6485 | 6.1008 | 1.3804 |
| a | Bernoulli model conditional on $\sum_{i=1}^{n} X_i = a$ for $m = 20$ | | | | | | | |
| 10 | 4.2962 | .8227 | 4.2504 | .8624 | 3.7172 | 1.3020 | 3.9890 | 1.2235 |
| 20 | 7.0983 | 1.0091 | 7.0427 | 1.0542 | 6.3180 | 1.7042 | 6.9212 | 1.3549 |

**Table 2.15:** Approximations for the expected value and standard deviation of one-dimensional discrete scan statistic for $N = 25$ and $m = 5$

| | Simulation | | Product-Type | | Poisson | | Compound Poisson | |
|---|---|---|---|---|---|---|---|---|
| | $E^*(S_{m,m})$ | $SD^*$ | $\hat{E}(S_{m,m})$ | $\hat{SD}$ | $\hat{E}(S_{m,m})$ | $\hat{SD}$ | $\hat{E}(S_{m,m})$ | $\hat{SD}$ |
| p | Bernoulli $(p)$ model | | | | | | | |
| .05 | 4.7072 | .9326 | 4.9714 | .8201 | 4.9951 | .8271 | 5.0281 | .8111 |
| .10 | 7.0991 | 1.0800 | 7.3211 | .9475 | 7.3490 | .9555 | 7.3961 | .9378 |
| p | Binomial $(5,p)$ model | | | | | | | |
| .05 | 13.5315 | 1.5710 | 13.8504 | 1.4043 | 13.8930 | 1.4170 | | |
| .10 | 22.1959 | 1.9725 | 22.5822 | 1.7540 | 22.6361 | 1.7698 | | |
| $\theta$ | Poisson $(\theta)$ model | | | | | | | |
| .25 | 13.7945 | 1.6514 | 14.1328 | 1.4790 | 14.1777 | 1.4926 | | |
| .50 | 22.8515 | 2.1594 | 23.3019 | 1.9238 | 23.3611 | 1.9415 | | |

# References


1. Aldous, D. (1989). *Probability Approximations via the Poisson Clumping Heuristic*, New York: Springer-Verlag.

2. Altschul, S. F. and Erickson, B. W. (1988). Significance levels for biological sequence comparison using non-linear similarity functions, *Bulletin of Mathematical Biology*, **50**, 77–92.

3. Arratia, R., Goldstein, L. and Gordon, L. (1989). Two moments suffice for Poisson approximations: The Chen-Stein method, *Annals of Applied Probability*, **17**, 9–25.

4. Arratia, R., Goldstein, L. and Gordon, L. (1990). Poisson approximation and the Chen-Stein method, *Statistical Science*, **5**, 403–434.

5. Arratia, R., Gordon, L. and Waterman, M. (1986). An extreme value theory for sequence matching, *Annals of Statistics*, **14**, 971–993.

6. Balakrishnan, N., Balasubramanian, K. and Viveros, R. (1993). On sampling inspection plans based on the theory of runs, *The Mathematical Scientist*, **18**, 113–126.

7. Balasubramanian, K., Viveros, R. and Balakrishnan, N. (1993). Sooner and later waiting time problems for Markovian Bernoulli trials, *Statistics & Probability Letters*, **18**, 153–161.

8. Banjevic, D. (1990). On order statistics in waiting time for runs in Markov chains, *Statistics & Probability Letters*, **9**, 125–127.

9. Barbour, A. D., Chryssaphinou, O. and Roos, M. (1995). Compound Poisson approximation in reliability theory, *IEEE Transactions on Reliability*, **44**, 398–402.

10. Barbour, A. D., Holst, L. and Janson, S. (1992). *Poisson Approximations*, Oxford, England: Oxford University Press.

11. Bogush, Jr., A. J. (1972). Correlated clutter and resultant properties of binary signals, *IEEE Transactions on Aerospace Electronic Systems*, **9**, 208–213.

12. Chao, M. T., Fu, J. C. and Koutras, M. V. (1995). Survey of reliability studies of consecutive-k-out-of-n: F and related systems, *IEEE Transactions on Reliability*, **44**, 120–127.

13. Chen, J. and Glaz, J. (1995). Two dimensional discrete scan statistics, *Technical Report No. 19*, Department of Statistics, University of Connecticut, Storrs, CT.

14. Chen, J. and Glaz, J. (1996). Two dimensional discrete scan statistics, *Statistics & Probability Letters*, **31**, 59–68.

15. Chen, J. and Glaz, J. (1997). Approximations and inequalities for the distribution of a scan statistic for 0-1 Bernoulli trials, In *Advances in the Theory and Practice of Statistics: A Volume in Honor of Samuel Kotz. Chapter 16* (Eds., N. L. Johnson and N. Balakrishnan), pp. 285–298 , New York: John Wiley & Sons.

16. Chen, J. and Glaz, J. (1997). Approximation for discrete scan statistics on the circle, *submitted for publication*.

17. Chryssaphinou, O. and Papastavridis, S. G. (1990). Limit distribution for a consecutive-k-out-of-n: F system, *Advances in Applied Probability*, **22**, 491–493.

18. Darling, R. W. R. and Waterman, M. S. (1986). Extreme value distributions for the largest cube in random lattice, *SIAM Journal of Applied Mathematics*, **46**, 118–132.

19. Fousler, D. E. and Karlin, S. (1987). Maximal success duration for a semi-Markov process, *Stochastic Processes and their Applications*, **24**, 203–224.

20. Fu, J. C. (1986). Reliability of consecutive-k-out-of-n: F system with (k-1)-step Markov dependence, *IEEE Transactions on Reliability*, **35**, 602–603.

21. Fu, J. C. and Hu, B. (1987). On reliability of a large consecutive-k-out-of-n: F stystem with (k-1)-step Markov dependence, *IEEE Transactions on Reliability*, **36**, 75–77.

22. Fu, J. C. and Koutras, M. V. (1994). Distribution theory of runs: A Markov chain approach, *Journal of the American Statistical Association*, **89**, 1050–1058.

23. Fu, J. C. and Koutras, M. V. (1994). Poisson approximation for 2-dimensional patterns, *Annals of the Institute of Statistical Mathematics*, **46**, 1979–1992.

24. Fu, Y. X. and Curnow, R. N. (1990). Locating a changed estimation of multiple change points, *Biometrika*, **77**, 295–304.

25. Glaz, J. (1983). Moving window detection for discrete data, *IEEE Transactions on Information Theory*, **29**, 457–462.

26. Glaz, J. (1995). Discrete scan statistics with applications to minefields detection, In *Proceedings of Conference SPIE*, **2765**, pp. 420–429, Orlando, FL.

27. Glaz, J. and Naus, J. (1983). Multiple cluster on the line, *Communications in Statistics—Theory and Methods*, **12**, 1961–1986.

28. Glaz, J. and Naus, J. I. (1991). Tight bounds and approximations for scan statistic probabilities for discrete data, *Annals of Applied Probability*, **1**, 306–318.

29. Glaz, J., Naus, J., Roos, M. and Wallenstein, S. (1994). Poisson approximations for the distribution and moments of ordered m-spacings, *Journal of Applied Probability*, **31**, 271–281.

30. Godbole, A. P. (1990). Specific formulae for some success runs distributions, *Statistics & Probability Letters*, **10**, 119–124.

31. Godbole, A. P. (1991). Poisson approximations for runs and patterns of rare events, *Advances in Applied Probability*, **23**, 851–865.

32. Godbole A. P. (1993). Approximate reliabilities of m-consecutive-k-out-of-n failure systems, *Statistica Sinica*, **3**, 321–327.

33. Goldstein, L. and Waterman, M. S. (1992). Poisson, compound Poisson and process approximations for testing statistical significance in sequence comparisons, *Bulletin of Mathematical Biology*, **54**, 785–812.

34. Gordon, L., Schilling, M. F. and Waterman, M. S. (1986). An extreme value theory for long head runs, *Probability Theory & Related Fields*, **72**, 279–288.

35. Gotoh, O. (1990). Optimal sequence alignments, *Bulletin of Mathematical Biology*, **52**, 509–525.

36. Greenberg, I. (1970). On sums of random variables defined on a two-state Markov chain, *Journal of Applied Probability*, **13**, 604–607.

37. Hirano, K. and Aki, S. (1993). On number of occurrences of success runs of specified length in a two-state Markov chain, *Statistica Sinica*, **3**, 313–320.

38. Karlin, S., Blaisdell, B., Mocarski, E. and Brendel, V. (1989). A method to identify distinctive charge configurations in protein sequences with applications to human Herpesvirus polypeptides, *Journal of Molecular Biology*, **205**, 165–177.

39. Karlin, S. and Ost, F. (1987). Counts of long aligned word matches among random letter sequences, *Advances in Applied Probability*, **19**, 293–351.

40. Karwe, V. and Naus, J. (1997). New recursive methods for scan statistic probabilities, *Computational Statistics & Data Analysis*, **23**, 389–404.

41. Koutras, M. V. and Alexandrou V. A. (1996). Runs, scans and urn model distributions: A unified Markov chain approach, *Annals of the Institute of Statistical Mathematics*, **47**, 743–766.

42. Koutras, M. V. and Alexandrou V. A. (1997). Non-parametric randomness test based on success runs of fixed length, *Statistics & Probability Letters*, **32**, 393–404.

43. Koutras, M. V. and Papastavridis, S. G. (1993). *New Trends in System Reliability Evaluation*, Elsevier Science Publ. B. V. pp. 228–248.

44. Koutras, M. V., Papadopoulos, G. K. and Papastavridis, S. G. (1993). Reliability of 2-dimensional consecutive-k-out-of-n: F systems, *IEEE Transactions on Reliability*, **42**, 658–661.

45. Krauth, J. (1992). Bounds for the upper-tail probabilities of the circular ratchet scan statistic, *Biometrics*, **48**, 1177–1185.

46. Lou, W. Y. W. (1997). An application of the method of finite Markov-chain into runs tests, *Statistics & Probability Letters*, **31**, 155–161.

47. Mosteller, F. (1941). Note on an application of runs to quality control charts, *Annals of Mathematical Statistics*, **12**, 228–232.

48. Mott, R. F., Kirkwood, T. B. L. and Curnow, R. N. (1990). An accurate approximation to the distribution of the length of longest matching word between two random DNA sequences, *Bulletin of Mathematical Biology*, **52**, 773–784

49. Naus, J. I. (1974). Probabilities for a generalized birthday problem, *Journal of the American Statistical Association*, **69**, 810–815.

50. Naus, J. I. (1982). Approximations for distributions of scan statistics, *Journal of the American Statistical Association*, **77**, 377–385.

51. Naus, J. I. and Sheng, K. N. (1996). Screening for unusual matched segments in multiple protein sequences, *Communications in Statistics— Simulation and Computation*, **25**, 937–952.

52. Naus, J. I. and Sheng, K. N. (1997). Matching among multiple random sequences, *Bulletin of Mathematical Biology*, **59**, 483–496.

53. Nelson, J. B. (1978). Minimal order models for false alarm calculations on sliding windows, *IEEE Transactions on Aerospace and Electronic System*, **15**, 352–363.

54. Patefield, W. M. (1981). An efficient method of generating random $R \times C$ tables with given row and column totals, *Applied Statistics*, **30**, 91–97.

55. Philippou, A. N. and Makri, F. S. (1986). Successes, runs and longest runs, *Statistics & Probability Letters*, **4**, 211–215.

56. Roos, M. (1993a). Compound Poisson approximations for the number of extreme spacings, *Advances in Applied Probability*, **25**, 847–874.

57. Roos, M. (1993b). Stein–Chen Method for compound Poisson Approximation, *Ph.D. Dissertation*, University of Zurich, Zurich, Switzerland.

58. Roos, M. (1994). Stein's method for compound Poisson approximation, *Annals of Applied Probability*, **4**, 1177–1187.

59. Saperstein, B. (1972). The generalized birthday problem, *Journal of the American Statistical Association*, **67**, 425–428.

60. Schwager, S. J. (1983). Run probabilities in sequences of Markov-dependent trials, *Journal of the American Statistical Association*, **78**, 168–175.

61. Sheng, K. N. and Naus, J. I. (1994). Pattern matching between two non-aligned random sequences, *Bulletin of Mathematical Biology*, **56**, 1143–1162.

62. Sheng, K. N. and Naus, J. I. (1996). Matching rectangles in 2-dimensions, *Statistics & Probability Letters*, **26**, 83–90.

63. Viveros, R. and Balakrishnan, N. (1993). Statistical inference from start-up demonstration test data, *Journal of Quality Technology*, **25**, 119–130.

64. Wallenstein, S., Naus, J. and Glaz J. (1994). Power of the scan statistic in detecting a changed segment in a Bernoulli sequence, *Biometrika*, **81**, 595–601.

65. Wallenstein, S. and Neff, N. (1987). An approximation for the distribution of the scan statistic, *Statistics in Medicine*, **6**, 197–207.

66. Wallenstein, S., Weinberg, C. R. and Gould, M. (1989). Testing for a pulse in seasonal event data, *Biometrics*, **45**, 817–830.

67. Waterman, M. S. (1995). *Introduction to Computational Biology*, London, England: Chapman & Hall.

# 3

## Ratchet Scan and Disjoint Statistics

**Joachim Krauth**

*Düsseldorf University, Düsseldorf, Germany*

**Abstract:** A general definition of ratchet scan and disjoint statistics is given. The known results for the disjoint statistic, the linear ratchet scan statistic, and the circular ratchet scan statistic are reviewed. This concerns the exact and asymptotic distributions as well as exact bounds for the upper tail probabilities of the test statistics under the null hypothesis of no clustering. Further, results concerning the power of the tests in comparison with other tests for clustering are reported. In addition, certain modifications and extensions, e.g., the EMM procedure, the Grimson models, and the test of Hewitt *et al.* (1971), are studied. Finally, a general approach to derive exact upper and lower bounds for the tail probabilities of the general ratchet scan statistic is described.

**Keywords and phrases:** Ratchet scan statistic, disjoint statistic, EMM procedure, Hewitt's test, bounds for tail probabilities

## 3.1  Introduction

The one-dimensional scan statistic for the continuous case in the conditional situation has been used by many authors to test for clusters in time in epidemiological data. This statistic is defined as the maximum number of events within a window of given length which is moved along the time axis. If this statistic is used, the exact time of occurrence for each event has to be known. This is no longer the case if only the number of events for certain disjoint time intervals, e.g., months or years, is available. Then, the scan statistic cannot be calculated and a discrete version of it, which was named *ratchet scan statistic* by Wallenstein, Weinberg, and Gould (1989b), has to be used. For the special case that the window has only the length of one of the given disjoint time intervals, Naus (1966) introduced the term disjoint test. Other authors used the term EMM, or Ederer–Myers–Mantel, procedure for both, the general and the special case, because this kind of statistic seems to have been studied by Ederer, Myers, and Mantel (1964) for the first time. The present statistic must not be confused with the scan statistics for the discrete case in the conditional situation, which are considered elsewhere in this volume.

In a more formal and more general way, we may introduce ratchet scan and disjoint statistics as follows:

Let $(N_1, \ldots, N_c)$ be a random vector which is multinomially distributed with parameters $N, p_1, \ldots, p_c$. Here, $N$ and $c$ with $1 \leq N < \infty$, $2 \leq c < \infty$, are integers, and the $p_i$ with $0 < p_i < 1$, $1 \leq i \leq c$, $p_1 + \ldots + p_c = 1$, are interpreted as probabilities assigned to certain cells $C_1, \ldots, C_c$.

It is assumed that a neighborhood structure exists for the cells $C_1, \ldots, C_c$, i.e., for each pair of cells, it is known whether or not the cells are neighbors. Two nonempty subsets of cells are called *isomorphic* if they contain the same number of cells and if there exists a one-to-one mapping from one subset to the other by which pairs of neighbors are transformed into pairs of neighbors and pairs of non-neighbors into pairs of non-neighbors. The two subsets may not be disjoint. A set of cells is called *connected* if no partition of the set into two nonempty subsets exists such that the cells of one subset have no neighbor in the other subset. If each cell has exactly two neighbors and if all $c$ cells are connected, the cells are said to form a *circle*. The cells are said to form a *line* if all $c$ cells are connected, if $c - 2$ of the cells have exactly two neighbors, and if each of the two remaining cells has exactly one neighbor. If at least one cell has at least three neighbors, the structure is said to have two or more dimensions.

A *window* is a connected subset of $m$ cells, $1 \leq m < c$. We consider the set of all possible subsets of $m$ cells of the $c$ cells which are isomorphic to the window and assume that each of the $c$ cells is contained in at least one of these subsets. For each subset, the sum of the $N_i$'s corresponding to the cells in this subset is calculated. The most important statistic to be considered here is the maximum of the sums in the set of subsets which are isomorphic to the window. This statistic is denoted by $M(m)$. Of interest are the exact and asymptotic distributions of $M(m)$, as well as approximations and bounds for the tail probabilities.

The statistic $M(m)$ for the circle is called *ratchet circular scan statistic* by Wallenstein, Weinberg, and Gould (1989b), the statistic $M(m)$ for the line is called *linear ratchet scan statistic* by Krauth (1992b), while the test based on the statistic $M(1)$ is called *disjoint test* by Naus (1966).

## 3.2 Disjoint Statistic

### 3.2.1 Definition

Consider $N$ independent identically distributed $d$-dimensional random vectors $X_1, \ldots, X_N$ with a distribution that is concentrated on a bounded Borel set $B$ in the $d$-dimensional Euclidean space. Under the null hypothesis $H_0$, we assume a uniform distribution, i.e., a distribution with constant density over $B$. This null hypothesis is to be tested against clustering alternatives. We assume a dissection of $B$ into $c \geq 2$ disjoint Borel subsets or cells $C_1, \ldots, C_c$, of which the probabilities $p_1, \ldots, p_c$, with $p_1 + \cdots + p_c = 1$, are known under $H_0$. Let $N_i$ denote the number of random vectors observed in cell $C_i$, $i = 1, \ldots, c$. Then, the *disjoint statistic* is defined by

$$M(1) = \max_{1 \leq i \leq c} N_i.$$

The distribution of the $c$ frequencies $N_1, \ldots, N_c$ under $H_0$ is given by the multinomial distribution

$$P\left( \bigcap_{i=1}^{c} \{N_i = n_i\} \right) = N! \prod_{i=1}^{c} \left( p_i^{n_i} / n_i! \right),$$

$$0 \leq n_i \leq N \ (i = 1, \ldots, c), \ \sum_{i=1}^{c} n_i = N.$$

Under $H_0$, most authors assume the equiprobable case with

$$p_1 = \cdots = p_c = 1/c.$$

Results for this case are of a simpler form and can be derived in an easier way than results for the general case with cell probabilities $p_1, \ldots, p_c$ which may be unequal.

### 3.2.2 Results for the equiprobable case

**Exact distribution**

Barton and David (1959) have given the formula for the cumulative distribution function of $M(1)$ as

$$P(M(1) \leq x) = c^{-N} \sum \left( N! / \prod_{i=1}^{c} n_i! \right),$$

where the summation is over $c$-compositions of $N$ with no component greater than $x$. Here and in the following, we assume without loss of generality $x$ to be integer-valued. They have stated that $P(M(1) \leq x)$ is the coefficient of $t^N$ in the expansion of

$$N! c^{-N} \left( \sum_{i=0}^{x} (t^i / i!) \right)^c .$$

This was also observed by Good (1957).

A FORTRAN algorithm for calculating $P(M(1) \geq x)$ is presented by Freeman (1979), while Grimson (1993) has derived the explicit formula

$$P(M(1) \geq x)$$
$$= \; c^{-N} \sum_{k \geq 1} (-1)^{k+1} \binom{c}{k} \sum_{j_1 + \ldots + j_k = x}^{N} (c-k)^{N - j_1 - \ldots - j_k} \frac{N!}{j_1! \ldots j_k!} .$$

A table of the exact distribution of $M(1)$ has been given by Ederer, Myers, and Mantel (1964) for $c = 5$ and $N = 2(1)15$.


## Approximations

Barton and David (1959) and David and Barton (1962, p. 239) have derived the approximation

$$P(M(1) \geq x) \doteq 1 - \exp\left( -c G_b\left( x; N, \frac{1}{c} \right) \right)$$

with the binomial tail $G(x; N, p)$.

Good (1957) has discussed the usage of saddle-point methods to approximate $P(M(1) \leq x)$.

Johnson and Young (1960) have derived three normal approximations for the upper percentage points of $NM(1)$. Using a normal approximation of the multinomial distribution and the upper Bonferroni bound of degree one, Kozelka (1956) has derived approximate upper percentage points for $M(1)$ of the form

$$x_\alpha \doteq \frac{N}{c} + \sqrt{\frac{(c-1)N}{c^2}} \, \Phi^{-1}\left( 1 - \frac{\alpha}{c} \right),$$

where $\Phi$ denotes the cumulative distribution function of the standard normal distribution. This approximation is identical to the third approximation of Johnson and Young (1960) if 0.5 is added to the right side of Kozelka's approximation as a correction for continuity.

Viktorova and Sevastyanov (1967) have derived the asymptotic distribution of $M(1)$ for $N, c \to \infty$ for the situations with $N/(c \ln c) \to 0$ and $N/(c \ln c) \to b > 0$, while Viktorova (1969) has considered the remaining situation with $N/(c \ln c) \to \infty$.

## Bounds

The computation of the exact distribution of the disjoint statistic is only feasible for small values of $N$ and $c$, while it is not known how good the result of an approximation or simulation will be in a concrete situation. Therefore, close exact bounds for $P(M(1) \geq x)$ which are easy to compute have advantages over the other approaches when statistical tests have to be performed.

By using Bonferroni's inequalities and the fact that $N_1, \ldots, N_c$ are nonpositively correlated, Mallows (1968) has derived the bounds

$$1 - \exp\left(-cG_b\left(x; N, \frac{1}{c}\right)\right) \leq 1 - \left(1 - G_b\left(x; N, \frac{1}{c}\right)\right)^c$$
$$\leq P(M(1) \geq x) \leq cG_b\left(x; N, \frac{1}{c}\right).$$

By similar arguments, Yusas (1972) has obtained

$$cG_b\left(x; N, \frac{1}{c}\right) - \frac{1}{2}c(c-1)\left(G_b\left(x; N, \frac{1}{c}\right)\right)^2$$
$$\leq P(M(1) \geq x) \leq cG_b\left(x; N, \frac{1}{c}\right).$$

Krauth (1991) has derived upper and lower bounds of degrees one and two for $P(M(1) \geq x)$. The upper bounds are given by

$$U_1 = \min\{1, cq_1\},$$
$$U_2 = \min\{1, cq_1 - (c-1)q_{12}\},$$

while the lower bounds are given by

$$L_1 = \frac{c}{k(k+1)}(2kq_1 - (c-1)q_1^2) \text{ with } k = 1 + \lfloor(c-1)q_1\rfloor,$$
$$L_2 = \frac{c}{k(k+1)}(2kq_1 - (c-1)q_{12}) \text{ with } k = 1 + \lfloor(c-1)q_{12}/q_1\rfloor.$$

Here, $\lfloor x \rfloor$ denotes the integer part of $x$ and

$$q_1 = G_b\left(x; N, \frac{1}{c}\right),$$
$$q_{12} = \sum_{s=x}^{N-x}\sum_{t=x}^{N-s} \frac{N!}{s!t!(N-s-t)!}\left(\frac{1}{c}\right)^{s+t}\left(1-\frac{2}{c}\right)^{N-s-t}$$

with $q_{12} = 0$ for $2x > N$. The bound $U_1$ is the best linear upper bound of degree one, the bound $U_2$ is the best linear upper bound of degree two, while the bound $L_2$ is the best linear lower bound of degree two. The inequality

$$L_1 \leq L_2 \leq P(M(1) \geq x) \leq U_2 \leq U_1$$

holds. As noted by Krauth (1996b), for $2x > N$ we get the exact result

$$
\begin{aligned}
P(M(1) \geq x) &= 1 && \text{for } x \geq 1 \\
&= cq_1 && \text{for } 2x > N \\
&= 0 && \text{for } x > N.
\end{aligned}
$$

The lower bound of Yusas (1972) is identical to $L_1$ if $k = 1$. For $k > 1$, the bound $L_1$ is better than Yusas' bound. Likewise, the classical lower Bonferroni bound of degree two, which is used by David and Barton (1962, pp. 238–240), is identical to $L_2$ if $k = 1$. For $k > 1$, the bound $L_2$ is closer than David and Barton's bound. The bound $U_1$ is identical to the upper bounds used by David and Barton, Mallows, and Yusas.

Considering the example used in Mallows (1968) with $N = 500$, $c = 50$, and $x = 20$, we find 0.1447 and 0.1449 for Mallows' lower bounds, 0.1443 for both Yusas' lower bound and $L_1$, 0.1470 for both David and Barton's lower bound and $L_2$, 0.1563 for $U_1$, and 0.1559 for $U_2$. In a situation with a tail probability of 0.05 or smaller, the bounds would have been considerably closer than in our case with a tail probability of about 0.15.

### 3.2.3   Results for the general case

In most applications, one has to assume that the assumption of equal cell probabilities under $H_0$ is not justified. Months as well as years differ in length and occasionally it is necessary to choose the probabilities in proportion to the total number of subjects in a cell.

**Exact distribution**

Good (1957) and Levin (1981) have noted that $P(M(1) \leq x)$ is the coefficient of $t^N$ in the expansion of

$$
N! \prod_{i=1}^{c} \sum_{j=0}^{x} \left( (p_i t)^j / j! \right).
$$

Levin (1981, 1983) has proposed to compute the cumulative distribution function of $M(1)$ by means of the relation

$$
P(M(1) \leq x) = \frac{N!}{s^N \exp(-s)} \left\{ \prod_{i=1}^{c} P(X_i \leq x) \right\} P\left( \sum_{j=1}^{c} Y_j = N \right), \qquad (3.1)
$$

where $s$ is an arbitrary positive real number, $X_i$'s are independent Poisson random variables with mean $sp_i$, for $i = 1, \ldots, c$, and $Y_j$'s are independent truncated Poisson random variables with mean $sp_j$ and range $0, 1, \ldots, x$, for $j = 1, \ldots, c$. For $s$, Levin (1981, 1983) has recommended to choose $s = N$.

## Approximations

Levin (1981, 1983) has proposed normal approximations of $P(M(1) \leq x)$ based on the formula given in (3.1). These approximations are of the form

$$P\left(\sum_{j=1}^{c} Y_j = N\right) \doteq f\left(\frac{N - \sum_{j=1}^{c} \mu_j}{\sqrt{\sum_{j=1}^{c} \sigma_j^2}}\right) \frac{1}{\sqrt{\sum_{j=1}^{c} \sigma_j^2}}$$

for the last term of the formula in (3.1). Here, $\mu_i = E[Y_i]$, $\sigma_i^2 = V[Y_i]$, for $i = 1, \ldots, c$, and $f(x)$ denotes an Edgeworth expansion, where the first order term is the density function of the standard normal distribution.

## Bounds

We use the following notation:

$$\begin{aligned}
q_i &= G_b(x; N; p_i) \text{ for } i = 1, \ldots, c, \\
q_{ij} &= 0 \text{ for } 2x > N, \\
q_{ij} &= \sum_{s=x}^{N-x} \sum_{t=x}^{N-s} \frac{N!}{s!t!(N-s-t)!} p_i^s p_j^t (1 - p_i - p_j)^{N-s-t} \\
&\qquad \text{for } i \neq j, \ i, j = 1, \ldots, c, \ 2x \leq N, \\
S_1 &= \sum_{i=1}^{c} q_i, \quad S_2 = \sum_{i=2}^{c} \sum_{j=1}^{i-1} q_{ij}, \quad S_2^\star = \sum_{i=2}^{c} \sum_{j=1}^{i-1} q_i q_j.
\end{aligned}$$

Mallows (1968) has shown that

$$1 - \exp(-S_1) \leq 1 - \prod_{i=1}^{s}(1 - q_i) \leq P\left(M(1) \geq x\right) \leq S_1$$

holds, while Yusas (1972) has proved that

$$S_1 - S_2^\star \leq P(M(1) \geq x) \leq S_1.$$

Krauth (1991) has derived the upper bounds

$$U_1 = \min\{1, S_1\}$$

of degree one and

$$U_2 = \min\left\{1, S_1 - \max_{1 \leq s \leq c} \sum_{\substack{i=1 \\ i \neq s}}^{c} q_{is}\right\}$$

of degree two, and the lower bounds

$$L_1 = \frac{2}{k(k+1)} (kS_1 - S_2^\star) \text{ with } k = 1 + \lfloor 2S_2^\star/S_1 \rfloor$$

of degree one and

$$L_2 = \frac{2}{k(k+1)} \left(kS_1 - S_2\right) \text{ with } k = 1 + \lfloor 2S_2/S_1 \rfloor$$

of degree two.

For these bounds, we have

$$L_1 \leq L_2 \leq P(M(1) \geq x) \leq U_2 \leq U_1.$$

Here, $U_1$ is the best linear upper bound of degree one, while $L_2$ is the best linear lower bound of degree two, if only the values of $S_1$ and $S_2$ but not the probabilities $q_i$ and $q_{ij}$ are known.

The upper bound $U_1$ is identical to the upper bound proposed by Mallows and Yusas. If we assume $k = 1$ for the lower bound $L_1$, it is identical to Yusas' lower bound, i.e., $L_1$ is always at least as close as Yusas' bound.

For $2x > N$, we have the exact result

$$P(M(1) \geq x) = S_1.$$

### 3.2.4  Power

Naus (1966) has compared the scan statistic and the disjoint statistic with respect to power against clustering alternatives. For accomplishing this, assume that $X_1, \ldots, X_N$ are independent identically distributed random variables which take only values in the unit interval $(0,1)$ with a cumulative distribution function $F(x)$ with a continuous density function $f(x)$. Under $H_0$, let $F(x) = x$ for $x \in (0, 1)$. The cells $C_1, \ldots, C_c$ for the disjoint statistic are defined by dissecting the unit interval into $c$ equal-sized intervals. Thus, under $H_0$, the same probabilitiy $(1/c)$ corresponds to each of these cells, i.e., we are in the equiprobable case.

Naus (1966) has derived

$$P(M(1) \geq x) = \binom{N}{x} \left(\frac{1}{c}\right)^{x-1} \int_0^1 (f(t))^x \, dt + o\left(\left(\frac{1}{c}\right)^{x-1}\right).$$

For comparing the two tests, Naus has determined a randomization probability $g$ for which the levels of the scan test and the disjoint test are the same. This yields the level

$$g\binom{N}{x} \left(\frac{1}{c}\right)^{x-1} + o\left(\left(\frac{1}{c}\right)^{x-1}\right) + \binom{N}{x+1}\left(\frac{1}{c}\right)^x + o\left(\left(\frac{1}{c}\right)^x\right)$$

and the power

$$g\binom{N}{x}\left(\frac{1}{c}\right)^{x-1} \int_0^1 (f(t))^x \, dt + o\left(\left(\frac{1}{c}\right)^{x-1}\right)$$

$$+ \binom{N}{x+1}\left(\frac{1}{c}\right)^x \int_0^1 (f(t))^{x+1} \, dt + o\left(\left(\frac{1}{c}\right)^x\right)$$

for the randomized disjoint test.

In his Theorem I Naus (1966) has proved that, for $c$ sufficiently large, the scan test is more powerful than the disjoint test against all alternative hypotheses with continuous density function.

In particular, Naus (1966) has considered the alternative with $F(x) = x^2$ for $x \in (0, 1)$. In this case, the power of the disjoint test is given by

$$(N-1)\left(\frac{1}{c}\right)^{2N-1}\sum_{i=1}^{c}(2i-1)^{N-1} + \left(1 - (N-1)\frac{1}{c}\right)\left(\frac{1}{c}\right)^{2N}\sum_{i=1}^{c}(2i-1)^{N}$$

which reduces to

$$\frac{1}{3c}\left(7 - \frac{4}{c} - \left(\frac{1}{c}\right)^2 + \left(\frac{1}{c}\right)^3\right)$$

for $N = 2$. In the latter case, the scan test is more powerful than the disjoint test. This is not necessarily true for $N > 2$. From Table 1 in Naus (1966), we learn that the disjoint test is more powerful, e.g., for $N = 6$, $c = 2$; $N = 7$, $c = 2, 3$; $N = 8$, $c = 2, 3$; $N = 9$, $c = 2, 3, 4$; $N = 10$, $c = 2, 3, 4, 5$. Generally speaking, the disjoint test is more powerful than the scan test of the same level for $N$ large relative to $c$.

Naus (1966) has also considered the generalizations of the disjoint test and scan test to more than one dimension by considering the unit $k$-dimensional cube. To define the disjoint statistic, this cube is dissected into $c$ equal-sized rectangular solids. The power of the disjoint test is then given by

$$\left((x+1)^k - 1\right)\binom{N}{x+1}\left(\frac{1}{c}\right)^x \int_0^1 \cdots \int_0^1 (f(t_1, \ldots, t_k))^x \, dt_1 \ldots dt_k$$

$$+ \binom{N}{x+1}\left(\frac{1}{c}\right)^x \int_0^1 \cdots \int_0^1 (f(t_1, \ldots, t_k))^{x+1} \, dt_1 \ldots dt_k + o\left(\left(\frac{1}{c}\right)^x\right).$$

In his Theorem II Naus (1966) has proved that, for $c$ sufficiently large, the scan test is more powerful than the disjoint test against all alternative hypotheses with continuous density functions on the unit $k$-dimensional cube.

Yusas (1972) has compared the disjoint test with the corresponding chi-square test with respect to power in the case that the alternative does not differ very much from the null hypothesis of equal probabilities, i.e., for

$$\rho^2 = \sum_{i=1}^{c}\left(p_i - \frac{1}{c}\right)^2 = o\left(\frac{1}{N}\right).$$

In this situation, Yusas (1972) has shown that

$$f(p_1, \ldots, p_c) \leq \alpha + \frac{1}{2}\frac{c}{c-1}z\left(1 - \frac{\alpha}{c}\right)\frac{1}{\sqrt{2\pi}}\exp\left(-\frac{1}{2}z^2\left(1 - \frac{\alpha}{c}\right)\right)$$

$$\times Nc\rho^2(1 + o(1)),$$

where $f$ denotes the power function of the disjoint test and $z(1 - \frac{\alpha}{c})$ is the quantile of the standard normal distribution at level $(1 - \frac{\alpha}{c})$. Yusas (1972) has proved

$$\frac{f(p_1, \ldots, p_c) - \alpha}{g(p_1, \ldots, p_c) - \alpha} \leq \frac{\chi^2\left(1 - \frac{2\alpha}{c}; 1\right)}{\chi^2(1 - \alpha; c - 1)}(1 + o(1)),$$

where $g$ denotes the power function of the chi-square test and $\chi^2(1 - \alpha; c)$ is the $(1 - \alpha)$th quantile of the chi-square distribution with $c$ degrees of freedom. For $c = 2$, the two tests are identical, i.e., we have $f(p_1, \ldots, p_c) \equiv g(p_1, \ldots, p_c)$. Then, the right side of the inequality becomes $(1 + o(1))$. If $c \geq 2$ and $0 < \alpha < 0.5$, the right side of the inequality is smaller than one and is monotonically decreasing with decreasing $\alpha$ or increasing $c$.

### 3.2.5   Modifications and extensions

**EMM statistic**

The EMM procedure is proposed in Ederer, Myers, and Mantel (1964) as a test for clusters in space and time with applications in epidemiology. The first step is the dissection of the geographical area $A$ of interest into $d$ subareas $A_1, \ldots, A_d$ of equal size. In the second step, the observation period $T$ is dissected into $c$ subperiods $T_1, \ldots, T_c$ of equal length. From this, altogether $d \times c$ three-dimensional temporal–spatial cells $C_{ij} = A_i \cap T_j$ result, for $i = 1, \ldots, d$, $j = 1, \ldots, c$, and for each cell the number of cases $(N_{ij})$ is recorded.

In the next step, the maximum of the numbers of cases is determined for those cells which belong to the same subarea and differ only with respect to the subperiods:

$$M_i(1) = \max_{1 \leq j \leq c} N_{ij} \text{ for } i = 1, \ldots, d.$$

In other words, for a fixed subarea $A_i$ the disjoint statistic is calculated with respect to the $c$ time periods $T_1, \ldots, T_c$. Ederer, Myers, and Mantel (1964) have derived the conditional distribution of the disjoint statistic $M_i(1)$ given the number $W_i = N_{i1} + \cdots + N_{ic}$ of cases in the considered subarea $A_i$ under the null hypothesis of equal probabilities (for $c = 5$ and $W_i = 2(1)15$). For this distribution, the mean $\mathrm{E}[M_i(1) \mid W_i]$ and the variance $\mathrm{V}[M_i(1) \mid W_i]$, for $i = 1, \ldots, d$, are calculated. Next, Ederer, Myers, and Mantel (1964) have considered the sum

$$M_1 = \sum_{i=1}^{d} M_i(1)$$

with

$$\mathrm{E}[M_1] = \sum_{i=1}^{d} \mathrm{E}[M_i(1) \mid W_i], \ \mathrm{V}[M_1] = \sum_{i=1}^{d} \mathrm{V}[M_i(1) \mid W_i]$$

and calculate

$$\chi_1^2 = (\mid M_1 - \mathrm{E}[M_1] \mid - 0.5)^2 / \mathrm{V}[M_1].$$

They have assumed asymptotic normality of $M_1$ and an approximate chi-square distribution with one degree of freedom of $\chi_1^2$ under $H_0$, where the term 0.5 in $\chi_1^2$ denotes the correction for continuity which was introduced by Yates (1934). Mantel, Kryscio, and Myers (1976) have tried to justify the assumption of asymptotic normality arguing that, though each individual $M_i(1)$ might follow some extreme-value distribution, $(M_1 - \mathrm{E}[M_1])/(\mathrm{V}[M_1])^{1/2}$ "would tend to be asymptotically normally distributed under the central limit theorem by virtue of being a total of many independent observations each following distributions with finite second moment."

Ederer, Myers, and Mantel (1964) have given the values of $\mathrm{E}[M_i(1) \mid W_i]$ and $\mathrm{V}[M_i(1) \mid W_i]$ for $c = 5$ and $W_i = 2(1)15$. Mantel, Kryscio, and Myers (1976) have corrected errors in this table and extended it to cover the cases $c = 2$, $W_i = 2(1)100(5)200(100)500$; $c = 3$, $W_i = 2(1)50(5)200(100)500$; $c = 4$, $W_i = 2(1)50(5)200$; $c = 5$, $W_i = 2(1)50(5)100(25)200$.

Since tables of this kind do not cover all situations met in practice, Stark and Mantel (1967a) have derived approximate values of $\mathrm{E}[M_i(1) \mid W_i]$ and $\mathrm{V}[M_i(1) \mid W_i]$ based on a normal approximation of the multinomial distribution. Mantel, Kryscio, and Myers (1976) have improved the approximation proposed by Stark and Mantel (1967a), introducing the exact values of the expectations and variances for $W_i = 100, 200$, and $500$ in the approximation formulas if $W_i > 100, W_i > 200$, or $W_i > 500$. Former results concerning normal approximations of $\mathrm{E}[M_i(1) \mid W_i]$ and $\mathrm{V}[M_i(1) \mid W_i]$ are due to Greenwood and Glasgow (1950) and Owen and Steck (1962).

Wartenberg and Greenberg (1990) have compared the power of the EMM procedure with that of Mantel's space–time regression procedure [Mantel (1967)] by means of a simulation study. They have concluded that the two procedures are specific to different hypotheses and that both show low power for small numbers of cases.

Wallenstein, Gould, and Kleinman (1989a) have replaced the disjoint statistic $M(1)$ in the EMM procedure by the scan statistic and have argued (where $S$ denotes the modified EMM statistic, and $d$ is replaced by $I$): "As for the EMM statistic, the Central Limit Theorem indicates that if $I$ is sufficiently large and no single unit is much larger than the rest, then $S$ in equation 2 will have approximately a normal distribution under $H_0$." For the application of the modified statistic, they have provided tables of means and variances of the scan statistic which are exact for $W_i \leq 19$ and based on simulations for $W_i \geq 20$.

## The Grimson models

Grimson (1979, 1993) and Grimson and Oden (1996) have derived the distribution of the disjoint statistic $M(1)$ in the equiprobable case in two situations where no multinomial distribution of the vector $(N_1, \ldots, N_c)$ is assumed.

In thermodynamics, three well-known probability models are discussed, which are described by Feller (1968, pp. 39–41); also see Kotz and Balakrishnan (1997). Assume $N$ balls are to be placed into $c$ cells resulting in the occupancy numbers $N_1, \ldots, N_c$. Assuming that all $c^N$ possible placements are equally probable, the probability to obtain the given vector $(n_1, \ldots, n_c)$ of occupancy numbers equals

$$\left( N! / \prod_{i=1}^{c} n_i! \right) c^{-N},$$

i.e., a multinomial probability. This model is called *Maxwell–Boltzmann statistics*.

Assuming that all possible distinguishable arrangements of the $N$ balls in the $c$ cells have the same probability, each arrangement has the probability

$$\binom{c + N - 1}{c - 1}^{-1}.$$

This model is called *Bose–Einstein statistics*. Here, in contrast to the Maxwell–Boltzmann model, we do not consider the different ways to generate a particular configuration of the numbers $n_1, \ldots, n_c$.

Finally, we assume that no cell can contain more than one ball and that all distinguishable arrangements with this property have equal probabilities. This probability is given by

$$\binom{c}{N}^{-1}.$$

Obviously, in this case we must assume $N \leq c$. This model is called *Fermi–Dirac statistics*.

So far, we have only dealt with the Maxwell–Boltzmann model. Grimson (1979) has derived for the Bose–Einstein model

$$P(M(1) \geq x) = 1 - \sum_{j=0}^{c} (-1)^j \binom{c}{j} \binom{c + N - jx - 1}{c - 1} \binom{c + N - 1}{c - 1}^{-1},$$

and has also given formulas for the moments of $M(1)$, in particular for the mean $\mathrm{E}[M(1)]$. The exact values of $\mathrm{E}[M(1)]$ and $\mathrm{V}[M(1)]$ for $c = 5$ and $N = 1(1)400$ are listed so that an EMM-like procedure can be performed.

Grimson (1993) and Grimson and Oden (1996) have considered a certain Fermi–Dirac model which they view as the "matrix occupancy" analogue of the usual occupancy model. In this model, $N$ balls are placed into the cells of $c$ columns where each column has $r$ cells and each cell may contain no more than one ball. Identifying the columns with our original cells, the statistic $M(1)$ corresponds to the maximum number of occupied cells in a column. They have derived

$$P(M(1) \geq x) \; = \; \binom{rc}{N}^{-1} \sum_{k \geq 1} (-1)^{k+1} \binom{c}{k} \sum_{j_1,\ldots,j_k = x}^{r} \binom{r(c-k)}{N - j_1 - \ldots - j_k} \times \binom{r}{j_1} \cdots \binom{r}{j_k}.$$

**Other approaches**

Krauth (1993) has considered an extension of the disjoint statistic $M(1)$, where not the maximum of $N_1, \ldots, N_c$ but the sum of the $v$ largest values of the $N_1, \ldots, N_c$ is studied, i.e.,

$$M_v(1) = \max_{j_1 < \cdots < j_v} \sum_{t=1}^{v} N_{j_t} \text{ for } v = 1, \ldots, c-1.$$

Obviously, $M(1) = M_1(1)$. For the equiprobable case, the upper Bonferroni bound of degree one for $P(M_v(1) \geq x)$ is given by

$$U_1^{(v)} = \min \left\{ 1, \binom{c}{v} q^{(v)} \right\}$$

with

$$q^{(v)} = G_b \left( x; N, \frac{v}{c} \right).$$

Krauth (1996a) has cosidered a Fermi–Dirac model (see the last subsection), where $c$ cells and in each cell $r_i$ subcells are given, for $i = 1, \ldots, c$, into each of which at most one ball of altogether $N$ balls may be placed in a random fashion. This kind of model is called *binary occupancy model* by Grimson (1993) and Grimson and Oden (1996). Under $H_0$, this results in a multivariate hypergeometric distribution of $(N_1, \ldots, N_c)$, i.e.,

$$P \left( \bigcap_{i=1}^{c} \{ N_i = n_i \} \right) = \binom{R}{N}^{-1} \prod_{i=1}^{c} \binom{r_i}{n_i}.$$

Here, $R = r_1 + \cdots + r_c$ and $N \leq R, n_i \leq r_i$, for $i = 1, \ldots, c$.

For this model, Krauth (1996a,b) has derived exact upper and lower bounds of degrees one and two for the tail of the distribution $M(1)$ under $H_0$, so that we have

$$L_1 \leq L_2 \leq P\left(M(1) \geq x\right) \leq U_2 \leq U_1,$$

where

$$U_1 \;\; = \;\; \min\{1, S_1\},$$

$$U_2 = \min\left\{1, S_1 - \max_{\substack{1 \le j \le c}} \sum_{\substack{i=1 \\ i \ne j}}^{c} q_{ij}\right\},$$

$$L_1 = \frac{2}{k(k+1)}\left(kS_1 - S_2^{\star}\right) \text{ with } k = 1 + \lfloor 2S_2^{\star}/S_1 \rfloor,$$

$$L_2 = \frac{2}{k(k+1)}\left(kS_1 - S_2\right) \text{ with } k = 1 + \lfloor 2S_2/S_1 \rfloor,$$

$$S_1 = \sum_{i=1}^{c} q_i, \quad S_2 = \sum_{i=2}^{c}\sum_{j=1}^{c-1} q_{ij}, \quad S_2^{\star} = \sum_{i=2}^{c}\sum_{j=1}^{c-1} q_i q_j,$$

$$q_i = 0 \text{ for } x > r_i,$$

$$q_i = \sum_{u=x}^{\min\{r_i,N\}} \binom{R}{N}^{-1}\binom{r_i}{u}\binom{R-r_i}{N-u} \text{ otherwise, for } i = 1, \dots, c,$$

$$q_{ij} = 0 \text{ for } x > \min\{r_i, r_j\},$$

$$q_{ij} = 0 \text{ for } 2x > N,$$

$$q_{ij} = \sum_{u=x}^{\min\{r_i,N\}}\sum_{v=x}^{\min\{r_j,N-u\}} \binom{R}{N}^{-1}\binom{r_i}{u}\binom{r_j}{v}\binom{R-r_i-r_j}{N-u-v}$$

$$\text{otherwise, for } i \ne j, \quad i, j = 1, \dots, c.$$

For $2x > N$, we get the exact result

$$P(M(1) \ge x) = \sum_{i=1}^{c} q_i.$$

The bound $U_1$ is the best linear upper bound of degree one, while the bound $L_2$ is the best linear lower bound of degree two for $P(M(1) \ge x)$ if only the values of $S_1$ and $S_2$ but not the probabilities $q_i$ and $q_{ij}$ are known.

Krauth (1996b) has also derived bounds for the special case of the matrix occupancy model [Grimson and Oden (1996)] with $r_1 = \cdots = r_c = r$. These follow immediately from the bounds for the general binary occupancy model given above by replacing $r_i$ and $r_j$ by $r$, $R$ by $rc$ and $r_i + r_j$ by $2r$.

### 3.2.6 Applications

Most applications of the disjoint statistic and its modifications can be found in epidemiology; for example, for detecting clusters of childhood leukemia [Ederer, Myers, and Mantel (1964), Fraumeni, Ederer, and Hardy (1966), Mantel, Kryscio, and Myers (1976), and Stark and Mantel (1967b)], poliomyelitis [Ederer, Myers, and Mantel (1964), and Mantel, Kryscio, and Myers (1976)], hepatitis [Ederer, Myers, and Mantel (1964), and Grimson (1979)], occurrences of bone fractures at an infirmary [Grimson (1993), and Krauth (1996b)], Trypanosoma cruzi seropositivity data [Grimson and Oden (1996)], Down's syndrome births [Mantel, Kryscio, and Myers (1976), and Stark and Mantel (1967a)], first-born anencephalics [Krauth (1991)], and Green Tobacco Sickness [McKnight *et al.* (1996)]. Other applications concern the clustering of uranium deposits [Krauth (1991)], seabirds [Krauth (1993)], species from a subtidal marsh creek [Krauth (1993)], and of neurons in chick embryos, rats, and humans [Krauth (1996a)].

## 3.3 Linear Ratchet Scan Statistic

### 3.3.1 Definition

Let $X_1, \ldots, X_N$ be independent identically distributed random variables. Under $H_0$, a continuous uniform distribution on a real-valued interval $[a, b]$ is assumed. The interval $[a, b]$ is dissected into $c$ subintervals or cells $C_1, \ldots, C_c$, i.e., the cells are pairwise disjoint and their union is $[a, b]$. The number of $X_j$'s observed in $C_i$ is denoted by $N_i$ for $i = 1, \ldots, c$, and

$$p_i = P(C_i) \text{ for } i = 1, \ldots, c, \quad p_1 + \ldots + p_c = 1$$

are the probabilities assigned to the $c$ cells. We set

$$T_i(m) = \sum_{t=i}^{i+m-1} N_t \text{ for } i = 1, \ldots, c - m + 1, \quad m = 1, \ldots, c - 1,$$

and define the *linear ratchet scan statistic* as the maximum number of $X_j$'s observed in $m$ consecutive cells:

$$M(m) = \max_{1 \leq i \leq c-m+1} T_i(m).$$

For $m = 1$, the sum $T_i(m)$ reduces to $T_i(1) = N_i$ and the statistic $M(m)$ to the disjoint statistic $M(1)$.

Under $H_0$, the vector $(N_1, \ldots, N_c)$ is multinomially distributed with parameters $N, p_1, \ldots, p_c$.

### 3.3.2 Results

Krauth (1992b) has derived exact upper and lower bounds for the upper tail probabilities of the linear ratchet scan statistic for the general and the equiprobable cases.

We use the notation $b(s; N, p)$ for the binomial probability, $G_b(s; N, p)$ for the binomial tail and

$$G_t(s; N, p, q) = \sum_{u=s}^{N-s} \sum_{v=s}^{N-u} \frac{N!}{u! v! (N - u - v)!} p^u q^v (1 - p - q)^{N-u-v}$$

$$\text{for } 0 < p, \quad q < 1, \quad 0 \le 2s \le N;$$

$$G_t(s; N, p, q) = 0 \text{ for } 2s > N;$$

$$G_t(s; N, p, q) = 1 \text{ for } s \le 0;$$

$$G_t(s; N, p, q) = G_b(s; N, p) \text{ for } p + q = 1;$$

$$q_i = G_b\left(x; N, \sum_{s=i}^{i+m-1} p_s\right) \text{ for } i = 1, \ldots, c - m + 1,$$

$$m = 1, \ldots, c - 1;$$

$$q_{ji} = q_{ij} \text{ for } i, j = 1, \ldots, c - m + 1, \quad m = 1, \ldots, c - 1;$$

$$q_{ij} = G_t\left(x; N, \sum_{s=i}^{i+m-1} p_s, \sum_{s=j}^{j+m-1} p_s\right) \text{ for } c \ge 2m,$$

$$i = 1, \ldots, c - 2m + 1,$$

$$j = i + m, \ldots, c - m + 1, \quad m = 1, \ldots, c - 1;$$

$$q_{ij} = \sum_{s=0}^{x-1} b\left(s; N, \sum_{r=i+u}^{i+m-1} p_r\right)$$

$$\times G_t\left(x - s; N - s, \frac{\sum_{r=i}^{i+u-1} p_r}{1 - \sum_{r=i+u}^{i+m-1} p_r}, \frac{\sum_{r=i+m}^{i+m+u-1} p_r}{1 - \sum_{r=i+u}^{i+m-1} p_r}\right)$$

$$+ G_b\left(x; N, \sum_{r=i+u}^{i+m-1} p_r\right)$$

$$\text{for } c \ge 2m, \quad u = 1, \ldots, m - 1, \quad i = 1, \ldots, c - m + 1 - u,$$

$$j = i + u$$

$$\text{or } c < 2m, \quad u = 1, \ldots, c - m, \quad i = 1, \ldots, c - m + 1 - u,$$

$$j = i + u;$$

$$S_1 = \sum_{i=1}^{c-m+1} q_i;$$

$$S_2 = \sum_{j=2}^{c-m+1} \sum_{i=1}^{j-1} q_{ij}.$$

Then, the exact results

$$P(M(1) \ge x) = S_1 \text{ for } x > N/2,$$

$$P(M(1) \ge x) = S_1 - S_2 \text{ for } x > N/3,$$

$$P(M(2) \ge x) = S_1 - S_2 \text{ for } x > N/2$$

hold.

Furthermore, the inequalities

$$L \le P(M(m) \ge x) \le U$$

with

$$U = \min \left\{ 1, S_1 - \max_{1 \le j \le c-m+1} \sum_{\substack{i=1 \\ i \ne j}}^{c-m+1} q_{ij} \right\}$$

and

$$L = \frac{2}{k(k+1)} (kS_1 - S_2) \text{ with } k = 1 + \lfloor 2S_2/S_1 \rfloor$$

are valid. Here, $L$ is the best linear lower bound for $P(M(m) \ge x)$ if only the values of $S_1$ and $S_2$ but not the probabilities $q_i$ and $q_{ij}$ are known.

The lower bound $L$ will be larger than $S_1 - S_2$ if $k \ge 2$. In view of the exact results for $M(1)$ and $M(2)$, we may speculate that the lower bound $L$ in general will be a better approximation of $P(M(m) \ge x)$ than the upper bound $U$.

In the equiprobable case with $p_1 = \cdots = p_c = 1/c$, all formulae are considerably simplified:

$$q_1 = G_b\left(x; N, \frac{m}{c}\right);$$

$$q_{1,m+1} = G_t\left(x; N, \frac{m}{c}, \frac{m}{c}\right) \text{ for } c \ge 2m;$$

$$q_{1,u+1} = \sum_{s=0}^{x-1} b\left(s; N, \frac{m-u}{c}\right) G_t\left(x - s; N - s, \frac{u}{c-m+u}, \frac{u}{c-m+u}\right)$$

$$+ G_b\left(x; N, \frac{m-u}{c}\right)$$

$$\text{for } c \ge 2m, \quad u = 1, \ldots, m-1$$

$$\text{or } c < 2m, \quad u = 1, \ldots, c-m;$$

$$S_1 = (c - m + 1)q_1;$$

$$S_2 = \sum_{u=1}^{m-1} (c - m + 1 - u)q_{1,u+1} + \frac{1}{2}(c - 2m + 1)(c - 2m + 2)q_{1,m+1}$$

$$\text{for } c \ge 2m;$$

$$S_2 = \sum_{u=1}^{c-m} (c - m + 1 - u)q_{1,u+1} \text{ for } c < 2m;$$

$$U = S_1 - 2\sum_{u=1}^{m-1} q_{1,u+1} - (c - 3m + 2)q_{1,m+1} \text{ for } c \ge 3m - 2;$$

$$U = S_1 - 2\sum_{u=1}^{w} q_{1,u+1} - (c - m - 2k)q_{1,k+2} \text{ for } c < 3m - 2,$$

$$k = \left\lfloor \frac{c-m}{2} \right\rfloor;$$

$$L = \frac{2}{k(k+1)} (kS_1 - S_2) \text{ with } k = 1 + \lfloor 2S_2/S_1 \rfloor.$$

### 3.3.3   The EMM procedure

The EMM procedure of Ederer, Myers, and Mantel (1964) has been described earlier in Section 3.2.5. There, for a fixed subarea $A_i$, the disjoint statistic $M_i(1)$ is calculated with respect to the $c$ time periods $T_1, \ldots, T_c$. They have suggested that, instead of $M_i(1)$, the linear ratchet scan statistic $M_i(2)$ or even $M_i(m)$ with a more general $m \geq 2$ may be calculated for each subarea $A_i$. For $m = 2$, they have derived the conditional distribution of $M(2)$ given the number $W_i = N_{i1} + \cdots + N_{ic}$ of cases in the respective subarea $A_i$ under the null hypothesis of equal probabilities (for $c = 5$, $W_i = 2(1)15$). For this distribution, the mean $E[M_i(2) \mid W_i]$ and the variance $V[M_i(2) \mid W_i]$, for $i = 1, \ldots, d$, have also been calculated. As for $M_i(1)$, they have calculated

$$M_2 \;=\; \sum_{i=1}^{d} M_i(2),$$

$$E[M_2] \;=\; \sum_{i=1}^{d} E[M_i(2) \mid W_i],$$

$$V[M_2] \;=\; \sum_{i=1}^{d} V[M_i(2) \mid W_i],$$

and

$$\chi_1^2 = (\mid M_2 - E[M_2] \mid -0.5)^2 / V[M_2],$$

where an approximate chi-square distribution with one degree of freedom is assumed for $\chi_1^2$.

Mantel, Kryscio, and Myers (1976) have presented an extended table of $E[M_i(2) \mid W_i]$ and $V[M_i(2) \mid W_i]$ for $c = 3$, $W_i = 2(1)50(5)200(100)500$; $c = 4$, $W_i = 2(1)50(5)200$; $c = 5$, $W_i = 2(1)50(5)100(25)200$.

### 3.3.4   Applications

The linear ratchet scan statistic and the corresponding EMM procedure are applied in epidemiology where these methods are used to detect clusters of childhood leukemia [Ederer, Myers, and Mantel (1964), Fraumeni, Ederer, and Handy (1966), and Mantel, Kryscio, and Myers (1976)], poliomyelitis [Ederer, Myers, and Mantel (1964), and Mantel, Kryscio, and Myers (1976)], infectious hepatitis [Ederer, Myers, and Mantel (1964), and Mantel, Kryscio, and Myers (1976)], and trisomies among karyotyped spontaneous abortions [Krauth (1992b)].

## 3.4  Circular Ratchet Scan Statistic

### 3.4.1  Definition

Let $X_1, \ldots, X_N$ be independent identically distributed random variables with a continuous uniform distribution on the circumference of a circle. This circumference is dissected into $c$ disjoint arcs $C_1, \ldots, C_c$ whose union is the circumference. The number of $X_j$'s observed in $C_i$ is denoted by $N_i$, for $i = 1, \ldots, c$, and

$$p_i = P(C_i) \text{ for } i = 1, \ldots, c, \ p_1 + \cdots + p_c = 1$$

are the probabilities assigned to the $c$ cells. We set

$$T_i(m) = \sum_{t=i}^{i+m-1} N_{t \bmod c} \text{ for } i = 1, \ldots, c, \ m = 1, \ldots, c-1,$$

where $c \bmod c := c$. The circular ratchet scan statistic is defined as the maximum number of $X_j$'s observed in $m$ consecutive cells:

$$M(m) = \max_{1 \leq i \leq c} T_i(m).$$

For $m = 1$, the sum $T_i(m)$ is reduced to $T_i(m) = N_i$ and the statistic $M(m)$ to the disjoint statistic $M(1)$.

Under $H_0$, the vector $(N_1, \ldots, N_c)$ is multinomially distributed with parameters $N, p_1, \ldots, p_c$.

### 3.4.2  Results

Wallenstein, Weinberg, and Gould (1989b) have given a table of the exact values of $P(M(m) \geq x)$ for $c = 12$, $m = 2$ and 3, $N = 8(1)25, 30, 35$ and "small" $p$-values for the equiprobable case with $p_1 = \cdots = p_c = 1/c$. The values of $x$ vary between 5 and 16.

For $c = 12$, they have considered approximate upper percentage points of the form

$$x_\alpha(m) \doteq \frac{1}{12}\left(Nm + \sqrt{Nm(12-m)}\, y_\alpha(m)\right),$$

where $y_\alpha(m)$ is the upper percentage point for the maximum of certain normally distributed random variables. The values of $y_\alpha(m)$ are estimated by means of a simulation.

Krauth (1992a) has derived exact upper and lower bounds for the upper tail probabilities of the circular ratchet scan statistic for the general and equiprobable case.

We use the notation, where $b(s; N, p)$, $G_b(s; N, p)$, and $G_t(s; N, p, q)$ are as defined in Section 3.3.2:

$$q_i = G_b\left(x; N; \sum_{r=i}^{i+m-1} p_r \bmod c\right) \text{ for } i = 1, \ldots, c, \quad m = 1, \ldots, c-1;$$

$$q_{ji} = q_{ij} \text{ for } i, j = 1, \ldots, c;$$

$$q_{ij} = G_t\left(x; N, \sum_{r=i}^{i+m-1} p_r \bmod c, \sum_{r=j}^{j+m-1} p_r \bmod c\right)$$
$$\text{for } i = 1, \ldots, c-m, \quad j = i+m, \ldots, i+c-m, \quad j \le c;$$

$$q_{ij} = \sum_{s=0}^{x-1} b\left(s; N, \sum_{r=i+u}^{i+m-1} p_r \bmod c\right) G_t\left(x-s; N-s, \frac{\sum_{r=i}^{i+u-1} p_r \bmod c}{1 - \sum_{r=i+u}^{i+m-1} p_r \bmod c},\right.$$
$$\left.\frac{\sum_{r=i+m}^{i+m+u-1} p_r \bmod c}{1 - \sum_{r=i+u}^{i+m-1} p_r \bmod c}\right) + G_b\left(x; N, \sum_{r=i+u}^{i+m-1} p_r \bmod c\right)$$
$$\text{for } u = 1, \ldots, m-1, \quad i = 1, \ldots, c-u, \quad j = i+u;$$

$$q_{ij} = \sum_{s=0}^{x-1} b\left(s; N, \sum_{r=i}^{i+m-u-1} p_r \bmod c\right)$$
$$\times G_t\left(x-s; N-s, \frac{\sum_{r=i+c-u}^{i+c-1} p_r \bmod c}{1 - \sum_{r=i}^{i+m-u-1} p_r \bmod c},\right.$$
$$\left.\frac{\sum_{r=i+m-u}^{i+m-1} p_r \bmod c}{1 - \sum_{r=i}^{i+m-u-1} p_r \bmod c}\right) + G_b\left(x; N, \sum_{r=1}^{i+m-u-1} p_r \bmod c\right)$$
$$\text{for } u = 1, \ldots, m-1, \quad i = 1, \ldots, u, \quad j = i+c-u;$$

$$S_1 = \sum_{i=1}^{c} q_i, \quad S_2 = \sum_{j=2}^{c} \sum_{i=1}^{j-1} q_{ij}.$$

The following exact results hold:

$$\begin{aligned}
P(M(1) \ge x) &= S_1 \text{ for } x > N/2, \\
P(M(1) \ge x) &= S_1 - S_2 \text{ for } x > N/3, \\
P(M(2) \ge x) &= S_1 - S_2 \text{ for } x > N/2.
\end{aligned}$$

The inequalities

$$L \le P(M(m) \ge x) \le U$$

are true with

$$U = \min\left\{1, S_1 - \max_{1 \le j \le c} \sum_{\substack{i=1 \\ i \ne j}}^{c} q_{ij}\right\}$$

and

$$L = \frac{2}{k(k+1)} (kS_1 - S_2) \text{ with } k = 1 + \lfloor 2S_2/S_1 \rfloor.$$

Here, $L$ is the best linear lower bound for $P(M(m) \geq x)$ if only the values of $S_1$ and $S_2$ are known but not the probabilities $q_i$ and $q_{ij}$. The lower bound $L$ will be larger than $S_1 - S_2$ for $k \geq 2$. In view of the exact results for $M(1)$ and $M(2)$, we may speculate that the lower bound $L$, in general, is a better approximation to $P(M(m) \geq x)$ than the upper bound $U$.

In the equiprobable case with $p_1 = \cdots = p_c = 1/c$, all formulae are considerably simplified:

$$q_1 = G_b\left(x; N, \frac{m}{c}\right);$$

$$q_{1,m+1} = G_t\left(x; N, \frac{m}{c}, \frac{m}{c}\right);$$

$$q_{i,u+1} = \sum_{s=0}^{x-1} b\left(s; N, \frac{m-u}{c}\right) G_t\left(x-s; N-s, \frac{u}{c-m+u}, \frac{u}{c-m+u}\right)$$
$$+ G_b\left(x; N, \frac{m-u}{c}\right) \quad \text{for } u = 1, \ldots, m-1;$$

$$S_1 = cq_1,$$

$$S_2 = \frac{1}{2} c(c - 2m + 1)q_{1,m+1} + c \sum_{u=1}^{m-1} q_{1,u+1},$$

$$L = \frac{2}{k(k+1)} (kS_1 - S_2) \quad \text{with } k = 1 + \lfloor 2S_2/S_1 \rfloor,$$

$$U = S_1 - \frac{2}{c} S_2.$$

Wallenstein, Weinberg, and Gould (1989b) have performed a simulation study in which they have compared the circular ratchet scan statistic $M(3)$ with four other statistics for five alternative distributions with respect to power. The circular ratchet scan statistic showed the best result of all competitors for a pure 3-month pulse alternative. As expected, this statistic, just as the continuous scan statistic, behaved in a less satisfactory way for sinusoidal alternatives.

### 3.4.3  Modifications

Edwards (1961) has considered to look for $m$ consecutive cells with values of the $N_j$'s which are all larger (or smaller) than the median of $N_1, \ldots, N_c$ as a simple but inefficient distribution-free test for cyclic trend. This idea is extended by Hewitt *et al.* (1971), who calculated the largest rank–sum for any 6-month segment, i.e., replaced the $N_j$'s in the circular ratchet scan statistic $M(6)$ by their ranks. The null distribution of this test statistic is derived by means of a simulation. Walter and Elwood (1975) have compared the test of Hewitt *et al.*, with four parametric competitors for a real data set and concluded that this test will probably have a low power to detect a seasonal trend.

Freedman (1979) has presented a simulation study where the test of Hewitt *et al.*, is compared with three other tests, and he has observed that the test of Hewitt *et al.*, is less powerful than the parametric test of Edwards (1961) and a distribution-free Kolmogorov–Smirnov-type statistic if a sinusoidal alternative is considered. Walter (1980) has provided exact and simulated null distributions for the statistic of Hewitt *et al.*, and has addressed the problem of ties.

In a simulation study, Marrero (1983) has compared the test of Hewitt *et al.*, to seven competitors with respect to power for three alternative distributions. However, the results for the test of Hewitt *et al.*, have not been included in the tables because of the weak performance of this test. Nevertheless, Marrero (1988) has performed a simulation study where the test of Hewitt *et al.*, has not been applied to the $N_j$'s but to the incidence rates, i.e., to the number of cases divided by the size of the population at risk. The power of the test has been simulated for a simple sinusoidal curve with one peak and one trough, for a sinusoidal curve with two peaks and two troughs and a one-pulse model. The test can be very powerful for the first alternative, and has, in general, a low power for the two remaining alternatives.

Marrero (1992) has given the exact null distributions for $m = 3, 4, 5$, and 6 months for the statistic of Hewitt *et al.* In a simulation study, he has found that this rank version of the circular ratchet scan statistic has a high power for one-pulse alternatives if the relative height of the pulse or the sample size is large. Average ranks are assigned to tied observations, and unequal month lengths are adjusted by enlarging the $N_j$'s corresponding to months with fewer than 31 days by assuming that the observations are equidistributed over the year. Obviously, not knowing Marrero's (1992) paper, Rogerson (1996) has derived simulated and exact null distributions for the test of Hewitt *et al.* (1971), for $m = 3, 4, 5$, and 6, and has compared the power of this test with that of the chi-square test and (for $m = 3$) the circular ratchet scan test. For one-pulse alternatives and a sample size of $N = 50$, the test of Hewitt *et al.*, seems to have more power than the chi-square test though the circular ratchet scan test is superior to both other tests.

Assume that $c = 2d$ holds, and consider

$$
\begin{aligned}
T_i(d) &= \sum_{t=i}^{i+d-1} N_{t \bmod c} \text{ for } i = 1, \ldots, c, \\
U_i &= (T_i(d) - T_{i+d}(d))\, N^{-1/2} = 2\left(T_i(d) - \frac{1}{2}N\right) N^{-1/2} \text{ for } i = 1, \ldots, c, \\
V_d &= \max_{1 \le i \le d} |U_i|.
\end{aligned}
$$

David and Newell (1965) have considered the statistic $V_d$, which, in an obvious way, is related to the circular ratchet scan statistic

$$
M(d) = \max_{1 \le i \le c} T_i(d).
$$

Assume that, under $H_0$, the cells $C_1, \ldots, C_c$ have equal probabilities and that the $U_i$, for $i = 1, \ldots, c$, approximately have a standard normal distribution. Under these assumptions, the approximate upper Bonferroni bound of degree one and the approximate lower Bonferroni bound of degree two are derived:

$$\sum_{i=1}^{d} P\left(|\,U_i\,| \geq x\right) - \sum_{i=2}^{d} \sum_{j=1}^{i} P\left(|\,U_i\,| \geq x, |\,U_j\,| \geq x\right)$$

$$\leq P(V_d \geq x) \leq \sum_{i=1}^{d} P\left(|\,U_i\,| \geq x\right).$$

Here,

$$\sum_{i=1}^{d} P\left(|\,U_i\,| \geq x\right) = dP\left(|\,U_i\,| \geq x\right).$$

### 3.4.4 Applications

The circular ratchet scan statistic and its modifications are mainly applied to problems in epidemiology. In particular, extrahepatic biliary atresia [Wallenstein, Weinberg, and Gould (1989b), Krauth (1992a), and Marrero (1992)], autism [Bolton *et al.* (1992)], duodenal ulcer [Cohen (1994, 1995)], Guillain–Barr syndrome [Ward (1992)], attempted suicides [Marrero (1992), Gould *et al.* (1994), and Rogerson (1996)], leukemia [David and Newell (1965), and Krauth (1992a)], anencephalus [Walter and Elwood (1975), and Krauth (1992a)], cardiac malformations [Hewitt *et al.* (1971)], and motor-vehicle-related fatalities [Marrero (1992)] are considered.

---

## 3.5 Exact Bounds for the Upper Tail Probabilities of the Statistic $M(m)$

In Section 3.1, we introduced the general statistic $M(m)$ and in Sections 3.2–3.4 we gave results for some special situations in which this statistic is considered. It is obvious that for most practical problems the computation of the exact distribution of $M(m)$ is not feasible and asymptotic results may not be trustworthy. Fortunately, we have the alternative option to derive exact bounds for the upper tail probabilities of $M(m)$, which on the one hand allow a rapid calculation and on the other hand are rather close for the extreme upper tails which alone are of practical interest. To perform statistical tests for clustering, it suffices to derive upper bounds. However, if the closeness of these bounds is to be evaluated it is of use to have additional lower bounds available.

The statistic $M(m)$ is defined as the maximum of the sums of frequencies in the cells of a moving window of size $m$:

$$M(m) = \max_{i \in I} T_i(m).$$

Here, $I$ is a finite index set and $T_i(m)$ is a sum of $m$ frequencies. Obviously, we have

$$P(M(m) \geq x) = P\left(\bigcup_{i \in I} \{T_i(m) \geq x\}\right) = P\left(\bigcup_{i \in I} A_i\right)$$

$$\text{with } A_i = \{T_i(m) \geq x\} \text{ for } i \in I.$$

Upper and lower bounds for the probability of a union are given by the classical Bonferroni bounds and the many improvements of these bounds which have been derived up to now. The bounds which we consider in the following are neither new nor are they the best available bounds known today. We looked for a compromise, where the bounds are of an acceptable accuracy and at the same time may be computed with low effort. Concerning the latter aspect, we must keep in mind that the probabilities $P(A_i)$ or $P(A_i \cap A_j)$ for $i, j \in I$ may be given by rather complicated expressions and it is often not advisable to consider intersections of more than two events. This is particularly true for $m \geq 2$ and unequal cell probabilities.

Kounias and Marin (1974) have proved that the best linear bounds of degree one are given by

$$\max_{i \in I} P(A_i) \leq P(M(m) \geq x) \leq \min\left\{1, \sum_{i \in I} P(A_i)\right\}.$$

Here, the right side is the classical upper Bonferroni bound of degree one.

If only the values

$$S_1 = \sum_{i \in I} P(A_i) \quad \text{and} \quad S_2 = \sum_{\substack{i, j \in I \\ i < j}} P(A_i \cap A_j)$$

are known, the best linear upper bound of degree two is given by

$$\min\left\{1, S_1 - \frac{2}{c} S_2\right\}$$

[Kwerel (1975, Corollary to Theorem 5)].

However, if we assume that the probabilities $P(A_i \cap A_j)$ are known for $i, j \in I$, we had better choose the superior upper bound of degree two given by

$$\min\left\{1, S_1 - \max_{j \in I} \sum_{\substack{i \in I \\ i \neq j}} P(A_i \cap A_j)\right\}$$

which is due to Kounias (1968).

Kwerel (1975) has proved that the bound

$$\frac{2}{k(k+1)}\,(kS_1 - S_2) \text{ with } k = 1 + \lfloor 2S_2/S_1 \rfloor,$$

which is equivalent to the bound of Dawson and Sankoff (1967), is the best linear lower bound of degree two if only the values of $S_1$ and $S_2$ are known. For $k = 1$, we get just the Bonferroni lower bound $S_1 - S_2$ of degree two. From Galambos (1977), we can conclude that the above expression will yield lower bounds of $P(M(m) \geq x)$ for each integer $k \geq 1$.

The lower bound of degree two proposed by Kwerel (1975) can be replaced in certain situations by an inferior lower bound of degree one assuming only the knowledge of $P(A_i)$ for $i \in I$. Jogdeo and Patil (1975) have proved that for certain multivariate discrete distributions, e.g., for the multinomial distribution, the Dirichlet distribution, and the multivariate hypergeometric distribution, the inequality

$$P(A_i \cap A_j) \leq P(A_i)P(A_j) \text{ for } i, j \in I$$

holds; see, for example, Johnson, Kotz, and Balakrishnan (1997). For the multinomial distribution, this result was also proved by Mallows (1968), Yusas (1972), and Proschan and Sethuraman (1975). If we introduce this inequality in the lower bound of degree two of Kwerel (1975), we get a weaker lower bound of degree one:

$$\frac{2}{k(k+1)}\,(kS_1 - S_2^{\star}) \text{ with } k = 1 + \lfloor 2S_2^{\star}/S_1 \rfloor, \quad S_2^{\star} = \sum_{\substack{i,j \in I \\ i < j}} P(A_i)P(A_j).$$

This nonlinear bound is superior to the best linear lower bound of degree one given above.

The upper and lower bounds described above can be used to get exact bounds for $P(M(m) \geq x)$ in the general case with a complex shape of the moving window, with unequal cell probabilities, and if not only the Maxwell–Boltzmann model but also the Bose–Einstein and the Fermi–Dirac models are considered. In many cases, the simple upper bound of degree one will be sufficient for performing a statistical test based on $M(m)$.

---

# References

1. Barton, D. E. and David, F. N. (1959). Combinatorial extreme value distributions, *Mathematika*, **6**, 63–76.

2. Bolton, P., Pickles, A., Harrington, R., Macdonald, H. and Rutter, M. (1992). Season of birth: issues, approaches and findings for autism, *Journal of Child Psychology and Psychiatry*, **33**, 509–530.

3. Cohen, P. (1994). Is duodenal ulcer a seasonal disease? Statistical considerations, *American Journal of Gastroenterology*, **89**, 1121–1122.

4. Cohen, P. (1995). Seasonality in duodenal ulcer disease: possible relationship with circannually cycling neurons enclosed in the biological clock, *American Journal of Gastroenterology*, **90**, 1189–1190.

5. David, F. N. and Barton, D. E. (1962). *Combinatorial Chance*, London, England: Charles Griffin.

6. David, H. A. and Newell, D. J. (1965). The identification of annual peak periods for a disease, *Biometrics*, **21**, 645–650.

7. Dawson, D. A. and Sankoff, D. (1967). An inequality for probabilities, *Proceedings of the American Mathematical Society*, **18**, 504–507.

8. Ederer, F., Myers, M. H. and Mantel, N. (1964). A statistical problem in space and time: Do leukemia cases come in clusters?, *Biometrics*, **20**, 626–638.

9. Edwards, J. H. (1961). The recognition and estimation of cyclic trends, *Annals of Human Genetics*, **25**, 83–86.

10. Feller, W. (1968). *An Introduction to Probability Theory and Its Applications*, Volume I, Third Edition, New York: John Wiley & Sons.

11. Fraumeni, J. F., Ederer, F. and Handy, V. H. (1966). Temporal–spatial distribution of childhood leukemia in New York State, special reference to case clustering by year of birth, *Cancer*, **19**, 996–1000.

12. Freedman, L. S. (1979). The use of a Kolmogorov–Smirnov type statistic in testing hypotheses about seasonal variation, *Journal of Epidemiology and Community Health*, **33**, 223–228.

13. Freeman, P. R. (1979). Algorithm AS 145: Exact distribution of the largest multinomial frequency, *Applied Statistics*, **28**, 333–336.

14. Galambos, J (1977). Bonferroni inequalities, *Annals of Probability*, **5**, 577–581.

15. Good, I. J. (1957). Saddle–point methods for the multinomial distribution, *Annals of Mathematical Statistics*, **28**, 861–881.

16. Gould, M. S., Petrie, K., Kleinman, M. H. and Wallenstein, S. (1994). Clustering of attempted suicide: New Zealand national data, *International Journal of Epidemiology*, **23**, 1185–1189.

17. Greenwood, R. E. and Glasgow, M. O. (1950). Distribution of maximum and minimum frequencies in a sample drawn from a multinomial distribution, *Annals of Mathematical Statistics*, **21**, 416–424.

18. Grimson, R. C. (1979). The clustering of disease, *Mathematical Biosciences*, **46**, 257–278.

19. Grimson, R. C. (1993). Disease clusters, exact distributions of maxima, and P–values, *Statistics in Medicine*, **12**, 1773–1794.

20. Grimson, R. C. and Oden, N. (1996). Disease clusters in structured environments, *Statistics in Medicine*, **15**, 851–871.

21. Hewitt, D., Milner, J., Csima, A. and Pakula, A. (1971). On Edwards' criterion of seasonality and a non-parametric alternative, *British Journal of Preventive and Social Medicine*, **25**, 174–176.

22. Jogdeo, K. and Patil, G. P. (1975). Probability inequalities for certain multivariate discrete distribution, *Sankhyā, Series B*, **37**, 158–164.

23. Johnson, N. L., Kotz, S. and Balakrishnan, N. (1997). *Discrete Multivariate Distributions*, New York: John Wiley & Sons.

24. Johnson, N. L. and Young, D. H. (1960). Some applications of two approximations to the multinomial distribution, *Biometrika*, **47**, 463–469.

25. Kotz, S. and Balakrishnan, N. (1997). Advances in urn models during the past two decades, In *Advances in Combinatorial Methods and Applications to Probability and Statistics* (Ed., N. Balakrishnan), pp. 203–256, Boston: Birkhaüser.

26. Kounias, E. G. (1968). Bounds for the probability of a union, with applications, *Annals of Mathematical Statistics*, **39**, 2154–2158.

27. Kounias, E. and Marin, J. (1974). Best linear Bonferroni bounds, In *Proceedings of the Prague Symposium on Asymptotic Statistics, Vol. II* (Ed., J. Hájek), pp. 179–213, Prague, Czech Republic: Charles University.

28. Kozelka, R. M. (1956). Approximate upper percentage points for extreme values in multinomial sampling, *Annals of Mathematical Statistics*, **27**, 507–512.

29. Krauth, J. (1991). Bounds for the upper tail probabilities of the multivariate disjoint test, *Biometrie und Informatik in Medizin und Biologie*, **22**, 147–155.

30. Krauth, J. (1992a). Bounds for the upper tail–probabilities of the circular ratchet scan statistic, *Biometrics*, **48**, 1177–1185.

31. Krauth, J. (1992b). Bounds for the tail–probabilities of the linear ratchet scan statistic, In *Analyzing and Modeling Data and Knowledge* (Ed., M. Schader), pp. 51–61, Berlin, Germany: Springer-Verlag.

32. Krauth, J. (1993). Spatial clustering of species based on quadrat sampling, In *Information and Classification* (Eds., O. Opitz, B. Lausen and E. Klar), pp. 17–23, Berlin, Germany: Springer-Verlag.

33. Krauth, J. (1996a). Spatial clustering of neurons by hypergeometric disjoint statistics, In *From Data to Knowledge* (Eds., W. Gaul and D. Pfeifer), pp. 253–261, Berlin, Germany: Springer-Verlag.

34. Krauth, J. (1996b). Bounds for p–values of combinatorial tests for clustering in epidemiology, In *Analysis and Information Systems* (Eds., H. H. Bock and W. Polasek), pp. 64–72, Berlin, Germany: Springer-Verlag.

35. Kwerel, S. M. (1975). Most stringent bounds on aggregated probabilities of partially specified dependent probability systems, *Journal of the American Statistical Association*, **70**, 472–479.

36. Levin, B. (1981). A representation for multinomial cumulative distribution functions, *Annals of Statistics*, **9**, 1123–1126.

37. Levin, B. (1983). On calculations involving the maximum cell frequency, *Communications in Statistics—Theory and Methods*, **12**, 1299–1327.

38. Mallows, C. L. (1968). An inequality involving multinomial probabilities, *Biometrika*, **55**, 422–424.

39. Mantel, N. (1967). The detection of disease clustering and a generalized regression approach, *Cancer Research*, **27**, 209–220.

40. Mantel, N., Kryscio, R. J. and Myers, M. H. (1976). Tables and formulas for extended use of the Ederer–Myers–Mantel disease-clustering procedure, *American Journal of Epidemiology*, **104**, 576–584.

41. Marrero, O. (1983). The performance of several statistical tests for seasonality in monthly data, *Journal of Statistical Computation and Simulation*, **17**, 275–296.

42. Marrero, O. (1988). The power of a nonparametric test for seasonality, *Biometrical Journal*, **30**, 495–502.

43. Marrero, O. (1992). A maximum rank–sum test for one-pulse variation in monthly data, *Biometrical Journal*, **34**, 485–500.

44. McKnight, R. H., Kryscio, R. J., Mays, J. R. and Rodgers, G. C. (1996). Spatial and temporal clustering of an occupational poisoning: the example of Green Tobacco Sickness, *Statistics in Medicine*, **15**, 747–757.

45. Naus, J. I. (1966). A power comparison of two tests of non-random clustering, *Technometrics*, **8**, 493–517.

46. Owen, D. B. and Steck, G. P. (1962). Moments of order statistics from the equicorrelated multivariate normal distribution, *Annals of Mathematical Statistics*, **33**, 1286–1291.

47. Proschan, F. and Sethuraman, J. (1975). Simple multivariate inequality using association, *Theory of Probability and Its Applications*, **20**, 193–195.

48. Rogerson, P. A. (1996). A generalization of Hewitt's test for seasonality, *International Journal of Epidemiology*, **25**, 644–648.

49. Stark, C. R. and Mantel, N. (1967a). Lack of seasonal– or temporal–spatial clustering of Down's syndrome births in Michigan, *American Journal of Epidemiology*, **86**, 199–213.

50. Stark, C. R. and Mantel, N. (1967b). Temporal–spatial distribution of birth dates for Michigan children with leukemia, *Cancer Research*, **27**, 1749–1775.

51. Viktorova, I. I. (1969). Asymptotic behavior of maximum of an equiprobable polynomial scheme, *Mathematical Notes*, **5**, 184–191.

52. Viktorova, I. I. and Sevastyanov, B. A. (1967). Limit behavior of the maximum in a polynomial representation, *Mathematical Notes*, **1**, 220–225.

53. Wallenstein, S., Gould, M. S. and Kleinman, M. (1989a). Use of the scan statistic to detect time–space clustering, *American Journal of Epidemiology*, **130**, 1057–1064.

54. Wallenstein, S., Weinberg, C. R. and Gould, M. (1989b). Testing for a pulse in seasonal event data, *Biometrics*, **45**, 817–830.

55. Walter, S. D. (1980). Exact significance levels for Hewitt's test for seasonality, *Journal of Epidemiology and Community Health*, **34**, 147–149.

56. Walter, S. D. and Elwood, J. M. (1975). A test for seasonality of events with a variable population at risk, *British Journal of Preventive and Social Medicine*, **29**, 18–21.

57. Ward, D. L. (1992). Guillan–Barré syndrome and influenza vaccination in the US army, 1980–1988, *American Journal of Epidemiology*, **136**, 374–375.

58. Wartenberg, D. and Greenberg, M. (1990). Detecting disease clusters: the importance of statistical power, *American Journal of Epidemiology*, **132**, 156–166.

59. Yates, F. (1934). Contingency tables involving small numbers and the $\chi^2$ test, *Journal of the Royal Statistical Society*, **1**, 217–235.

60. Yusas, I. S. (1972). On the distribution of the maximum frequency of a multinomial distribution, *Theory of Probability and Its Applications*, **17**, 712–717.

# 4

## Scanning Multiple Sequences

**Joseph I. Naus**

*Rutgers University, New Brunswick, NJ*

**Abstract:** Much of the scanning literature focuses on unusual clusters of a given type of event in a single sequence of trials or time period. In this chapter, we discuss approaches to simultaneously scan multiple series. In one set of problems, there are multiple series corresponding to the occurrence of different types of events over the same period of time; the researcher looks for multiple-type clusters allowing for lagged effects between the different types of events. In the second set of problems, one scans multiple series looking for the largest common perfect or almost perfect match between all or most of the series. This second set of problems is of importance to molecular biologists searching for strong homologies in DNA sequences. Some related problems in two-dimensional scanning are mentioned.

**Keywords and phrases:** Matching, DNA, scan statistic, double scan statistic

## 4.1 Discrete Scan Statistic and Its Generalization to Multiple Sequences

There is a large body of literature that describes properties of tests that scan for unusually large clusters, or multiple large clusters of events over time or space; see Naus (1988). Scan statistics and their distributions have been developed to study clustering in time (viewed as continuous), or in sequences of trials (the discrete case). In this chapter, we focus on the discrete case, but several of the problems have continuous counterparts, so it is useful to briefly distinguish among the cases. Furthermore, in certain applications the continuous case can serve as an approximation to the discrete case, and conversely.

For the continuous case, the times of occurrence of events are assumed to have a specified random distribution over a time interval $(0, T)$. The continuous scan statistic $S_w$ is the maximum number of points in any subinterval of $(0, T)$ of length $w$. The name follows from the fact that we are "scanning" $(0, T)$ with a window of length $w$, and looking for the largest number of points in the window.

For the discrete case, we start by considering a sequence $X_1, X_2, \ldots, X_N$ of integer-valued random variables. The scan statistic is the maximum moving sum of $m$ of the $X_i$'s. Formally, for $m$ an integer, and $t = 1, 2, \ldots, N - m + 1$, define the random variables $Y_t = \sum_{t \leq i \leq t+m-1} X_i$. The scan statistic is $S_m = \max_{1 \leq t \leq N-m+1} \{Y_t\}$.

Glaz and Naus (1991) have given accurate approximations and tight bounds for the distribution of $S_m$ for the case where the $X_i$'s are i.i.d. variables that take integer values. Various asymptotic results have been derived for this case under a variety of names: Erdös-Rényi Laws [Deheuvels (1985)], increments of partial sums. For the general case, the tight bounds and sharp approximations can sometimes be computationally complex, and the asymptotic results can be slow to converge.

An important special case is where the $X_i$'s only take the values 0 and 1. Many researchers deal with data that can be viewed as a series of trials, each with two possible outcomes. We will arbitrarily label the two alternative possible outcomes of a trial as "success" and "failure." $S_m$ is the maximum number of "successes" within any $m$ contiguous trials within the $N$ trials. Many results and applications exist for the distribution of the scan statistic for the special case of a sequence of 0-1 variates. These appear under a variety of names: Generalized Birthday Problem, Erdös-Rényi Laws, quotas, generalized runs, and $k$-in-$n$ in $m$ reliability sequences. For the case where the $X_i$'s are i.i.d. Bernoulli variables, exact results and readily computable highly accurate approximations and tight bounds exist for the distribution of $S_m$; see, for example, Naus (1974, 1982) and Glaz and Naus (1991).

Our generalizations deal with the simultaneous scanning of multiple sequences with a window, or windows of $m$ consecutive integers or letters. There are many possible ways in which this can be done, and they result in flavors of scan-type statistics of practical interest. Many of these generalizations lead to open combinatorial and probabilistic problems. Interesting applications have helped focus some of the directions of generalization, and in this chapter we will describe some of these. In the next two sections, we discuss two directions of generalization and the results derived. In the final section, we will mention some other directions of generalization of interest.

## 4.2 Clusters in Multiple Sequences Over the Same Time Period

Researchers studying series of multiple outcomes sometimes seek to determine whether the observed clustering of different types of outcomes is likely to arise under a given chance model. Most of the scan statistic literature describes properties of a statistic that scans for unusually large clusters, or multiple large clusters of events over time. However, almost all of these results deal with one type of event within the cluster. In this section, we discuss generalization to several types of events. Two early results and the applications that motivated them are as follows.

The first result grew out of a quality control setting. Quality control and acceptance sampling had led to some early results on the classic scan statistic. In some acceptance sampling plans, one looks at overlapping sets of items or batches of items, rather than just individual batches before passing sentence on individual batches. Anscombe, Godwin, and Plackett (1947) considered such types of deferred sentencing acceptance sampling plans. Ahn and Kuo (1994) considered the event that $k$ successes out of $m$ consecutive trials, occur for first time at $j$th trial, and note (p. 135) "In view of the control chart and acceptance sampling system, we recall that most of signaling and switching rules in classical statistical quality control (SQC) procedures take the form of signal or switch as soon as $k$ out of $m$ consecutive trials result in the occurrence of an event of interest." Page (1955), Roberts (1958) and some others have studied control chart procedures with warning lines in addition to the usual out-of-control lines. Some of the procedures flag the process if one point is out-of-control, or if $k$ out of $m$ consecutive points fall outside warning zones. Page (1955) has outlined (but does not use) an approach to find the expected waiting time until a cluster of the two types of events. The method generates all state combinations for the related discrete Markov Chain, and Page (1955) noted that even for the one type of event "$k$ out of $m$ consecutive points" it is complex except for some very simple cases.

For recent advances on the Markov-chain embedding approach for the one type of event scan that may be useful more generally, see Koutras and Alexandrou (1995). These authors have been motivated by examples in reliability of systems. In reliability theory, researchers evaluate the reliability of configurations of components, or design a system to have a certain reliability. Papastavridis and Koutras (1993) and others have derived results for "$k$ within consecutive $m$ in $N$ systems." These systems are viewed as a linearly ordered set of $N$ independent components with possibly different individual probabilities of being defective; the system fails if there are $k$ defectives within any $m$ consecutive components in the system.

Huntington (1976) was motivated by a problem of breaks in transoceanic cables. Such breaks were relatively rare, and the company had a specialized ship to service and repair such breaks. The ship had to service cables in two different distant regions. If there were two breaks in the same region within a few days of each other, that would not pose any problem. However, if there was a break in both of the regions' cables within a few days of each other, there would be an unacceptable delay in repairing the cables. This led Huntington to derive general results for the expected waiting time until a cluster that contained at least one of each of the two types of events (a break in region one, and a break in region two) within a few days of each other. Huntington then generalized this to a cluster of at least $k$ events within $d$ days, that satisfy various constraints on the number of types of events.

Naus and Wartenberg (1997) have developed a scan-type statistic called the *double scan statistic* based on the number of "declumped" (a type of non-overlapping) clusters that contain at least one of each of two types of event. They derived the expectation and approximate distribution of the number of declumped clusters for this test statistic for two chance models. They were analyzing a data set of different causes of death (homicide, suicide, accidents) among Americans aged 15–25 for a 7-year period. The data were broken down by day, type of death, race, and gender, and by county. However, the data did not give names or other identifiers that might relate one death to another.

The goal of the analysis was to identify interesting clusters that could then be checked with police departments or other outside sources. An earlier article by Greenberg *et al.* (1991) had used the scan statistic to focus on unusual homicide clusters, and on unusual suicide clusters. Naus and Wartenberg (1997) were interested in techniques for focusing on unusual clustering of the two different types of events. Unfortunately, it happens that someone murders another person and then kills themself. Several recent homicide/suicides in Kentucky, in Quebec, and elsewhere received wide publicity and concern. A different motivating example was one in which the two different events were male suicide and female suicide. While many of the statistical scan methods deal with the clustering of one type of event over time, one can readily apply occupancy theory to the clustering of two types of events on the same day. In other cases, there may be a lagged relation. If we anticipated a delayed effect, we would want to look at cases where two types of events might occur within the same $d$-day interval. It is for these cases that Naus and Wartenberg (1997) developed the double scan statistic.

*The double-scan statistic:* Given that at least one of each of the two types of events occur within a $d$-day period, we say that a "2-type $d$-day cluster" has occurred. Over a long period of $D$ days, a scientist may observe several such clusters. The scientist seeks to determine whether the observed number of clusters is unusually greater than what would be expected under certain chance models.

For the case $d = 1$, the number of 2-type 1-day clusters, $N_1$, can be counted simply as the number of days in the $D$-day period that contain at least one of each of the two types of events. For the case $d > 1$, there are many alternative ways to count the number of 2-type $d$-day clusters. Naus and Wartenberg (1997) have used an approach that avoids multiple counting of the same, or too closely overlapping clusters.

Define the event $E_i$ to have occurred if anywhere within the $d$ consecutive days $i, i+1, \ldots, i+d-1$ there are at least one of each of the two types of events. The event $E_i$ indicates the occurrence of a 2-type $d$-day cluster. Let $Z_i = 1$, if $E_i$ occurs and none of $E_{i-1}, E_{i-2}, \ldots, E_{i-d+1}$ occur. Let $Z_i = 0$, otherwise. Let $N_d = \sum_{1 \leq i \leq D-d+1} Z_i$. Naus and Wartenberg (1997) have termed $N_d$, the *double scan statistic*.

This method counts the number of times that an $E_i$ occurs with no previously overlapping $E_j$'s. This particular method of counting has the advantage that when the events are relatively rare and distributed according to certain chance models, the number of such "declumped" clusters is approximately Poisson distributed. Further, the method of declumping used allows one to estimate the error in the Poisson approximation through the Chen–Stein method. For a more detailed discussion of Poisson approximation and the declumping approach, see Aldous (1989), Arratia, Goldstein, and Gordon (1990), and Barbour, Holst, and Janson (1992).

Results for the expectation and variance of $N_d$ are derived for two chance models. In the retrospective model, there are exactly $A$ of the $D$ days where a type-one event occurs, and exactly $B$ of the $D$ days where a type-two event occurs. All $\binom{D}{A}$ ways of picking the $A$ type-one days, and all $\binom{D}{B}$ ways of picking the $B$ type-two days are equally likely, and the occurrence of the two types of days are independent. For this model, the exact expectation and approximate variance of $N_d$ are derived.

In the *prospective model*, let $A_{ki}$ denote the event that the $i$th day contains a type $k$ event, for $k = 1, 2$ and $i = 1, 2, \ldots, D$. Let $\alpha_{ki} = P(A_{ki})$, and all $A_{ki}$ are mutually independent. For this model, the exact expectation and variance of $N_d$ are derived. The Chen–Stein error bounds for the goodness-of-fit of the Poisson approximation are illustrated for the simple case where $\alpha_{ki} = \alpha_k$ for $k = 1, 2$ and all $i$.

Naus and Wartenberg (1997) have also considered a mixed model where some of the two types of events are linked in time, but many are not, and have illustrated how to evaluate the power of the double scan statistic against this alternative.

For $d > 1$, the double scan statistic $N_d$ measures for a possibly delayed relation between two types of events, but no order is prespecified for the two events. For the events in the quality control application, this might be reasonable. However, for homicide/suicide clusters one would be looking for unusual clusters where a person first commits a homicide and then kills themself. That is, one would anticipate that the day of the homicide would be the same or earlier than the day of the suicide. This led Naus and Wartenberg (1997) to develop a *directional double scan statistic*, and illustrate its application. Define an $E_i^*$ to have occurred if anywhere within the $d$ consecutive days $i, i+1, \ldots, i+d-1$ there are at least one of each of the two types of events with a type-one event on the same or previous day as a type-two event. The statistic declumps (counts the nonoverlapping $E^*$'s) as before.

The discrete double scan statistics have more general forms as well as continuous counterparts. The discrete double scan statistic can be generalized for the case of two types of events to cases where there are at least $r$ type-one and $s$ type-two events within a $d$-day period. For other applications, the statistics can be generalized to more than two types of events, and the distribution of the number of declumped clusters derived.

The following is an example of the continuous version of the double scan statistic under a simple model. Consider two independent Poisson processes on $(0, T)$, where the first process generates type-one points, and the second process generates type-two points. Scan $(0, T)$ with a window of length $w$ and count the number of non-overlapping clumps $C_w$ of times that there are at least one each of a type-one and type-two point in the window. For the case where $d \ll D$, the distribution of $N_d$ can frequently approximate the distribution of $C_w$, where $w = (d/D)T$, and conversely. To get a standardized form for the distribution of $C_w$, without loss of generality, choose the units so that $T = 1$. The continuous double scan can be generalized to cases where there are at least $r$ type-one and $s$ type-two events within a window of length $w$.

---

## 4.3 Matching in Multiple Random Letter Sequences

Given an alphabet consisting of $B$ different letters, there are $B^m$ possible different $m$-letter words. Consider $R$ sequences each consisting of $N$ letters chosen from the $B$-letter alphabet. If all the $R$ sequences share a common $m$-letter word, we say there is a *perfect $m$-word match* between the $R$ sequences. The common word could appear in different positions in the $R$ sequences; this is called the *nonaligned case*. Sometimes, we are looking for a common $m$-letter word in the same position in all $R$ sequences; this is called the *aligned case*. Researchers are also interested in almost-perfect $m$-word matches. There may be a common $m$-letter word that appears with up to $s$ letters changed in any match word. This is referred to as the *almost perfectly matching word allowing $s$ mismatches*. There are several other variations in how the number of mismatches can be counted.

The classic scan statistic is directly related to perfect or almost perfect $m$-

word matches in the aligned case. View the $R$ sequences of $N$ letters as $R$ rows aligned one above the other. Below the last row, add an $(R+1)$st row consisting of the sequence $X_1, X_2, \ldots, X_N$ where $X_i = 1$ if the $R$ letters above it are identical, and $X_i = 0$ otherwise. Now, scan the sequence of $X$'s with a window of $m$ consecutive letters. If the window contains $m$ 1's, this is equivalent to there being an aligned perfectly matching $m$-word. If the window contains $k$ 1's, this is equivalent to there being an aligned almost perfectly matching $m$-word with at most $m - k$ mismatches. For the case where the $R$ sequences are mutually independent sequences of independent (but not necessarily identically distributed) letters, the results for the classic 0-1 scan statistic give the probability of aligned matching words allowing for mismatches. Exact results, highly accurate approximations, and tight bounds exist for this aligned case.

The generalization that we discuss in this section involves the length of the longest common words in all $R$ (nonaligned) sequences. Here, we are scanning each of the $R$ sequences looking for any common word. Our motivation comes from an application in molecular biology.

Scientists and researchers compare sequences of DNA from several biological sources. The DNA can be viewed as a sequence of letters from a four-letter alphabet (the nucleotides A,C,G,T), or a 20-letter amino acid alphabet (certain sets of the triplets of nucleotides), or in other ways. Similarity between different sequences suggests commonality of functions or ancestry.

There are several large data banks of DNA sequences. Researchers with newly sequenced segments search these data looking for homologies. Computer algorithms have been developed to scan two long DNA sequences, searching for matching words. In comparing two long sequences, one would find, purely by chance, some matching words. The researcher seeks to determine, for various chance models, what an unusual match is.

In some applications, two sequences are aligned by some overall criteria, and one looks for positionally matching words. Piterbarg (1992) has noted that a standard method starts with a given alignment of the two sequences, and compares each pair of aligned letters to get a sequence of similarity scores. The researcher often focuses on whether the letters match (a 0,1 similarity scoring). The method then computes the scan statistic $S_m$ for the sequence of similarity scores. Under the usual null model, it is assumed that the sequence of similarity scores is mutually independent. Arratia, Gordon, and Waterman (1990) have noted that even though the letters within a DNA sequence are not independent, the proportion of time that long common words were observed in unrelated sequences of DNA, is close to that estimated from the independence model. For this model, 0-1 scoring, two or more aligned sequences, and perfectly matching words, or for two aligned sequences and almost perfectly matching words, the classic (0-1) scan statistic results provide the necessary measures of unusualness. For the case of more than two aligned squences and almost perfectly matching words, a variety of similarity or consensus systems can be used to score letters in the same position in the sequences. Naus and Sheng (1996) describe several integer scoring systems and have given distributional results for the scan statistic that measures the most similar (aligned) $m$-word.

In other applications, the researcher is looking for common words in two or more (nonaligned) sequences. Sometimes, the researcher is using long common words as a consensus-based approach for multiple sequence alignment; see, Waterman (1986) and Leung *et al.* (1991) for a discussion of such algorithmic alignment approaches. Other times, the researcher is looking for homologies in and between certain proteins. The researchers seek help to distinguish unusual matching, and significance levels are built in as part of the search and match engines. These measures of unusualness are based on the distribution of the length of the longest matching common word in $R$ sequences, under various chance models.

For the case of a (perfect) common word in two (nonaligned) sequences, and the independence model, Mott, Kirkwood, and Curnow (1990) and Sheng and Naus (1994) have presented excellent approximations. Karlin and Ost (1988) have given asymptotic results both for the independence and more general models that gives (in a modified form) excellent approximations for even moderate size sequences.

For the case of a perfect common word in three or more (nonaligned) sequences, Karlin and Ost (1988) have given asymptotic results, but these can converge very slowly, particularly as the number of sequences gets larger. For six or more moderate size sequences, the approximations can be off. Naus and Sheng (1997) have presented several approximations that are more accurate for this case. The following table illustrates this for a few examples. The probabilities are for an $m$ letter word in common to $R$ independent sequences each of $N$ i.i.d. letters drawn from an equally likely four-letter alphabet.

**Table 4.1:** Probability of an $m$ letter word in common to $R$ sequence of $N$ letters

| $m$ | $R$ | $N$ | K&O(4) | N&S(8) | 100,000 Simulations | Upper Bonferroni Bound |
|-----|-----|------|--------|--------|---------------------|------------------------|
| 4 | 6 | 74 | .110 | .050 | .050 | |
| 4 | 8 | 111 | .227 | .049 | .051 | |
| 4 | 10 | 118 | .082 | .010 | .010 | |
| 6 | 6 | 682 | .080 | .050 | | .056 |
| 6 | 10 | 1605 | .287 | .050 | | .062 |

Karlin and Ost (K&O) and Naus and Sheng (N&S) both have given a Poisson-type approximation (with some declumping) for this case. We will describe the approximations in a heuristic way for the last case in Table 4.1.

Karlin and Ost's approximation can be interpreted for this case as follows: For a 4-letter equally likely alphabet, the probability that the second sequence has the same first letter as the first sequence is $1/4$. The probability that the first letter of each of the 10-sequences is the same is $\lambda = p^9$, where $p = 1/4$. The probability that the first six letters of all 10-sequences are a common word $= (p^9)^6 = \lambda^6$. There are $(N-m+1) = 1600$ positions where a six-letter word could start in the first sequence, and similarly for all the sequences. Since the common word could be anywhere in each of the 10-sequences, we have to consider a very large number $(N-m+1)^R = 1600^{10}$ of possible combinations of positions. The probability of a 10-sequence match in any particular one of the combinations of positions is very small and equals $0.25^{54}$. The expected number of such 10-sequence matches is $(N-m+1)^R \lambda^m = 1600^{10}(0.25^{54}) = 0.338131$. Using a Poisson-type approximation for the probability that at least one of the possible sets of combinations leads to a common word, we get $1 - \exp(-0.338131) = 0.287$. Karlin and Ost used a declumping factor of $(1-\lambda)$ to multiply the expectation to adjust for the fact that a 7-letter common word would include two overlapping 6-letter common words. For this example, the $(1-\lambda)$ factor makes no difference, the approximation is still $0.287$.

There is a strong dependence of multiple matching that is not taken into account in the K&O approximation. Consider two sets of positions for the 6-letter words in the 10-sequences. The first set consists of the first six positions in all 10-sequences. The second set consists of the first six positions in the first 9-sequences, and the last six positions in the last sequence. These two sets are highly dependent.

Naus and Sheng (1997) have taken into account this type of dependence in the following Poisson-type approximation. Let $\delta$ denote the probability that the first $m = 6$ letters of sequence one appears somewhere in sequence two. Since there are $(N-m+1) = 1600$ positions for the matching word in sequence two, and the probability of a match in any one position is $p^6 = (0.25)^6$, the expected number of such matches is $1600(0.25)^6 = 0.390625$. A Poisson approximation gives $\delta \approx 1 - \exp\{-0.390625\} = 0.3233661$. The probability that the first 6 letters in sequence one find a matching word somewhere in all 9 other sequences is approximately $\delta^9$. This is the first stage of the approximation.

In sequence one, there are 1600 possible positions for a 6-letter word, and we could argue that the expected number of 10-sequence common matches would be $\delta^9 \times 1600$. However, one does not expect all the 1600 words in sequence one to be distinct. Given a 4-letter alphabet, there are $4^6 = 4096$ distinct 6-letter words. Viewing the 1600 positions as 1600 balls distributed at random into the 4096 distinct cells, we expect there to be $N^* = 4096\{1 - (4095/4096)^{1600}\} = 1324.64$ distinct 6-letter words in sequence one. The expected number of (declumped) 10-sequence matches is $1324.64(\delta^9) = 0.05121$. The probability of at least one 6-letter word is common to all 10-sequences is by the second stage Poisson approximation as $1 - \exp\{-.05121\} = 0.050$.

Karlin and Ost (1988) have given general asymptotic results for a common word in $R$ out of $S$ independent sequences, where the letters within an individual sequence are (a) i.i.d., or (b) independent but not identically distributed, or (c) alternative preasymptotic approximations. Currently, research is being carried out for the case of a matching word in multiple sequences allowing some mismatches within the common word.

## 4.4   Lattice Problems Related to Common Words in Multiple Sequences

The matching problem and the two event problem have various generalized problems on the lattice. In Section 4.3, we discussed the problem of scanning for common words in $R$ multiple aligned sequences of $N$ letters. There have been some results on the generalized problem of scanning $R$ $N \times N$ lattices, looking for a common rectangular set of letters. View the $R$ lattices as being placed on top of each other. Underneath all the $R$ lattices, place another (indicator) $N \times N$ lattice from the 0-1 alphabet. If there is a common letter in the $(i, j)$th position in all the $R$ lattices, place a 1 in the $(i, j)$th position in the bottom lattice, otherwise a 0. Darling and Waterman (1985) and Sheng and Naus (1996) have derived results for the probability of finding an $s \times t$ rectangular sub-lattice all of 1's within the indicator lattice. This is equivalent to finding a common generalized (2-dimensional word) in the same position in all lattices. Chen and Glaz (1996) have generalized this to allow for some mismatches in the 0-1 sub-lattice.

A different generalization of the scan statistic to the lattice is motivated by a recent approach for analysis of codes in text; see, Witztum, Rips, and Rosenberg (1994) and Drosnin (1997). In this approach, a continuous text is written as an alphabetical lattice, and the rows are scanned for a particular word, and the columns scanned for a related word, and the closesness of the two words is measured. Simulation approaches are used to evaluate the unusualness of such matches. This application suggests a range of interesting and useful combinatorial problems.

## References

1. Ahn, H. and Kuo, W. (1994). Applications of consecutive system reliability in selecting acceptance sampling strategies, In *Runs and Patterns in Probability* (Eds., A. P. Godbole and S. G. Papastavridis), pp. 131–162, Dordrecht, The Netherlands: Kluwer Academic Publishers.

2. Aldous, D. (1989). *Probability Approximations via the Poisson Clumping*

*Heuristic*, New York: Springer-Verlag.

3. Anscombe, F. J., Godwin, H. J. and Plackett, R. L. (1947). Methods of deferred sentencing in testing, *Journal of the Royal Statistical Society, Series B*, **7**, 198–217.

4. Arratia, R., Goldstein, L. and Gordon, L. (1990). Poisson approximation and the Chen-Stein method, *Statistical Science*, **5**, 403–434.

5. Arratia, R., Gordon, L. and Waterman, M. S. (1990). The Erdös-Rényi law in distribution, for coin tossing and sequence matching, *Annals of Statistics*, **18**, 539–570.

6. Barbour, A. D., Holst, L. and Janson, S. (1992). *Poisson Approximation*, Oxford, England: Clarendon Press.

7. Chen, J. and Glaz, J. (1996). Two-dimensional discrete scan statistics, *Statistics & Probability Letters*, **31**, 59–68.

8. Darling, R. W. R. and Waterman, M. S. (1985). Matching rectangles in *d*-dimensions: Algorithms and laws of large numbers, *Advances in Mathematics*, **55**, 1–12.

9. Deheuvels, P. (1985). On the Erdös-Rényi theorem for random fields and sequences and its relationships with the theory of runs and spacings, *Zeitschrift Wahrscheinlichkeitstheorie*, **70**, 91–115.

10. Drosnin, M. (1997). *The Bible Code*, New York: Simon & Schuster.

11. Glaz, J. and Naus J. (1991). Tight bounds and approximations for scan statistic probabilities for discrete data, *Annals of Applied Probability*, **1**, 306–318.

12. Greenberg, M., Naus, J., Schneider, D. and Wartenberg, D. (1991). Temporal clustering of homicide and suicide among 15-24 year old white and black Americans, *Ethnicity and Disease*, **1**, 342–350.

13. Huntington, R. J. (1976). Expected waiting time till a constrained quota, *Technical Report*, AT&T.

14. Karlin, S. and Ost, F. (1987). Counts of long aligned word matches among random letter sequences, *Advances in Applied Probability*, **19**, 293–351.

15. Karlin, S. and Ost, F. (1988). Maximal length of common words among random letter sequences, *Annals of Probability*, **16**, 535–563.

16. Koutras, M. V. and Alexandrou, V. A. (1995). Runs, scans and urn model distributions: A unified Markov chain approach, *Annals of the Institute of Statistical Mathematics*, **47**, 743–766.

17. Leung, M. Y., Blaisdell, B. E., Burge, C. and Karlin, S. (1991). An efficient algorithm for identifying matches with errors in multiple long molecular sequences, *Journal of Molecular Biology*, **221**, 1367–1378.

18. Mott, R. F., Kirkwood, T. B. L. and Curnow, R. N. (1990). An accurate approximation to the distribution of the length of the longest matching word between two random DNA sequences, *Bulletin of Mathematical Biology*, **52**, 773–784.

19. Naus, J. I. (1974). Probabilities for a generalized birthday problem, *Journal of the American Statistical Association*, **69**, 810–815.

20. Naus, J. I. (1982). Approximations for distributions of scan statistics, *Journal of the American Statistical Association*, **77**, 177–183.

21. Naus, J. I. (1988). Scan statistics, In *Encyclopedia of Statistical Sciences* Volume 8 (Eds., N. L. Johnson and S. Kotz), pp. 281–284, New York: John Wiley & Sons.

22. Naus, J. I. and Sheng, K. N. (1996). Screening for unusual matched segments in multiple protein sequences, *Communications in Statistics— Simulation and Computation*, **25**, 937–952.

23. Naus, J.I. and Sheng, K.N. (1997). Matching among multiple random sequences, *Bulletin of Mathematical Biology*, **59**, 483–496.

24. Naus, J. I. and Wartenberg, D. (1997). A double scan statistic for clusters of two types of events, *Journal of the American Statistical Association*, **92**, 1105–1113.

25. Page, E. S. (1955). Control charts with warning lines, *Biometrika*, **42**, 243–257.

26. Papastavridis, S. G. and Koutras, M. V. (1993). Bounds for reliability of consecutive $k$-within-$m$-out-of-$n$: $F$ system, *IEEE Transactions on Reliability*, **42**, 156–160.

27. Piterbarg, V. I. (1992). On the distribution of the maximum similarity score for fragments of two random sequences, In *Mathematical Methods of Analysis of Biopolymer Sequences* (Ed., Simon Gindikin), pp. 11–18, DIMACS series in Discrete Mathematics and Theoretical Computer Science, Volume 8, Providence, RI: American Mathematical Society.

28. Roberts, S. W. (1958). Properties of control chart zone tests, *Bell System Technical Journal*, **37**, 83–114.

29. Sheng, K. N. and Naus, J. I. (1994). Pattern matching between two non-aligned random sequences, *Bulletin of Mathematical Biology*, **56**, 1143–1162.

30. Sheng, K. N. and Naus, J. I. (1996). Matching fixed rectangles in 2-dimensions, *Statistics & Probability Letters*, **26**, 83–90.

31. Waterman, M. S. (1986). Multiple sequence alignment by consensus, *Nucleic Acids Research*, **14**, 9095–9102.

32. Witztum, D., Rips, E. and Rosenberg, Y. (1994). Equidistant letter sequences in the book of Genesis, *Statistical Science*, **9**, 429–438.

# PART III
## CONTINUOUS SCAN STATISTICS

# 5

# Approximations of the Distributions of Scan Statistics of Poisson Processes

**Sven Erick Alm**

*Uppsala University, Uppsala, Sweden*

**Abstract:** We study scan statistics of Poisson processes in one and higher dimensions. First, a very accurate approximation is established in one dimension. This is done by studying upcrossings of the scanning process and noting that these occur in clusters. The clusters appear more or less independently and the cluster size is estimated by a random walk argument. This idea is then used repeatedly to obtain approximations in higher dimensions. Simulation is used to check the accuracy of the approximations in two and three dimensions. A discussion of these simulations is included, as they are by no means trivial to perform.

**Keywords and phrases:** Scan statistic, Poisson process, higher dimensions, unconditional, simulation

## 5.1  Introduction

Clustering of points in Poisson processes, in one or more dimensions, is of interest in many applications, including risk analysis, telecommunication, epidemiology, reliability, and traffic theory.

It is, therefore, of interest to find the distribution (or an approximation) of the maximum number of points in such a cluster. This study originates from the following teletraffic problem: Calls arrive according to a Poisson process to a computer controlled exchange. The computer is protected against overload through an overload control system. This system may fail if there is an extremely large number of calls during a short time interval (1–5 seconds). The interesting quantity is the maximum number of calls in such an interval during the busy hour of a day, when the number of calls per second is typically 10–50.

The problem is also of interest for Poisson processes in more than one dimension. As an example, suppose that, in a certain material, microcracks appear at random, according to a Poisson process, on the surface (or in the body) of the material. One way to explain cracks in the material is that they develop if sufficiently many microcracks appear in a small area (volume). It is then of interest to study the distribution of the maximum number of microcracks in any translate of such an area (volume). This maximum is known as the scan statistic.

To derive approximations of the distribution of the scan statistic, the following notation will be used.

In one dimension, consider a Poisson process on $\mathbf{R}^+$, $\{X_t, t \geq 0\}$, with intensity $\lambda$, and the associated *scanning process* $\{Y_t(w), t \geq 0\}$, where $Y_t = X_{t+w} - X_t$. Then,

$$S_w = S_w(\lambda, T) = \max_{0 \leq t \leq T - w} Y_t(w)$$

is the *scan statistic* with a *scanning window* of length $w$. Several authors [e.g., Naus (1982), Alm (1983), Janson (1984), and Loader (1991)] have suggested approximations for the distribution of $S_w$.

In two dimensions, consider a Poisson process $X$, with intensity $\lambda$, on a rectangular area $A = [0, T_1] \times [0, T_2]$, and a fixed *scanning set* $W$. Let $W(x)$ be the translate of $W$ by $x \in \mathbf{R}^2$, and define the *scan statistic*

$$S_W = S_W(\lambda, A) = \max_{x \in \mathbf{R}^2} X(W(x) \cap A), \tag{5.1}$$

i.e., the maximum number of points in $A$ that can be covered by a translate of the scanning set $W$.

When it is necessary to distinguish between different scanning sets, we will use $S_R$ to denote the scan statistic for rectangular scanning sets, $S_T$ for triangular, and $S_C$ for circular.

Alm (1997) derived an accurate approximation for the distribution of the scan statistic in two dimensions. The arguments are based on the previous work in one dimension by Alm (1983), which is reviewed in Section 5.2. A summary of the approximations in two dimensions is presented in Section 5.3. Simulation is used to compare two suggested approximations for rectangular scanning sets. Based on these, we also give heuristic approximations for general convex scanning sets. The approximations are compared with simulations for circular and triangular scanning sets. Approximations for the distribution of $S_W$ have also been suggested by Aldous (1989) (rectangular and circular $W$).

The conditional case, where a fixed number of points are distributed uniformly over $A$, i.e., where $X(A) = N$, has been studied by Mack (1949) ("general" $W$), Loader (1991) (rectangular $W$), and Månsson (1994) (general convex $W$). In a rather different setting, two-dimensional scan statistics are used by Kulldorff and Nagarwalla (1995) and Hjalmars *et al.* (1996) to study epidemiological problems.

In Section 5.4 we discuss approximations in three dimensions, with a Poisson process $X$ on a rectangular volume, $A = [0, T_1] \times [0, T_2] \times [0, T_3]$, and a fixed rectangular scanning set $R$. In analogy with (5.1), let $R(x)$ be the translate of $R$ by $x \in \mathbf{R}^3$, and define the *scan statistic*

$$S_R = S_R(\lambda, A) = \max_{x \in \mathbf{R}^3} X(R(x) \cap A),$$

i.e., the maximum number of points in $A$ that can be covered by a translate of the rectangular scanning set $R$. This technique is generalized to $d$ dimensions in Section 5.5.

The agreement between the approximation and simulations is good also in three dimensions, although simulations only have been possible to perform for moderate parameter values.

The simulations are discussed in greater detail in Section 5.6. They are very time consuming in more than one dimension, so it is of great interest to perform them in an efficient way, which may include using some variance reducing techniques. The difficulty of simulating in higher dimensions, of course, make, the approximations more interesting.

Limit theorems for the distribution of the $d$-dimensional rectangular scan statistic are studied by Auer, Hornik, and Révśz (1991).

Throughout this chapter, we study scan statistics of homogeneous Poisson processes. In many applications, it would be of interest to generalize the results to the nonhomogeneous case, but this is far from a trivial task.

---

## 5.2 Approximation in One Dimension

Although the object of this chapter is to give approximations in higher dimensions, we will start by discussing an approximation in one dimension in some detail, as this forms the basis for the approximations in higher dimension. The results of this section were first presented by Alm (1983).

Let $\{X_t, t \geq 0\}$ be a Poisson process with intensity $\lambda$, and $Y = \{Y_t, t \geq 0\}$, where $Y_t = Y_t(w) = X_{t+w} - X_t$, be the associated scanning process. Then,

$$S_w = S_w(\lambda, T) = \max_{0 \leq t \leq T - w} Y_t$$

is called the scan statistic.

**Remark 5.2.1** For convenience, we use too many parameters above. $T' = T/w$ and $\lambda' = \lambda w$ are sufficient.

$Y$ is an integer valued stationary process that changes value through jumps of size $\pm 1$. A traditional technique for analyzing extreme values of stationary processes is to study upcrossings. When $Y$ changes value from $n-1$ to $n$, we say that an *upcrossing of level $n$* occurs. Let $M_n$ denote the number of upcrossings of level $n$ in the interval $(0, T - w)$.

The scan statistic $S_w = \max\limits_{0 \le t \le T-w} Y_t$ is related to $M_n$ through

$$
\begin{aligned}
P(S_w < n) &= P(Y_0 < n \cap M_n = 0) \\
&\approx P(Y_0 < n)P(M_n = 0) = F_p(n - 1; \lambda w)P(M_n = 0),
\end{aligned} \tag{5.2}
$$

where $F_p(n; \mu)$ is the distribution function of a Poisson distribution with mean $\mu$. Let $p(n; \mu)$ denote the corresponding probability function.

To approximate $P(M_n = 0)$, we first note the following.

**Lemma 5.2.1** $E(M_n) = \lambda(T - w)p(n - 1; \lambda w)$.

PROOF. An upcrossing in $Y$ at $t$, $0 \le t \le T - w$, can only occur if an event occurs in the Poisson process at $t + w$. The number of such $t$-values is $X_T - X_w$, with mean $\lambda(T - w)$, and such a $t$ is an upcrossing if the number of points in $(t, t + w)$ is exactly $n - 1$, which occurs with probability $p(n - 1; \lambda w)$. ∎

For very high levels $n$, $M_n$ can be approximated by a Poisson distribution, so that

$$
P(M_n = 0) \approx e^{-E(M_n)}. \tag{5.3}
$$

This well-known approximation gives unsatisfactory results for moderately high levels $n$, because of the dependence between upcrossings that occur close to each other.

An example of this can be seen in Figure 5.1, where a simulation of the scanning process $Y$ is plotted around the maximum. There were two upcrossings of the maximum level $n = 70$, and these occurred very close to each other.

A way to improve the approximation (5.3) is to partition the upcrossings into *primary*, that occur (almost) independently, and *secondary*, that follow close after a primary.

Let $U_n$ be the number of primary and $V_n$ be the number of secondary upcrossings of level $n$. Then,

$$
M_n = U_n + V_n = U_n + \sum_{i=1}^{U_n} Z_{i,n} = \sum_{i=1}^{U_n}(1 + Z_{i,n}), \tag{5.4}
$$

where $Z_{i,n}$ is the number of secondary upcrossings between primary upcrossings $i$ and $i + 1$.

While $Y$ is at a high level $n$, it behaves approximately as a random walk with negative drift and jump probabilities

$$
P(+1) = \frac{\lambda w}{\lambda w + n} \quad \text{and} \quad P(-1) = \frac{n}{\lambda w + n}.
$$

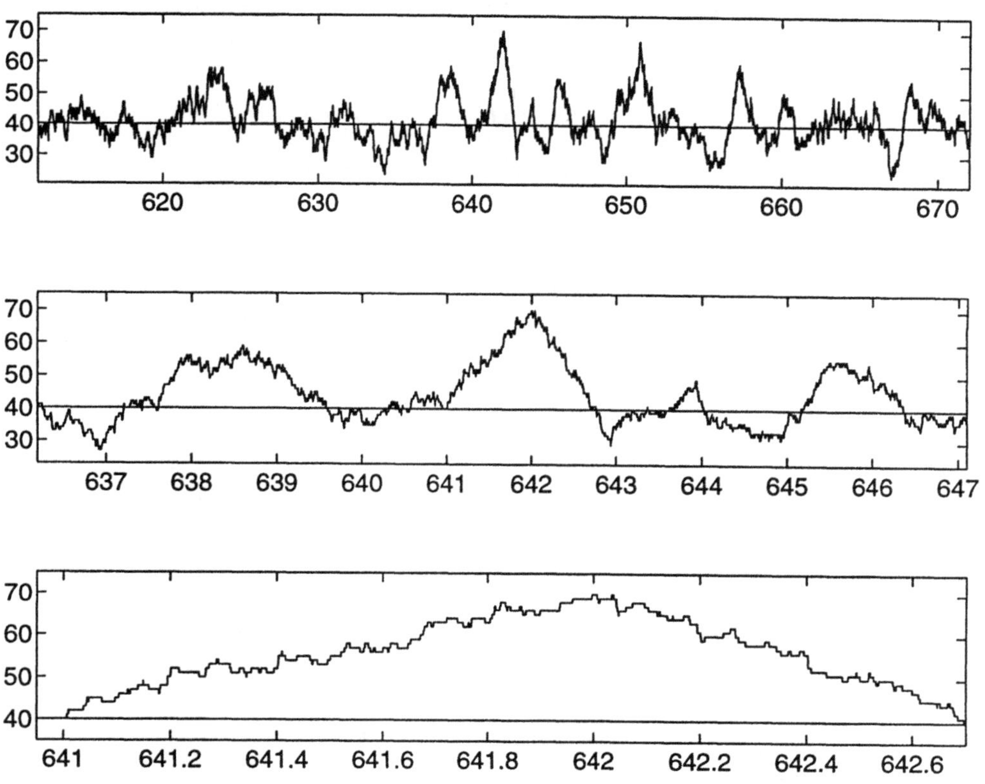

**Figure 5.1:** The scanning process plotted around its maxima

This random walk, starting at level $n - 1$, will have $Z'_n$ upcrossings of level $n$, where

$$E(Z'_n) = \frac{P(+1)}{P(-1) - P(+1)} = \frac{\lambda w}{n - \lambda w},$$

so that

$$E(Z_{i,n}) \approx E(Z'_n) = \frac{\lambda w}{n - \lambda w}.$$

This gives, using Wald's Lemma,

$$E(M_n) \approx E(U_n)\{1 + E(Z'_n)\} = E(U_n)\left(1 + \frac{\lambda w}{n - \lambda w}\right) = E(U_n)\frac{n}{n - \lambda w}.$$

Combining this with Lemma 5.2.1, we get

$$
\begin{aligned}
\mu_n = E(U_n) &\approx \frac{n - \lambda w}{n} E(M_n) = \left(1 - \frac{\lambda w}{n}\right)\lambda(T - w)p(n - 1; \lambda w) \\
&= \left(1 - \frac{\lambda w}{n}\right)\lambda w \left(\frac{T}{w} - 1\right)p(n - 1; \lambda w) \\
&= \left(1 - \frac{\lambda'}{n}\right)\lambda'(T' - 1)p(n - 1; \lambda').
\end{aligned}
$$

$$(5.5)$$

As primary upcrossings of high levels are rare and occur (almost) independently, $U_n$ can be approximated by a Poisson distributed random variable, so that

$$P(M_n = 0) = P(U_n = 0) \approx e^{-\mu_n}\,.$$

Combining this with (5.2) and (5.5), we get the approximation

$$
\begin{aligned}
F_{S_w}(n) &= P(S_w \le n) = P(S_w < n+1) \\
&\approx F_p(n; \lambda w)\, e^{-\mu_{n+1}} \\
&\approx F_p(n; \lambda w)\, e^{-(1-\frac{\lambda w}{n+1})\lambda w(T/w-1)p(n;\lambda w)} \\
&= F_p(n; \lambda')\, e^{-(1-\frac{\lambda'}{n+1})\lambda'(T'-1)p(n;\lambda')}\,.
\end{aligned}
\tag{5.6}
$$

This approximation is very accurate, as can be seen by comparing it with the exact lower and upper bounds given by Janson (1984). Computationally, approximation (5.6) is simpler than Janson's bounds as well as Naus' (1982) approximation.

Table 5.1: Approximation (5.6) compared with lower and upper bounds for $T' = 3600$ and $\lambda' = 40$

| $n$ | Lower bound | Appr. (5.6) | Upper bound |
|---|---|---|---|
| 63 | 0.0001 | 0.0001 | 0.0001 |
| 64 | 0.0018 | 0.0018 | 0.0018 |
| 65 | 0.0186 | 0.0188 | 0.0187 |
| 66 | 0.0847 | 0.0850 | 0.0849 |
| 67 | 0.2220 | 0.2223 | 0.2221 |
| 68 | 0.4052 | 0.4054 | 0.4053 |
| 69 | 0.5863 | 0.5864 | 0.5864 |
| 70 | 0.7329 | 0.7329 | 0.7329 |
| 71 | 0.8368 | 0.8368 | 0.8368 |
| 72 | 0.9042 | 0.9042 | 0.9042 |
| 73 | 0.9455 | 0.9455 | 0.9455 |
| 74 | 0.9697 | 0.9697 | 0.9697 |
| 75 | 0.9835 | 0.9835 | 0.9835 |
| 76 | 0.9911 | 0.9911 | 0.9911 |
| 77 | 0.9953 | 0.9953 | 0.9953 |
| 78 | 0.9976 | 0.9976 | 0.9976 |
| 79 | 0.9988 | 0.9988 | 0.9988 |
| 80 | 0.9994 | 0.9994 | 0.9994 |
| 81 | 0.9997 | 0.9997 | 0.9997 |
| 82 | 0.9998 | 0.9998 | 0.9998 |
| 83 | 0.9999 | 0.9999 | 0.9999 |
| 84 | 1.0000 | 1.0000 | 1.0000 |

When both $T'$ and $\lambda'$ are large, the precision is very high even for smaller $n$, as can be seen in Table 5.1, where we have $T' = 3600$ and $\lambda' = 40$. The approximation gives a slight overestimate.

**Table 5.2:** Approximations and Janson's bounds for $T' = 10$ and $\lambda' = 3$

| $n$ | Lower bound | Appr. (5.6) | Naus' appr. | Upper bound |
|---|---|---|---|---|
| 4 | 0.0050 | 0.1328 | 0.0202 | 0.0758 |
| 5 | 0.0960 | 0.2349 | 0.1341 | 0.1891 |
| 6 | 0.3621 | 0.4440 | 0.3883 | 0.4234 |
| 7 | 0.6604 | 0.6862 | 0.6684 | 0.6806 |
| 8 | 0.8558 | 0.8610 | 0.8573 | 0.8595 |
| 9 | 0.9484 | 0.9492 | 0.9486 | 0.9488 |
| 10 | 0.9838 | 0.9839 | 0.9838 | 0.9839 |
| 11 | 0.9954 | 0.9955 | 0.9954 | 0.9954 |
| 12 | 0.9988 | 0.9988 | 0.9988 | 0.9988 |
| 13 | 0.9997 | 0.9997 | 0.9997 | 0.9997 |
| 14 | 0.9999 | 0.9999 | 0.9999 | 0.9999 |
| 15 | 1.0000 | 1.0000 | 1.0000 | 1.0000 |

From Table 5.2, where $T' = 10$ and $\lambda' = 3$, we observe that the approximation is accurate even for moderate parameter values, at least for large $n$. As a comparison, Naus' (1982) approximation is included.

**Remark 5.2.2** The idea of partitioning the upcrossings into "independent clumps" is an example of the Poisson clumping technique of Aldous (1989).

**Remark 5.2.3** We could define the scanning process for all $t, 0 \le t \le T$, by either defining $X$ on $[0, T + w]$ or connecting the endpoints of the interval $[0, T]$ to form a torus. In both cases, the only effect on (5.6) is that $(T/w - 1)$ is replaced by $T/w$ (or $T' - 1$ by $T'$).

**Remark 5.2.4** By noting that the $Z$ variables in (5.4) are approximately independent and geometrically distributed, we actually get a compound Poisson approximation of the distribution of $M_n$, and not only an approximation of the probability $P(M_n = 0)$.

In the next section, we will also need the joint distribution of $S_w$ and $U_{S_w}$. This is of independent interest, as it could also be used to obtain an approximation for the distribution of the so-called *multiple scan statistic*.

Consider a process $X'$, where the number of primary upcrossings of level $n$, $U'_n$, is exactly Poisson distributed with mean $\mu'_n$, and where, at each primary $n$-upcrossing, there is a fixed probability $p'_{n+1}$ of this causing a $(n+1)$-upcrossing. Let the maximum of this process be $S'$. Then

$$\mu'_{n+1} = \mu'_n p'_{n+1}$$

and, for $k > 0$,

$$
\begin{aligned}
P(U'_{S'} = k, S' = n) &= P(U'_n = k, S' = n) = P(U'_n = k, U'_{n+1} = 0) \\
&= P(U'_n = k)P(U'_{n+1} = 0 \mid U'_n = k) \\
&= \frac{(\mu'_n)^k}{k!} e^{-\mu'_n}(1 - p'_{n+1})^k \\
&= \frac{(\mu'_n - \mu'_{n+1})^k}{k!} e^{-(\mu'_n - \mu'_{n+1})}e^{-\mu'_{n+1}} .
\end{aligned}
$$

For $n$, such that $\mu_n > \mu_{n+1}$, and $k > 0$, this gives the approximation

$$
P(U_{S_w} = k, S_w = n) \approx \frac{(\mu_n - \mu_{n+1})^k}{k!}e^{-(\mu_n - \mu_{n+1})}e^{-\mu_{n+1}} . \tag{5.7}
$$

---

## 5.3  Approximation in Two Dimensions

Let $X$ be a two-dimensional Poisson process with intensity $\lambda$, and define the scan statistic

$$
S_W = S_W(\lambda, A) = \max_{x \in \mathbf{R}^2} X(W(x) \cap A),
$$

where $A = [0, T_1] \times [0, T_2]$, and $W$ is a general scanning set.

For a rectangular scanning set $R = [0, w_1] \times [0, w_2]$, this gives the scan statistic

$$
S_R = S_R(\lambda, A) = \max_{\substack{0 \leq t_1 \leq T_1 - w_1 \\ 0 \leq t_2 \leq T_2 - w_2}} X\left([t_1, t_1 + w_1] \times [t_2, t_2 + w_2]\right) .
$$

It should be noted that, just as in the one-dimensional case, we have unnecessarily many parameters. $\lambda' = \lambda w_1 w_2$, $T'_1 = T_1/w_1$, and $T'_2 = T_2/w_2$ are sufficient, and will, at times, be used to simplify the notation.

**Remark 5.3.1** When comparing the effect of the shapes of different scanning sets, we will use the torus convention to avoid boundary effects. The scan statistic will then be denoted $S_W(\lambda, A_t)$, where $A_t$ denotes the torus version of $A$; see, for example, Alm (1997) for details.

To approximate the distribution of $S_R$, we can not simply copy the technique from the previous section. First, it is not trivial to define upcrossings in two dimensions, as we have only a partial ordering in $\mathbf{R}^2$. This problem can be overcome by instead considering $E_n = \#$ *extreme $n$-rectangles*, defined to be those $w_1 \times w_2$ rectangles in $A$ that contain exactly $n$ points, but which can not be translated to the left or down without losing points. On the torus, $A_t$, the corresponding number is denoted $E_{n,t}$.

**Remark 5.3.2** In one dimension, an upcrossing of level $n$ obviously corresponds to an *extreme $n$-interval*, i.e., one that can not be moved to the left without losing a point.

There are two types of extreme $n$-rectangles:

(i) Those with one point in the top right corner and $n - 1$ in the interior.

(ii) Those with one point on the top side, one point on the right side, and $n - 2$ in the interior.

By a similar argument to that of Lemma 5.2.1, we get the following.

**Lemma 5.3.1** $E(E_{n,t}) = n\lambda T_1 T_2 p(n - 1; \lambda w_1 w_2)$.

PROOF. The contribution from rectangles of type (i) is $\lambda T_1 T_2 p(n - 1; \lambda w_1 w_2)$, since for each of the points in $[0, T_1] \times [0, T_2]$ the probability is $p(n - 1; \lambda w_1 w_2)$ that it is the top right corner of an $n$-rectangle, and the mean number of such points is $\lambda T_1 T_2$.

The mean number of pairs of points that are close enough to be the two points that define a rectangle of type (ii) is $\lambda T_1 T_2 \lambda w_1 w_2$, and the probability of the correct number of interior points is $p(n - 2; \lambda w_1 w_2)$.

Collecting terms, we now get

$$
\begin{aligned}
E(E_{n,t}) &= \lambda T_1 T_2 p(n - 1; \lambda w_1 w_2) + \lambda T_1 T_2 \lambda w_1 w_2 p(n - 2; \lambda w_1 w_2) \\
&= \lambda T_1 T_2 (1 + (n - 1)) p(n - 1; \lambda w_1 w_2) = n\lambda T_1 T_2 p(n - 1; \lambda w_1 w_2).
\end{aligned}
$$

■

By a similar but slightly more complicated argument, we get the following.

**Lemma 5.3.2** $E(E_n) = n\lambda (T_1 - w_1)(T_2 - w_2) p(n - 1; \lambda w_1 w_2)$.

Lemma 5.3.2 gives the possible, although poor, approximation

$$
P(S_R < n) \approx P(E_n = 0) \approx e^{-E(E_n)}. \tag{5.8}
$$

**Remark 5.3.3** Although the concept of extreme rectangles (or boxes in three dimensions) does not lead to any usable approximation, it is fundamental in the simulations; see Section 5.6.

Another approach is to study $G_n$, *the number of subsets of $n$ points that can be covered by a translate of the scanning set $W$.* Using Månsson's (1994) expression for $E(G_n)$, we get an approximation

$$
P(S_W < n) \approx P(G_n = 0) \approx e^{-E(G_n)},
$$

which, as well as the related approximation (5.8), is even worse than the corresponding (5.3) in one dimension, due to the strong dependence between different $n$-subsets with common points. Note that, if $G_n = 1$, then $G_{n-1} \geq n$.

Following the one-dimensional approach, we would like to partition the $n$-subsets (or extreme $n$-rectangles), $E_n$, into primary and secondary ones. Again, this is not easily done because of the partial ordering in $\mathbf{R}^2$. However, with a suitable definition of $U_n$ as the number *of primary n-rectangles*, we would expect a reasonable approximation by using that

$$P(S_R < n) \approx P(E_n = 0) = P(U_n = 0) \approx e^{-E(U_n)},$$

as in the one-dimensional case.

This idea, combined with the results of Månsson (1994), will be used in Section 5.3.2 to get approximations for general scanning sets, given the approximation for rectangular scanning sets derived in the next section.

For rectangular scanning sets, we use a different approach which uses the one-dimensional technique in a more direct way.

### 5.3.1  Rectangular scanning sets

The idea behind the one-dimensional approximation was to form the stationary scanning process $Y$, and approximate the maximum of this by a random walk argument.

To approximate the scan statistic in two dimensions, we will scan the rectangle $A$ by first, for fixed $t_2$, doing a one-dimensional scanning in the strip $[0, T_1] \times [t_2, t_2 + w_2]$, only considering the first coordinate. These one-dimensional scan statistics, which can be approximated using (5.6), form a new one-dimensional stationary process, corresponding to the scanning process $Y$ in one dimension, whose maximum is the two-dimensional scan statistic. Thus, the distribution of the two-dimensional scan statistic can be approximated by using a slight variation of the one-dimensional technique.

Consider, for fixed $t_2$, the Poisson process on a vertical strip of width $w_2$, $[0, T_1] \times [t_2, t_2 + w_2]$. If we only consider the first coordinate, we get a one-dimensional Poisson process, $X' = X'(t_2) = \{X'_{t_1}(t_2), 0 \leq t_1 \leq T_1\}$, with intensity $\lambda w_2$.

For each $t_2$, define a one-dimensional scan statistic corresponding to a scanning interval of length $w_1$ [see Figure 5.2],

$$S_{w_1} = S_{w_1}(t_2) = \max_{0 \leq t_1 \leq T_1 - w_1} (X'_{t_1 + w_1}(t_2) - X'_{t_1}(t_2))$$

as before. The distribution of $S_{w_1}$ can, for fixed $t_2$, be approximated using (5.6) as

$$\begin{aligned}
P(S_{w_1} \leq n) &\approx F_p(n; \lambda w_1 w_2) e^{-(1 - \frac{\lambda w_1 w_2}{n+1})\lambda w_1 w_2 (T_1/w_1 - 1)p(n;\lambda w_1 w_2)} \\
&= F_p(n; \lambda') e^{-(1 - \frac{\lambda'}{n+1})\lambda'(T_1' - 1)p(n;\lambda')},
\end{aligned} \tag{5.9}$$

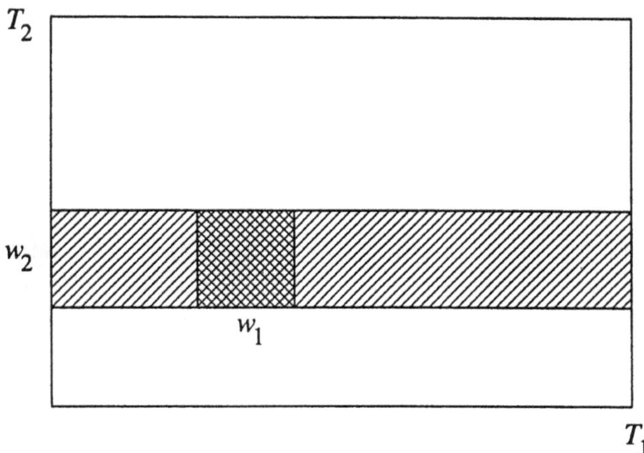


**Figure 5.2:** One-dimensional scanning in a strip

using the notation $T_1' = T_1/w_1$, $T_2' = T_2/w_2$, and $\lambda' = \lambda w_1 w_2$.

Now, $Y = \{S_{w_1}(t_2),\, 0 \le t_2 \le T_2 - w_2\}$ is a new one-dimensional stationary stochastic process with maximum

$$
\begin{aligned}
\max_{0 \le t_2 \le T_2 - w_2} S_{w_1}(t_2) &= \max_{0 \le t_2 \le T_2 - w_2} \left( \max_{0 \le t_1 \le T_1' - w_1} \left( X_{t_1 + w_1}'(t_2) - X_{t_1}'(t_2) \right) \right) \\
&= \max_{0 \le t_2 \le T_2 - w_2} \left( \max_{0 \le t_1 \le T_1' - w_1} X\left([t_1, t_1 + w_1] \times [t_2, t_2 + w_2]\right) \right) \\
&= S_R = S_R(\lambda, A) ,
\end{aligned}
$$

so that $S_R$ is both the two-dimensional scan statistic of the Poisson process $X$ and the maximum of the one-dimensional scanning process $Y$.

Repeating the one-dimensional argument for $Y$, we get [see (5.6)]

$$F_{S_R}(n) \approx F_{Y_0}(n)\, e^{-\gamma_{n+1}} , \tag{5.10}$$

where $F_{Y_0}(n) = P(S_{w_1} \le n)$ is approximated by (5.9), and

$$\gamma_n = E(\text{number of primary } n\text{-upcrossings of } Y = \{S_{w_1}(t_2),\, 0 \le t_2 \le T_2 - w_2\}).$$

$\gamma_n$ is approximated by a random walk argument similar to that in one dimension. A complication is that, if $S_{w_1}(t_2) = n$, we may have several primary and/or secondary upcrossings of level $n$ within the strip $[t_2, t_2 + w_2]$. Disregarding this possibility, we can copy the argument from the preceding section, replacing $\lambda$ with $\lambda w_1$ and $p(n-1; \lambda w_2)$ by $P(S_{w_1} = n-1)$ in (5.5) to get

$$
\begin{aligned}
\gamma_n \approx \gamma_n^{(1)} &= \left(1 - \frac{\lambda w_1 w_2}{n}\right) \lambda w_1 w_2 \left(\frac{T_2}{w_2} - 1\right) P(S_{w_1} = n-1) \\
&= \left(1 - \frac{\lambda'}{n}\right) \lambda' (T_2' - 1) P(S_{w_1} = n-1) ,
\end{aligned}
\tag{5.11}
$$

which, combined with (5.10), gives the approximation

$$F_{S_R}(n) \approx F_{S_R}^{(1)}(n) = P(S_{w_1} \le n)e^{-\gamma_{n+1}^{(1)}} . \tag{5.12}$$

The effect of secondary upcrossings is a slight increase in the intensity, $\lambda w_1$, of upward jumps in the random walk, which may be estimated. There is also a more complicated increase in the probability of downward jumps. Fortunately, the total effect seems to be of small significance, so it will be neglected.

By using (5.7), we can take into account the possibility of multiple primary upcrossings in a strip. This gives a second approximation for $\gamma_{n+1}$ [see Alm (1997) for details] as

$$\gamma_{n+1} \approx \gamma_{n+1}^{(2)} = \left(1 - \frac{\lambda w_1 w_2}{n+1}\right) \lambda w_1 w_2 \left(\frac{T_2}{w_2} - 1\right) (\mu_n - \mu_{n+1})e^{-\mu_{n+1}}$$

$$= \left(1 - \frac{\lambda'}{n+1}\right) \lambda' (T_2' - 1)(\mu_n - \mu_{n+1})e^{-\mu_{n+1}} , \tag{5.13}$$

where, as in (5.5), we have

$$\mu_n \approx \left(1 - \frac{\lambda w_1 w_2}{n}\right) \lambda w_1 w_2 \left(\frac{T_1}{w_1} - 1\right) p(n-1; \lambda w_1 w_2)$$

$$\approx \left(1 - \frac{\lambda'}{n}\right) \lambda' (T_1' - 1)p(n-1; \lambda') .$$

This gives a second approximation

$$F_{S_R}(n) \approx F_{S_R}^{(2)}(n) = P(S_{w_1} \le n)e^{-\gamma_{n+1}^{(2)}} . \tag{5.14}$$

**Remark 5.3.4** To use the approximations on the torus $A_t$ we only need to replace $T_1/w_1 - 1$ by $T_1/w_1$ and $T_2/w_2 - 1$ by $T_2/w_2$ (or $T_1' - 1$ by $T_1'$ and $T_2' - 1$ by $T_2'$).

To compare the two approximations, (5.12) and (5.14), we have simulated the Poisson process for various values of the parameters $T_1'$, $T_2'$, and $\lambda'$.

In each simulation, we have calculated the value of the scan statistic $S_R$ by studying all extreme $n$-rectangles. To get a reasonably accurate empirical distribution, 10,000 simulations were performed for each parameter combination.

Alm (1997) has presented numerous tables comparing the empirical distribution function with the approximations (5.12) and (5.14). We give two of these as examples. Table 5.3 gives the comparison for $T_1' = T_2' = 30$, while Table 5.4 uses $T_1' = T_2' = 20$ on a torus. Both use $\lambda' = 5$. As can be seen from the tables, both approximations give a good agreement with the simulations.

Here, we summarize the results of the simulations by measuring the distance between the empirical probability function $p_e$ and the approximate $p_a$ with the total variation distance

$$d(p_e, p_a) = \sum_k |p_e(k) - p_a(k)| . \tag{5.15}$$

**Table 5.3:** Comparison of approximations (5.12) and (5.14) with simulations for $T'_1 = T'_2 = 30$ and $\lambda' = 5$

| $n$ | Empirical | Appr. (5.12) | Appr. (5.14) |
|---|---|---|---|
| 14 | 0.0016 | 0.0009 | 0.0007 |
| 15 | 0.0571 | 0.0537 | 0.0522 |
| 16 | 0.3346 | 0.3368 | 0.3365 |
| 17 | 0.6906 | 0.6907 | 0.6913 |
| 18 | 0.8945 | 0.8899 | 0.8902 |
| 19 | 0.9693 | 0.9662 | 0.9663 |
| 20 | 0.9907 | 0.9905 | 0.9905 |
| 21 | 0.9979 | 0.9975 | 0.9975 |
| 22 | 0.9994 | 0.9994 | 0.9994 |
| 23 | 0.9998 | 0.9999 | 0.9999 |
| 24 | 1.0000 | 1.0000 | 1.0000 |

**Table 5.4:** Comparison of approximations (5.12) and (5.14) with simulations on a torus for $T'_1 = T'_2 = 20$ and $\lambda' = 5$

| $n$ | Empirical | Appr. (5.12) | Appr. (5.14) |
|---|---|---|---|
| 13 | 0.0004 | 0.0008 | 0.0005 |
| 14 | 0.0333 | 0.0326 | 0.0306 |
| 15 | 0.2535 | 0.2440 | 0.2430 |
| 16 | 0.6024 | 0.5936 | 0.5945 |
| 17 | 0.8442 | 0.8379 | 0.8386 |
| 18 | 0.9446 | 0.9459 | 0.9461 |
| 19 | 0.9825 | 0.9837 | 0.9838 |
| 20 | 0.9940 | 0.9955 | 0.9955 |
| 21 | 0.9981 | 0.9988 | 0.9988 |
| 22 | 0.9999 | 0.9997 | 0.9997 |
| 23 | 1.0000 | 0.9999 | 0.9999 |

**Remark 5.3.5** Alm (1998) has used several measures to compare the approximations. Fortunately, the result of the comparison does not seem to depend too much on the choice of measure. We have, therefore, chosen to use the variation distance $d$, as it is both easy to calculate and to interpret.

The comparison is summarized in Table 5.5 for $T'_1 = T'_2$, and in Table 5.6 for $T'_1 \neq T'_2$.

Note that we would really like to compare the approximations with the true distribution. Replacing this with the empirical distribution introduces a random error. A way to measure this error would be through $d(p, p_e)$, where $p$ is the true distribution. Here we face two complications, since not only is the true distribution unknown, but the distance $d(p, p_e)$ is a random variable. To get an idea about the distribution of $d(p, p_e)$, we have used bootstrap. Repeated resampling (1000 samples of size 10,000) from the empirical distribution gives approximate estimates of the mean and standard deviation of $d(p, p_e)$. These are denoted by $\mu_e^*$ and $\sigma_e^*$ and are included in the tables for comparison.

**Table 5.5:** Comparison of approximations (5.12) and (5.14) when $T_1' = T_2'$

| $T_1'$ | $T_2'$ | $\lambda'$ | Torus | $d(5.12)$ | $d(5.14)$ | $\mu_e^*$ | $\sigma_e^*$ |
|---|---|---|---|---|---|---|---|
| 30 | 30 | 5 | $-$ | 0.0208 | 0.0227 | 0.0173 | 0.0060 |
| 20 | 20 | 5 | $+$ | 0.0231 | 0.0249 | 0.0176 | 0.0059 |
| 20 | 20 | 2 | $-$ | 0.0529 | 0.0433 | 0.0150 | 0.0057 |
| 10 | 10 | 5 | $-$ | 0.0656 | 0.0608 | 0.0190 | 0.0058 |
| 10 | 10 | 5 | $+$ | 0.0383 | 0.0268 | 0.0190 | 0.0058 |
| 5 | 5 | 5 | $+$ | 0.1715 | 0.1454 | 0.0209 | 0.0060 |

**Table 5.6:** Comparison of approximations (5.12) and (5.14) when $T_1' \neq T_2'$

| $T_1'$ | $T_2'$ | $\lambda'$ | Torus | $d(5.12)$ | $d(5.14)$ | $\mu_e^*$ | $\sigma_e^*$ |
|---|---|---|---|---|---|---|---|
| 20 | 20 | 2 | $-$ | 0.0529 | 0.0433 | 0.0150 | 0.0057 |
| 10 | 40 | 2 | $-$ | 0.0480 | 0.0509 | 0.0149 | 0.0056 |
| 40 | 10 | 2 | $-$ | 0.0987 | 0.0619 | 0.0149 | 0.0056 |
| 5 | 80 | 2 | $-$ | 0.0453 | 0.0711 | 0.0148 | 0.0056 |
| 80 | 5 | 2 | $-$ | 0.1813 | 0.0963 | 0.0148 | 0.0056 |
| 2 | 200 | 2 | $-$ | 0.0652 | 0.2056 | 0.0149 | 0.0060 |
| 200 | 2 | 2 | $-$ | 0.4310 | 0.2126 | 0.0149 | 0.0060 |

From Table 5.5 we see that when $T_1' = T_2'$, the approximations give similar results, although (5.14), as might be expected, seems to be slightly better.

When $T_1' \neq T_2'$, we actually have four different approximations to consider, as we can freely interchange $T_1'$ and $T_2'$. From Table 5.6, we see that (5.12) with $T_1' < T_2'$ is preferable. As a general recommendation: *Use (5.12) with $T_1' \leq T_2'$.*

### 5.3.2 General scanning sets

In one dimension, the only convex scanning sets are intervals. In two dimensions, we obtained good approximations for rectangular scanning sets in the last section. The technique that was used relied heavily on the scanning sets being rectangular, and can not, in any obvious way, be extended to general scanning sets.

Aldous (1989), has given an approximation for a circular scanning set of radius $r$, which in our notation reads as

$$P(S_C < n) \approx \exp\left(-\frac{n^2 T_1 T_2}{\pi r^2} p(n; \lambda \pi r^2)(1 - 2\frac{r}{T_1})(1 - 2\frac{r}{T_2})\right). \qquad (5.16)$$

Aldous technique can be used for various shapes of the scanning set, but requires detailed, nontrivial, analysis for each shape.

There is no method available that works for general scanning sets, but we will use the results of Section 5.3.1, combined with a result by Månsson (1994), to get reasonable approximations for arbitrary convex scanning sets $W$.

Let, as before, $G_n$ be the number of subsets of $n$ points that can be covered by a translate of $W$, and $G_n^N$ be the corresponding quantity when the number of points is fixed as $N$ instead of being Poisson. Under the torus convention, we will use the notation $G_{n,t}$ and $G_{n,t}^N$.

Månsson (1994) has given the following result for $E(G_{n,t}^N(W))$:

$$E(G_{n,t}^N(W)) = \binom{N}{n}\left(n + n(n-1)\frac{\nu(W,\breve{W})}{|W|}\right)\left(\frac{|W|}{|A|}\right)^{n-1}. \tag{5.17}$$

Here, $W$ can be any convex set. $|W|$ denotes the area of $W$, and $\nu(W,\breve{W})$ denotes the mixed area of $W$ and $\breve{W}$, the reflection of $W$.

Suppose that we can partition the subsets as

$$G_n = U_n + V_n,$$

where $U_n$ denotes primary and $V_n$ secondary subsets, in such a way that the primary subsets occur (almost) independently and [see (5.4)],

$$V_n = \sum_{k=1}^{U_n} Z_k,$$

so that

$$E(G_n) = E(U_n) + E(V_n) \approx E(U_n)(1 + E(Z)).$$

A reasonable approximation would then be

$$P(S_W < n) = P(G_n = 0) = P(U_n = 0) \approx e^{-E(U_n)}, \tag{5.18}$$

but $E(U_n)$ is not easily calculated for general $W$.

For rectangular scanning sets, we have obtained approximations of $\gamma_n = E(U_n)$ in (5.11) and (5.13). Using (5.17) both for $W$ and a rectangular scanning set $R$, we can calculate

$$E(G_{n,t}^N(W)) \quad \text{and} \quad E(G_{n,t}^N(R)),$$

and use the heuristic approximation

$$R_n(W,R) := \frac{E(G_{n,t}^N(W))}{E(G_{n,t}^N(R))} \approx \frac{E(G_n(W))}{E(G_n(R))} \approx \frac{E(U_n(W))}{E(U_n(R))}\frac{1 + E(Z_W)}{1 + E(Z_R)}. \tag{5.19}$$

Using (5.17), choosing $R$ so that $|R| = |W|$, and noting that $\nu(R, \check{R}) = |R|$, we get

$$R_n(W, R) = \frac{n + n(n-1)\frac{\nu(W,\check{W})}{|W|}}{n + n(n-1)\frac{\nu(R,\check{R})}{|R|}} \frac{|W|^{n-1}}{|R|^{n-1}} = \frac{1 + (n-1)\frac{\nu(W,\check{W})}{|W|}}{n}.$$

By (5.19), the value of this ratio depends on the two factors

$$\frac{E(U_n(W))}{E(U_n(R))} \quad \text{and} \quad \frac{1 + E(Z_W)}{1 + E(Z_R)},$$

and we need to estimate one of them to get an approximation for $E(U_n(W))$.

A rough estimate is obtained by assuming that $E(Z_W)$ does not depend (too much) on the shape of the set $W$. This gives a first approximation

$$E(U_n(W)) \approx E(U_n(R))R_n(W, R) = \gamma_n R_n(W, R), \tag{5.20}$$

probably overestimating $E(U_n(W))$.

Assuming the other extreme, that $E(U_n(W))$ does not depend (too much) on the shape of the set $W$, gives the trivial approximation

$$E(U_n(W)) \approx \gamma_n, \tag{5.21}$$

which probably underestimates $E(U_n(W))$.

To improve the approximation (5.20), a more careful study of how $E(Z_W)$ depends on the shape of W is needed. Here, we will use a simple heuristic argument to try to improve the approximation.

Assuming that the relative number of primary and secondary subsets are affected similarly by the shape of $W$, we could approximate

$$\frac{E(U_n(W))}{E(U_n(R))} \approx \frac{E(Z_W)}{E(Z_R)}, \tag{5.22}$$

or simpler

$$\frac{E(U_n(W))}{E(U_n(R))} \approx \frac{1 + E(Z_W)}{1 + E(Z_R)}. \tag{5.23}$$

(5.23) gives the approximation

$$E(U_n(W)) \approx \gamma_n \sqrt{R_n(W, R)}, \tag{5.24}$$

and (5.22) gives the slightly more complicated

$$E(U_n(W)) \approx \sqrt{\frac{1}{4} + R_n(W, R)\gamma_n(1 + \gamma_n)} - \frac{1}{2}. \tag{5.25}$$

Combining (5.20), (5.21), (5.24), or (5.25) with (5.18) gives four possible approximations. In the following sections, these will be compared with simulations for circular scanning sets (where the approximations coincide) and for a triangular scanning set (where they differ).

**Circular scanning sets**

As $\nu(W, \check{W}) = |W|$ for all centrally symmetric sets $W$, e.g., rectangles and circles, we get the same approximation, $E(U_n) \approx \gamma_n$, for all such sets, and we can use either (5.11) or (5.13) to approximate $\gamma_n$, which gives approximation (5.12) or (5.14).

Using both a circular scanning set $C$, with radius $r$, and a square scanning set, with $w_1 = w_2 = \sqrt{\pi}r$, in the same simulation supports this result; see Tables 5.7 and 5.8, where we compare the empirical distribution functions of the scan statistics with approximations (5.12) and (5.14). We have also included Aldous' (1989) approximation (5.16) for comparison. The empirical distribution functions are denoted $\hat{F}_S$ for a square scanning set and $\hat{F}_C$ for a circular.

In Table 5.7, we show the results of 10,000 simulations with $\lambda' = 5$ and $T_1' = T_2' = 10$. The empirical distributions are quite close, and in reasonable agreement with both (5.12) and (5.14), but differ significantly from (5.16).

**Table 5.7:** Square and circular scanning sets, $T_1' = T_2' = 10$ and $\lambda' = 5$

| $n$ | $\hat{F}_S$ | $\hat{F}_C$ | (5.12) | (5.14) | (5.16) |
|---|---|---|---|---|---|
| 11 | 0.0006 | 0.0007 | 0.0095 | 0.0064 | 0.0000 |
| 12 | 0.0222 | 0.0232 | 0.0477 | 0.0428 | 0.0000 |
| 13 | 0.1687 | 0.1726 | 0.1994 | 0.1973 | 0.0007 |
| 14 | 0.4663 | 0,4750 | 0.4814 | 0.4838 | 0.0618 |
| 15 | 0.7389 | 0.7464 | 0.7450 | 0.7476 | 0.3715 |
| 16 | 0.8976 | 0.9016 | 0.8979 | 0.8991 | 0.7198 |
| 17 | 0.9650 | 0.9648 | 0.9643 | 0.9647 | 0.9027 |
| 18 | 0.9873 | 0.9880 | 0.9886 | 0.9888 | 0.9704 |
| 19 | 0.9958 | 0.9967 | 0.9966 | 0.9967 | 0.9917 |
| 20 | 0.9984 | 0.9988 | 0.9991 | 0.9991 | 1.0000 |
| 21 | 0.9995 | 0.9995 | 0.9998 | 0.9998 | 1.0000 |
| 22 | 1.0000 | 1.0000 | 0.9999 | 0.9999 | 1.0000 |

As square and circular scanning sets are affected differently by the boundary, we have included Table 5.8, where the simulation is performed on a torus with $\lambda' = T_1' = T_2' = 5$. Although $T_1'$ and $T_2'$ are smaller than in the previous table, we get a good agreement, except for (5.16), in particular, for large $n$.

**Triangular scanning sets**

The maximum value of $\nu(W, \check{W})/|W|$ is 2 for triangles, so that for a triangular scanning set $T$,

$$R_n(T, R) \approx \frac{1 + 2(n-1)}{n} = \left(2 - \frac{1}{n}\right). \tag{5.26}$$

**Table 5.8:** Square and circular scanning sets, $T'_1 = T'_2 = 5$ and $\lambda' = 5$ on a torus

| $n$ | $\hat{F}_S$ | $\hat{F}_C$ | (5.12) | (5.14) | (5.16) |
|---|---|---|---|---|---|
| 9 | 0.0018 | 0.0007 | 0.0791 | 0.0709 | 0.0000 |
| 10 | 0.0195 | 0.0180 | 0.0931 | 0.0858 | 0.0000 |
| 11 | 0.1176 | 0.1120 | 0.1730 | 0.1705 | 0.0006 |
| 12 | 0.3407 | 0.3360 | 0.3498 | 0.3542 | 0.0352 |
| 13 | 0.5942 | 0.5968 | 0.5879 | 0.5950 | 0.2501 |
| 14 | 0.7981 | 0.7961 | 0.7903 | 0.7952 | 0.5884 |
| 15 | 0.9146 | 0.9132 | 0.9106 | 0.9128 | 0.8281 |
| 16 | 0.9682 | 0.9679 | 0.9665 | 0.9673 | 0.9393 |
| 17 | 0.9884 | 0.9887 | 0.9886 | 0.9889 | 0.9807 |
| 18 | 0.9961 | 0.9956 | 0.9964 | 0.9965 | 0.9943 |
| 19 | 0.9984 | 0.9987 | 0.9989 | 0.9990 | 0.9984 |
| 20 | 0.9997 | 0.9998 | 0.9997 | 0.9997 | 0.9996 |
| 21 | 0.9998 | 0.9999 | 0.9999 | 0.9999 | 0.9999 |
| 22 | 0.9999 | 0.9999 | 1.0000 | 1.0000 | 1.0000 |
| 23 | 1.0000 | 1.0000 | 1.0000 | 1.0000 | 1.0000 |

It is quite remarkable that triangular scanning sets give such a dramatic increase in $R_n(T, R)$, almost double that of centrally symmetric scanning sets! Using (5.20), (5.24), or (5.25) gives the approximations

$$E(U_n) \approx \gamma_n \left(2 - \frac{1}{n}\right), \tag{5.27}$$

$$E(U_n) \approx \gamma_n \sqrt{\left(2 - \frac{1}{n}\right)}, \tag{5.28}$$

and

$$E(U_n) \approx \sqrt{\frac{1}{4} + \left(2 - \frac{1}{n}\right)\gamma_n(1 + \gamma_n)}. \tag{5.29}$$

In all of these, we have used (5.14) to approximate $\gamma_n$.

The scanning sets that have been used in this section are a square, with $w_1 = w_2$, and a right-angled isosceles triangle with the same area. $\hat{F}_T$ denotes the empirical distribution function for a triangular scanning set.

We give two comparisons, each based on 10,000 simulations on a torus. Table 5.9 uses $T'_1 = T'_2 = 20$ and $\lambda' = 10$. In Table 5.10, the area is larger ($T'_1 = T'_2 = 30$) but the intensity is smaller ($\lambda' = 5$).

As expected, (5.27) underestimates the distribution function, whereas (5.14) overestimates it. Both (5.28) and (5.29) give good approximations, with a slightly better agreement for (5.28) in Table 5.9 and for (5.29) in Table 5.10. Note that (5.28) tends to overestimate and that (5.29) tends to underestimate the distribution function.

**Table 5.9:** Square and triangular scanning sets, $T_1' = T_2' = 20$ and $\lambda' = 10$ on a torus

| $n$ | $\hat{F}_S$ | (5.14) | $\hat{F}_T$ | (5.27) | (5.28) | (5.29) |
|---|---|---|---|---|---|---|
| 22 | 0.0037 | 0.0038 | 0.0004 | 0.0000 | 0.0004 | 0.0003 |
| 23 | 0.0611 | 0.0551 | 0.0166 | 0.0035 | 0.0175 | 0.0147 |
| 24 | 0.2592 | 0.2462 | 0.1273 | 0.0647 | 0.1411 | 0.1209 |
| 25 | 0.5338 | 0.5280 | 0.3814 | 0.2867 | 0.4094 | 0.3621 |
| 26 | 0.7541 | 0.7587 | 0.6568 | 0.5823 | 0.6795 | 0.6252 |
| 27 | 0.8918 | 0.8924 | 0.8375 | 0.8001 | 0.8527 | 0.8138 |
| 28 | 0.9563 | 0.9561 | 0.9335 | 0.9158 | 0.9391 | 0.9187 |
| 29 | 0.9827 | 0.9832 | 0.9756 | 0.9673 | 0.9765 | 0.9678 |
| 30 | 0.9939 | 0.9939 | 0.9902 | 0.9880 | 0.9914 | 0.9880 |
| 31 | 0.9975 | 0.9978 | 0.9965 | 0.9958 | 0.9970 | 0.9958 |
| 32 | 0.9991 | 0.9993 | 0.9984 | 0.9986 | 0.9990 | 0.9986 |
| 33 | 0.9998 | 0.9998 | 0.9996 | 0.9995 | 0.9997 | 0.9995 |
| 34 | 0.9999 | 0.9999 | 0.9999 | 0.9999 | 0.9999 | 0.9999 |
| 35 | 0.9999 | 1.0000 | 0.9999 | 1.0000 | 1.0000 | 1.0000 |
| 36 | 1.0000 | 1.0000 | 1.0000 | 1.0000 | 1.0000 | 1.0000 |

**Table 5.10:** Square and triangular scanning sets, $T_1' = T_2' = 30$ and $\lambda' = 5$ on a torus

| $n$ | $\hat{F}_S$ | (5.14) | $\hat{F}_T$ | (5.27) | (5.28) | (5.29) |
|---|---|---|---|---|---|---|
| 14 | 0.0007 | 0.0005 | 0.0000 | 0.0000 | 0.0000 | 0.0000 |
| 15 | 0.0396 | 0.0425 | 0.0092 | 0.0022 | 0.0124 | 0.0104 |
| 16 | 0.3075 | 0.3118 | 0.1543 | 0.1047 | 0.1976 | 0.1709 |
| 17 | 0.6725 | 0.6737 | 0.5274 | 0.4646 | 0.5769 | 0.5227 |
| 18 | 0.8823 | 0.8830 | 0.8160 | 0.7852 | 0.8408 | 0.8005 |
| 19 | 0.9660 | 0.9640 | 0.9409 | 0.9311 | 0.9501 | 0.9331 |
| 20 | 0.9923 | 0.9899 | 0.9840 | 0.9804 | 0.9859 | 0.9806 |
| 21 | 0.9977 | 0.9973 | 0.9961 | 0.9948 | 0.9963 | 0.9948 |
| 22 | 0.9996 | 0.9993 | 0.9992 | 0.9987 | 0.9991 | 0.9987 |
| 23 | 1.0000 | 0.9998 | 0.9998 | 0.9997 | 0.9998 | 0.9997 |
| 24 | 1.0000 | 1.0000 | 1.0000 | 0.9999 | 1.0000 | 0.9999 |

## 5.4  Approximation in Three Dimensions

The same techniques that were used in the last section to go from one to two dimensions can be used to get approximations in three dimensions based on the two-dimensional approximations. For simplicity, we only use the technique leading to approximation (5.12).

Consider a Poisson process $X$, with intensity $\lambda$ in three dimensions and let

$$S_{R_3} = S_{R_3}(\lambda, A_3) = \max_{\substack{0 \leq t_1 \leq T_1 - w_1 \\ 0 \leq t_2 \leq T_2 - w_2 \\ 0 \leq t_3 \leq T_3 - w_3}} X\left([t_1, t_1 + w_1] \times [t_2, t_2 + w_2] \times [t_3, t_3 + w_3]\right)$$

be the scan statistic corresponding to the rectangular scanning set $R_3 = [0, w_1] \times [0, w_2] \times [0, w_3]$, translated over the (larger) box $A_3 = [0, T_1] \times [0, T_2] \times [0, T_3]$.

Note that we have used too many parameters above. We only need $T_1' = T_1/w_1$, $T_2' = T_2/w_2$, $T_3' = T_3/w_3$, and $\lambda' = \lambda w_1 w_2 w_3$.

Consider, for fixed $t_3$, a layer of width $w_3$, $[0, T_1] \times [0, T_2] \times [t_3, t_3 + w_3]$, where, disregarding the last coordinate, we have a two-dimensional Poisson process $X'$ with intensity $\lambda w_3$ on $A_2 = [0, T_1] \times [0, T_2]$.

Then, [see (5.10)]

$$P(S_{R_3} \leq n) = P(S_{R_2}(\lambda w_3, A_2) \leq n)e^{-\varphi_{n+1}} = P(S_{R_2'}(\lambda', A_2') \leq n)e^{-\varphi_{n+1}},$$
(5.30)

where $R_2' = [0, 1] \times [0, 1]$, and

$$\begin{aligned}
\varphi_{n+1} &\approx \left(1 - \frac{\lambda w_1 w_2 w_3}{n + 1}\right) \lambda w_1 w_2 w_3 \left(\frac{T_1}{w_1} - 1\right) P(S_{R_2}(\lambda w_3, A_2) = n) \\
&= \left(1 - \frac{\lambda'}{n + 1}\right) \lambda'(T_1' - 1) P(S_{R_2'}(\lambda', A_2') = n),
\end{aligned}$$
(5.31)

in analogy with (5.11).

As $S_{R_2}(\lambda w_3, A_2)$ is a two-dimensional scan statistic, we can use (5.12) or (5.14) to approximate $P(S_{R_2}(\lambda w_3, A_2) \leq n) = P(S_{R_2'}(\lambda', A_2') \leq n)$, which gives the two approximations

$$P(S_{R_3} \leq n) \approx F_{S_{R_3}}^{(1)}(n) = F_{S_{R_2}}^{(1)}(n)e^{-\varphi_{n+1}}$$
(5.32)

and

$$P(S_{R_3} \leq n) \approx F_{S_{R_3}}^{(2)}(n) = F_{S_{R_2}}^{(2)}(n)e^{-\varphi_{n+1}}.$$
(5.33)

**Remark 5.4.1** In the next section, we will use this argument to get an approximation in $d$ dimensions given one in $d - 1$ dimensions.

**Remark 5.4.2** To use the approximations on a torus, simply substitute $T_1' - 1$ with $T_1'$, $T_2' - 1$ with $T_2'$, and $T_3' - 1$ with $T_3'$.

**Remark 5.4.3** Månsson (1995) gives a formula that, if combined with (5.30), can be used to obtain approximations for general convex scanning sets $W$, by the same procedure that was used in Section 5.3.2 in the two-dimensional case. The formula is more complicated in three dimensions, so the generalization from rectangular to general scanning sets is omitted here.

To estimate the precision of the approximations, we have simulated the process for a number of different parameter combinations. The simulations are much more time consuming in three dimensions than in two dimensions, and so we have only been able to use moderate values of $T_1'$, $T_2'$, $T_3'$, and $\lambda'$. To get a reasonably accurate empirical distribution, 10,000 simulations were performed for each parameter combination. All simulations were performed on a torus. Some aspects of the simulations are discussed in Section 5.6.

**Table 5.11:** Comparison with simulations for $T_1' = T_2' = T_3' = 10$ and $\lambda' = 5$

| $n$ | Empirical | (5.32) | (5.33) |
|---|---|---|---|
| 16 | 0.0003 | 0.0016 | 0.0016 |
| 17 | 0.0465 | 0.0525 | 0.0541 |
| 18 | 0.3400 | 0.3287 | 0.3327 |
| 19 | 0.6945 | 0.6885 | 0.6913 |
| 20 | 0.8910 | 0.8915 | 0.8925 |
| 21 | 0.9687 | 0.9676 | 0.9678 |
| 22 | 0.9902 | 0.9911 | 0.9912 |
| 23 | 0.9979 | 0.9977 | 0.9978 |
| 24 | 0.9996 | 0.9995 | 0.9995 |
| 25 | 0.9999 | 0.9999 | 0.9999 |
| 26 | 0.9999 | 1.0000 | 1.0000 |
| 27 | 1.0000 | 1.0000 | 1.0000 |

**Table 5.12:** Comparison with simulations for $T_1' = 4, T_2' = 8, T_3' = 16$ and $\lambda' = 2$

| $n$ | Empirical | (5.32) | (5.33) |
|---|---|---|---|
| 8 | 0.0000 | 0.0009 | 0.0011 |
| 9 | 0.0162 | 0.0343 | 0.0393 |
| 10 | 0.3010 | 0.3339 | 0.3494 |
| 11 | 0.7382 | 0.7530 | 0.7610 |
| 12 | 0.9357 | 0.9389 | 0.9409 |
| 13 | 0.9879 | 0.9876 | 0.9880 |
| 14 | 0.9974 | 0.9978 | 0.9978 |
| 15 | 0.9997 | 0.9996 | 0.9996 |
| 16 | 1.0000 | 0.9999 | 0.9999 |

Examples from the simulations, where the empirical distribution is compared with the approximations (5.32) and (5.33) are given in Table 5.11, where $T_1' = T_2' = T_3' = 10$ and $\lambda' = 5$ and in Table 5.12, where $T_1' = 4$, $T_2' = 8$, $T_3' = 16$, and $\lambda' = 2$.

From the tables, we see that both approximations give good agreement with the simulations, even for moderate values of $n$.

**Table 5.13:** Comparison of approximations (5.32) and (5.33) when $T_1'$, $T_2'$, and $T_3'$ are equal

| $T_1'$ | $T_2'$ | $T_3'$ | $\lambda'$ | Torus | $d(5.32)$ | $d(5.33)$ | $\mu_e^*$ | $\sigma_e^*$ |
|---|---|---|---|---|---|---|---|---|
| 4 | 4 | 4 | 8 | + | 0.2431 | 0.2198 | 0.0211 | 0.0058 |
| 5 | 5 | 5 | 2 | + | 0.1768 | 0.1820 | 0.0162 | 0.0057 |
| 5 | 5 | 5 | 5 | + | 0.1368 | 0.1206 | 0.0187 | 0.0060 |
| 10 | 10 | 10 | 1 | + | 0.2484 | 0.2560 | 0.0128 | 0.0054 |
| 10 | 10 | 10 | 2 | + | 0.1084 | 0.1144 | 0.0143 | 0.0060 |
| 10 | 10 | 10 | 3 | + | 0.0592 | 0.0668 | 0.0159 | 0.0057 |
| 10 | 10 | 10 | 5 | + | 0.0402 | 0.0369 | 0.0169 | 0.0061 |

**Table 5.14:** Comparison of approximations (5.32) and (5.33) when $T_1'$, $T_2'$, and $T_3'$ are unequal

| $T_1'$ | $T_2'$ | $T_3'$ | $\lambda'$ | Torus | $d(5.32)$ | $d(5.33)$ | $\mu_e^*$ | $\sigma_e^*$ |
|---|---|---|---|---|---|---|---|---|
| 4 | 8 | 16 | 2 | + | 0.0674 | 0.0976 | 0.0145 | 0.0058 |
| 4 | 16 | 8 | 2 | + | 0.0914 | 0.1210 | 0.0145 | 0.0058 |
| 8 | 4 | 16 | 2 | + | 0.0702 | 0.0824 | 0.0145 | 0.0058 |
| 8 | 16 | 4 | 2 | + | 0.1484 | 0.1500 | 0.0145 | 0.0058 |
| 16 | 4 | 8 | 2 | + | 0.0994 | 0.1002 | 0.0145 | 0.0058 |
| 16 | 8 | 4 | 2 | + | 0.1578 | 0.1502 | 0.0145 | 0.0058 |

To compare the approximations, we have, for each simulation, calculated the total variation distance $d(p_e, p_a)$ of (5.15) as a measure of the distance between the approximate and the empirical distributions, and bootstrap estimates, $\mu_e^*$ and $\sigma_e^*$, of the mean and standard deviation of the distance $d(p_e, p)$ between the empirical and the true distribution; see Section 5.3.1.

In Table 5.13, the parameters $T_1'$, $T_2'$ and $T_3'$ are equal, and the two approximations give similar results, with (5.33) doing slightly better for larger values of $\lambda'$. In Table 5.14, $T_1'$, $T_2'$ and $T_3'$ are unequal. Here, the precision depends heavily on the choice of $T_1'$, $T_2'$ and $T_3'$, with $T_1' < T_2' < T_3'$ and approximation (5.32) giving the best fit. As a general recommendation: *Use approximation (5.32) with $T_1' \leq T_2' \leq T_3'$.*

## 5.5 Approximation in Higher Dimensions

In $d$ dimensions, study a Poisson process with intensity $\lambda$ on the box $A_d = [0, T_1] \times [0, T_2] \times \ldots \times [0, T_d]$, with scanning set $R_d = [0, w_1] \times [0, w_2] \times \ldots \times [0, w_d]$. Introduce the notation

$$T_i' = T_i/w_i\,, i = 1, \ldots, d\,, \quad \lambda' = \lambda \prod_{i=1}^{d} w_i\,,$$

$$A_d' = [0, T_1'] \times \ldots \times [0, T_d']\,, \text{ and } R_d' = [0, 1] \times \ldots \times [0, 1]\,.$$

Then, the scan statistics $S_{R_d}(\lambda, A_d)$ and $S_{R_d'}(\lambda', A_d')$ have the same distribution. Repeating the argument that resulted in (5.32) gives

$$\begin{aligned} P(S_{R_d}(\lambda, A_d) \leq n) &= P(S_{R_d'}(\lambda', A_d') \leq n) \\ &= P(S_{R_{d-1}'}(\lambda', A_{d-1}') \leq n)e^{-\alpha_{d-1}(n+1)}\,, \end{aligned} \qquad (5.34)$$

where $\alpha_{d-1}(n+1)$ is approximated as in (5.11) and (5.31) by

$$\alpha_{d-1}(n+1) \approx \left(1 - \frac{\lambda'}{n+1}\right) \lambda'(T_d' - 1)P(S_{R_{d-1}'}(\lambda', A_{d-1}') = n)\,. \qquad (5.35)$$

Note that the $(d-1)$-dimensional scan statistic $S_{R_{d-1}'}(\lambda', A_{d-1}')$, in (5.34) and (5.35), has the same distribution as $S_{R_{d-1}}(\lambda w_d, A_{d-1})$.

**Remark 5.5.1** Formula (5.34) holds also for $d = 1$, if $S_{R_0'}(\lambda', A_0')$ is identified as a Poisson distributed random variable with mean $\lambda'$. Then, (5.34) reduces to (5.6). For $d = 2$ we get formula (5.12), and for $d = 3$ we get (5.30).

## 5.6 Comments on Simulations

For simplicity, we will throughout this section, with a slight abuse of notation, write $\lambda$, $T_1$, $T_2$, and $T_3$ instead of $\lambda'$, $T_1'$, $T_2'$, and $T_3'$.

The accuracy of the one-dimensional approximation (5.6) can be evaluated by comparing it with the lower and upper bounds given by Janson (1984). In two and more dimensions, there are no known corresponding bounds and so we have to use simulation to evaluate the different approximations.

The key idea behind the simulations, in all dimensions, is that, to determine the value of the scan statistic for a certain realization of the Poisson process, it is sufficient to study the number of points in a *finite* number of sets, *the extreme sets.*

In one dimension, the extreme sets are the intervals where one endpoint (e.g., the left) coincides with a point in the realization.

In two dimensions, with rectangular scanning sets, the extreme sets are two types of rectangles; see Figure 5.3.

(i) Those where one corner (e.g., the lower left) coincides with a point.

(ii) Those where two sides (e.g., the lower and the left) contain one point each.

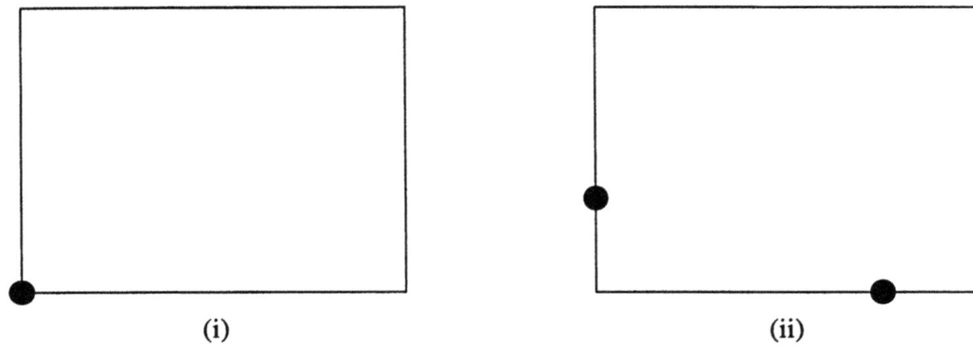


**Figure 5.3:** Extreme sets in two dimensions

To study the extreme rectangles of type (ii), we need to consider all pairs of points (at least those where the points lie sufficiently close), which makes the simulations much more time consuming in two dimensions than in one.

In three dimensions, with rectangular scanning sets, the extreme boxes are those that can not be translated by increasing any of the coordinates without losing a point. Consider a certain box. If it is extreme, it can not be fixed by interior points, but must have points in a corner (which fixes all three coordinates), on an edge (which fixes two coordinates), or on a side (which fixes one coordinate).

This gives three different types (one with three cases) to check:

(i) Those which are fixed by one point (in the corner where all coordinates are minimal).

(ii) Those which are fixed by two points, one on an edge and one on a side. There are three possible cases, depending on which coordinate is fixed by the point on the side.

(iii) Those which are fixed by three points, on three different sides.

From the above, one would expect the time required for the simulation to be of the order $\lambda T_1$, $(\lambda T_1 T_2)^2$ and $(\lambda T_1 T_2 T_3)^3$ in one, two, and three dimensions, respectively.

By using the possibility of ordering the points after one of the coordinates (for example, the first), it is possible to reduce these orders to

$$(\lambda T_2)^2 T_1 \quad \text{in two dimensions} \tag{5.36}$$

and

$$\lambda^3 (T_2 T_3)^2 T_1 \quad \text{in three dimensions.} \tag{5.37}$$

To indicate how the value of the scan statistic was determined, we will give a brief description of the procedure in two dimensions. The realization of the Poisson process was obtained by first generating the horizontal coordinate according to a one-dimensional Poisson process with intensity $\lambda T_2$, using the well-known fact that the gaps are exponentially distributed and independent. This gives an automatic ordering of the points with regard to this coordinate. The vertical coordinate is then, independently for different points, chosen according to a uniform distribution on $(0, T_2)$.

Going through the points from left to right, we then check for extreme rectangles with this point in the lower left corner, or on the left side with another point on the lower side. As the points are ordered, we need only check new points until we come across one whose horizontal coordinate is too large.

The idea in three dimensions is similar, although somewhat more complicated.

From (5.36) and (5.37), we see that it is important to sort the points according to the coordinate that corresponds to the largest of $T_1$, $T_2$ and $T_3$.

The simulations are very time consuming, especially in three dimensions, where some of the simulations required more than a hundred hours of CPU time. In contrast, to calculate the approximations, it took typically only a few seconds on a PC. The calculations were performed in UBASIC to guarantee sufficient numerical precision, but could be performed with any tool admitting at least double precision.

Because of the considerable execution time, it is tempting to try some variance reducing technique in order to obtain the same precision with fewer simulations.

One possibility is to study not only the scan statistics $S_R$, but also the related statistics $S_R'$, which are obtained by maximizing only over disjoint rectangles/boxes. As the latter statistics are the maximum of a number of independent random variables, their distributions can easily be calculated exactly.

This method is discussed in greater detail by Alm (1998).

For this technique to give a usable reduction of the variance, the correlation between $S_R$ and $S_R'$, $\rho = \rho(S_R, S_R')$, must not be too small. From experience, we can expect a reduction in variance by a factor of order $1 - \rho^2$.

Unfortunately, the correlations are decreasing with the dimension, and in three dimensions, where we really need a reduction of the variance, $1-\rho^2 > 0.85$, which unfortunately is of little practical use.

Simulations with non-rectangular scanning sets are more complicated, but can be performed in a similar way. As an example, with a circular scanning set in two dimensions, we need to check all pairs of points that are sufficiently close to lie on the circumference of the same circle. For such a pair there are two possible circles. It is sufficient to check one of them, e.g., the one whose center is closest to the origin. This method was used to perform simulation with circular scanning sets by Alm (1997). A similar technique can be used for any convex scanning set in two dimensions, and also for spherical scanning sets in three dimensions. The time required for the simulation increases markedly when non-rectangular scanning sets are used, increasing the need for good algorithms and approximations!

**Acknowledgements.** This paper is part of the joint research project *Risk Analysis for Building Structures* involving researchers from Structural Mechanics and Risk Analysis/Mathematical Statistics. The project is financially supported by *The Axel and Margaret Ax:son Johnson Foundation*.

I would like to thank Lennart Bondesson for valuable suggestions and inspiring discussions concerning variance reducing techniques and one of the referees for suggesting the inclusion of the general $d$-dimensional approximation of Section 5.4.

# References

1. Aldous, D. (1989). *Probability Approximations via the Poisson Clumping Heuristic*, New York: Springer-Verlag.

2. Alm, S. E. (1983). On the distribution of the scan statistic of a Poisson process, In *Probability and Mathematical Statistics, Essays in Honour of Carl-Gustav Esseen*, 1–10.

3. Alm, S. E. (1997). On the distribution of scan statistics of a two-dimensional Poisson process, *Advances in Applied Probability*, **29**, 1–18.

4. Alm, S. E. (1998). Approximation and simulation of the distributions of scan statistics for Poisson process in higher dimensions, *Extremes*, **1**, 111–126.

5. Auer, P., Hornik, K. and Révész, P. (1991). Some limit theorems for the homogeneous Poisson process, *Statistics and Probability Letters*, **12**, 91–96.

6. Hjalmars, U., Kulldorff, M., Gustafsson, G. and Nagarwalla, N. (1996). Childhood leukaemia in Sweden: Using GIS and a spatial scan statistic for cluster detection, *Statistics in Medicine*, **15**, 707–715.

7. Janson, S. (1984). Bounds on the distributions of extremal values of a scanning process, *Stochastic Processes and Their Applications*, **18**, 313–328.

8. Kulldorff, M. and Nagarwalla, N. (1995). Spatial disease clusters: Detection and inference, *Statistics in Medicine*, **14**, 799–810.

9. Loader, C. R. (1991). Large-deviation approximations to the distribution of scan statistics, *Advances in Applied Probability*, **23**, 751–771.

10. Mack, C. (1949). The expected number of aggregates in a random distribution of $n$ points, *Proceedings of the Cambridge Philosophical Society*, **46**, 285–292.

11. Månsson, M. (1994). Covering uniformly distributed points by convex scanning sets, *Preprint* 1994:17/ISSN 0347-2809, Department of Mathematics, Chalmers University of Technology, Göteborg, Sweden.

12. Månsson, M. (1995). Intersections of uniformly distributed translations of convex sets in two and three dimensions, *Preprint* 1995:11/ISSN 0347-2809, Department of Mathematics, Chalmers University of Technology, Göteborg, Sweden.

13. Naus, J. I. (1982). Approximations for distributions of scan statistics, *Journal of the American Statistical Association*, **77**, 177–183.

# 6

# An Approach to Computations Involving Spacings With Applications to the Scan Statistic


**Fred W. Huffer and Chien-Tai Lin**

*Florida State University, Tallahassee, FL*
*Tamkang University, Taiwan*



**Abstract:** Consider the order statistics from $N$ i.i.d. random variables uniformly distributed on the interval $(0, 1]$. We present a general method for computing probabilities involving differences of the order statistics or linear combinations of the spacings between the order statistics. This method is based on repeated use of a basic recursion to break up the joint distribution of linear combinations of spacings into simpler components which are easily evaluated. Let $S_w$ denote the (continuous conditional) scan statistic with window length $w$. Let $C_w$ denote the number of $m : w$ clumps among the $N$ random points, where an $m : w$ clump is defined as $m$ points falling within an interval of length $w$. We apply our general method to compute the distribution of $S_w$ (for small $N$) and the lower-order moments of $C_w$. The final answers produced by our approach are piecewise polynomials (in $w$) whose coefficients are computed exactly. These expressions can be stored and later used to rapidly compute numerical answers which are accurate to any required degree of precision.

**Keywords and phrases:** Uniform distribution, number of clumps, order statistics, $m$-spacings, symbolic computation


## 6.1 Introduction

We begin by giving a general description of our work. In order to do this, we need some notation. Let $X_1, X_2, \ldots, X_N$ be i.i.d. random variables which are uniformly distributed on the interval $(0, 1]$. Let $X_{(1)} < X_{(2)} < \cdots < X_{(N)}$ be the corresponding order statistics. The spacings $G_1, G_2, \ldots, G_{N+1}$ are the lengths of the spaces (or gaps) between consecutive order statistics, that is, $G_i = X_{(i)} - X_{(i-1)}$ for $i = 1, 2, \ldots, N + 1$ where, for convenience, we take

$X_{(0)} = 0$ and $X_{(N+1)}=1$. Let $\boldsymbol{G} = (G_1, G_2, \ldots, G_{N+1})'$ denote the column vector of spacings.

We have developed general methods for computing probabilities involving linear combinations of spacings. In particular, we are able to compute quantities of the form

$$P(\boldsymbol{\Gamma G} > w) \quad \text{or} \quad P(\boldsymbol{\Gamma G} < w) \tag{6.1}$$

for a wide variety of matrices $\Gamma$. (For a vector $\boldsymbol{U} = (u_1, \ldots, u_k)$ and a real value $w$, we say that $\boldsymbol{U} > w$ if $u_i > w$ for all $i$; $\boldsymbol{U} < w$ is similarly defined.) Most of our attention has been directed to linear combinations which are simple sums of the spacings. To represent these sums, we use the following notation: for any set $\Delta \subset \{1, 2, \ldots, N + 1\}$, define

$$G(\Delta) = \sum_{i \in \Delta} G_i . \tag{6.2}$$

We are able to compute

$$P\left(\bigcap_{i=1}^{r} \{G(\Delta_i) > w\}\right) \quad \text{or} \quad P\left(\bigcap_{i=1}^{r} \{G(\Delta_i) < w\}\right) \tag{6.3}$$

for a fairly broad class of configurations of the sets $\Delta_1, \ldots, \Delta_r$. Differences between the order statistics $X_{(j)}$ are sums of consecutive spacings (that is, $X_{(k)} - X_{(j)} = G_{j+1} + G_{j+2} + \cdots + G_k$ for $j < k$) and an important special class of (6.3) is to evaluate the quantities

$$P\left(\bigcap_{i=1}^{r} \left\{X_{(k_i)} - X_{(j_i)} > w\right\}\right) \quad (\text{or replace} > \text{by} <) \tag{6.4}$$

where $j_i < k_i$ for $i = 1, \ldots, r$.

The work we have done in finding probabilities in (6.1), (6.3), and (6.4) is described in Huffer (1988), Lin (1993), and Huffer and Lin (1995, 1996, 1997a, 1997b). The current status of our work is as follows: We are now able to find arbitrary probabilities of the form (6.4), but are not yet able to find probabilities having the more general forms (6.1) and (6.3). We are in the process of developing algorithms and software which can find probability in (6.1) for arbitrary matrices $\Gamma$ with rational entries.

The main application of our methods so far has been to compute the distribution of the scan statistic $S_w$ and the moments of the number of clumps $C_w$. We now formally define these quantities. The scan statistic $S_w$ is simply the maximum number of points $X_1, X_2, \ldots, X_N$ contained in a scanning interval (window) of length $w$. If we let $Y_t(w)$ denote the number of these points in the interval $(t, t + w]$, then we can write $S_w = \max_{0 < t < 1-w} Y_t(w)$. We say that $m$ points form an $m : w$ clump if these points are all contained in some interval of length $w$. There is an $m:w$ clump beginning at $X_{(i)}$ if $X_{(i+m-1)} - X_{(i)} < w$. We define the total number of clumps $C_w$ to be

$$C_w = \sum_{i=1}^{N-m+1} I(X_{(i+m-1)} - X_{(i)} < w). \qquad (6.5)$$

It is clear that $S_w \geq m$ if and only if there exists at least one $m:w$ clump. This implies

$$P(S_w < m) = 1 - P(S_w \geq m) = P\left( \bigcap_{i=1}^{N-m+1} \left\{ X_{(i+m-1)} - X_{(i)} > w \right\} \right) \qquad (6.6)$$

which is a probability of the type (6.4). Similarly, we show in Section 6.8 that the moments of $C_w$ can be expressed as sums of probabilities all having the form in (6.4).

The remainder of this chapter is organized as follows. Section 6.2 lists some results about spacings that we shall need. Section 6.3 presents the recursion which is the basis of our approach; we evaluate quantities such as (6.1) and (6.3) by repeated use of this recursion. This section also presents some notation and properties which are used in conjunction with the recursion. Section 6.4 extends the recursion to i.i.d. exponential random variables. Section 6.5 illustrates the use of the recursion by applying it in a simple, but important special case: the evaluation of (6.1) when $\Gamma$ is a binary matrix with two rows. In Section 6.6, we show how our methods may be used to evaluate the distribution of the scan statistic $S_w$. In Section 6.7, we present an algorithm for systematically applying the recursion to evaluate any quantity of the form (6.4). Finally, in Section 6.8 we illustrate the use of our methods to evaluate the moments of the number of clumps $C_w$.

## 6.2  Properties of Spacings

We now review some well-known properties of spacings that we shall need later. The vector of spacings $G$ is uniformly distributed on the simplex

$$\{ G \in \mathcal{R}^{N+1} : G_i > 0 \text{ for all } i \text{ and } G_1 + G_2 + \cdots + G_{N+1} = 1 \}.$$

This is equivalent to saying that the spacings have a joint Dirichlet distribution with parameters $1, 1, \ldots, 1$ which we abbreviate as $\mathcal{D}(1, 1, \ldots, 1)$; see Wilks (1962) and Johnson, Kotz, and Balakrishnan (1999) for a review of the Dirichlet distribution. With this notation, we may write

$$(G_1, G_2, \ldots, G_{N+1}) \sim \mathcal{D}(1, 1, \ldots, 1). \qquad (6.7)$$

Many facts about spacings can be easily derived from the following formula which may be found in Chapter I, Exercise 23 of Feller (1971). For arbitrary $a_1 \geq 0, \ldots, a_{N+1} \geq 0$, we have

$$P(G_1 > a_1, G_2 > a_2, \ldots, G_{N+1} > a_{N+1}) = \left(1 - \sum_{i=1}^{N+1} a_i\right)_+^N. \qquad (6.8)$$

Here, we use $(x)_+$ to denote the positive part, that is, $(x)_+ = \max(x, 0)$. From this formula, it is immediate that the spacings $G_1, G_2, \ldots, G_{N+1}$ are exchangeable random variables, each with a beta$(1, N)$ distribution.

The joint distribution of nonoverlapping sums of spacings also has a Dirichlet distribution. This is an easy consequence of (6.7) and standard properties of the Dirichlet distribution. Here is a precise statement of this fact. Let $\Delta_1, \ldots, \Delta_r$ be disjoint nonempty subsets of $\{1, 2, \ldots, N+1\}$ with cardinalities $|\Delta_i| = p_i$ for $1 \leq i \leq r$. Then

$$(G(\Delta_1), \ldots, G(\Delta_r)) \sim \mathcal{D}(p_1, \ldots, p_r, N+1 - \textstyle\sum_i p_i).$$

Starting from this joint distribution, it can be shown that

$$P\left(\bigcap_{i=1}^{r}\{G(\Delta_i) > a_i\}\right) = \sum_{(k_1,\ldots,k_r)} \binom{N}{k_1, \ldots, k_r} \left(1 - \sum_{i=1}^{r} a_i\right)_+^{N-\sum_i k_i} \prod_{i=1}^{r} a_i^{k_i},$$

$$(6.9)$$

where $a_1, \ldots, a_r$ are nonnegative, and the sum is over all $r$-tuples of integers $(k_1, k_2, \ldots, k_r)$ satisfying $0 \leq k_i \leq p_i - 1$ for all $i$. This formula may be found as Theorem 2.1 of Khatri and Mitra (1969). If $p_1 = p_2 = \cdots = p_r = 1$, then (6.9) reduces to (6.8).

We rely on (6.9) for most of the explicit formulas in our work. This formula gives the solution to problems of the form (6.3) in the case where the sets $\Delta_i$ are disjoint. When the sets $\Delta_1, \ldots, \Delta_r$ overlap, we can use the recursion (6.15) presented in the next section to re-express the probability (6.3) as a sum of similar probabilities which involve only disjoint sets; these probabilities can then be evaluated using (6.9).

We can express probabilities of the form (6.3) in a much more compact form by introducing the following notation. For integers $j \geq 0$ and real values $\lambda \geq 0$, define

$$R(j, \lambda) = \begin{cases} \binom{N}{j} w^j (1 - \lambda w)^{N-j} & \text{for } \lambda w < 1, \\ 0 & \text{for } \lambda w \geq 1. \end{cases} \qquad (6.10)$$

The dependence of $R$ on $N$ and $w$ can be left implicit because these values are fixed in any given application of our methods. Viewed as a function of $w$ (taking values in (0,1)) with $N$, $j$ and $\lambda$ fixed, R is a piecewise polynomial with two pieces: $w \in (0, 1/\lambda)$ and $w \in (1/\lambda, 1)$. In terms of $R$, when $a_1 = \cdots = a_r = w$, formula (6.9) becomes

$$P\left(\bigcap_{i=1}^{r}\{G(\Delta_i) > w\}\right) = \sum_{(k_1,\ldots,k_r)} \binom{\sum_i k_i}{k_1,\ldots,k_r} R(\textstyle\sum_i k_i, r), \qquad (6.11)$$

where again the summation is over all $r$-tuples of integers $(k_1, k_2, \ldots, k_r)$ satisfying $0 \le k_i \le p_i - 1$ for all $i$. Suppressing the dependence on $N$ and $w$ reduces the length of the answers we obtain and makes it easier to state answers which are valid for all $N$ and $w$.

---

## 6.3  The Basic Recursion

In this section, we describe the recursion which is the basis of our approach. This recursion was first obtained by Micchelli (1980) as a result about multivariate B-splines. The result was rediscovered in a probabilistic setting by Huffer (1988). We also introduce some matrix notation and simplify properties that we find very useful when applying this recursion.

Let $\boldsymbol{\Gamma}$ be an $r \times (N + 1)$ real matrix. Let $\boldsymbol{G} = (G_1, G_2, \ldots, G_{N+1})'$ be the vector of spacings between uniform random variables as defined in Section 6.1. For any $\boldsymbol{\xi} \in \mathcal{R}^r$, define $\boldsymbol{\Gamma}_{i,\xi}$ to be the $r \times (N + 1)$ matrix obtained by replacing the $i$th column of $\boldsymbol{\Gamma}$ by $\boldsymbol{\xi}$. The basic recursion is the following.

**Theorem 6.3.1** *Suppose* $\boldsymbol{c} = (c_1, c_2, \ldots, c_{N+1})'$ *satisfies* $\sum_{i=1}^{N+1} c_i = 1$. *Let* $\boldsymbol{\xi} = \boldsymbol{\Gamma} \boldsymbol{c}$. *Then,*

$$P(\boldsymbol{\Gamma}\boldsymbol{G} \in B) = \sum_{i=1}^{N+1} c_i\, P(\boldsymbol{\Gamma}_{i,\xi}\boldsymbol{G} \in B) \qquad (6.12)$$

*for any measurable set* $B \subset \mathcal{R}^r$.

For a proof, see Huffer (1988). The proof given there uses the moment generating function of $\boldsymbol{\Gamma}\boldsymbol{G}$ and properties of divided differences. In some special cases, the recursion can be proved by very elementary means. See Chapter 6 of Huffer (1982) for an example of this kind.

### Matrix notation for spacings problems

We now introduce some matrix notation which allows us to re-express the recursion (6.12) in a form convenient for presenting examples of its use.

Let $\boldsymbol{A}$ be any matrix having $r$ rows and at most $N + 1$ columns. Take $\boldsymbol{\Gamma}$ to be the matrix with $N + 1$ columns obtained by padding $\boldsymbol{A}$ with columns of zeros; $\boldsymbol{\Gamma} = (\boldsymbol{A}|\,\boldsymbol{0})$. Define $\boldsymbol{Y} = (Y_1, Y_2, \ldots, Y_r)$ by $\boldsymbol{Y} = \boldsymbol{\Gamma G}$. For any value of $w$, we define

$$\{\boldsymbol{A}\}_w^1 = P(\boldsymbol{\Gamma G} > w) = P(\min Y_i > w), \qquad (6.13)$$

and

$$\{\boldsymbol{A}\}_w^2 = P(\boldsymbol{\Gamma G} < w) = P(\max Y_i < w). \qquad (6.14)$$

When the value of $w$ is held fixed in an argument, we delete the subscript and just write $\{\boldsymbol{A}\}^1$ or $\{\boldsymbol{A}\}^2$. We also omit the superscript when convenient.

The quantity $\{\boldsymbol{A}\}$ is well defined so long as the number of spacings $N + 1$ is greater than or equal to the number of columns in $\boldsymbol{A}$. The value of $\{\boldsymbol{A}\}$ depends, of course, on $N$, but we do not indicate this in the notation because the value of $N$ is not important in most of our manipulations.

When specialized to probabilities of the type (6.13) and (6.14), the basic recursion (6.12) becomes the following:

*Let $\boldsymbol{A}$ be a matrix having $k$ columns with $k \leq N+1$. Suppose $\boldsymbol{c} = (c_1, c_2, \ldots, c_k)'$ satisfies $\sum_{i=1}^k c_i = 1$. Let $\boldsymbol{\xi} = \boldsymbol{Ac}$. Then,*

$$\{\boldsymbol{A}\} = \sum_{i=1}^k c_i \{\boldsymbol{A}_{i,\xi}\}. \qquad (6.15)$$

It is understood that all of the braces $\{\cdot\}$ appearing in (6.15) have a common superscript of 1 or 2 and a common subscript of $w$.

### Simplification properties

Properties S1–S4 given below allow us to rearrange and simplify the matrix $\boldsymbol{A}$ of (6.13) or (6.14) in various ways. We state these properties in terms of evaluating $\{\boldsymbol{A}\}^1$.

**(S1)** If the $i$th row of $\boldsymbol{A}$ dominates (is componentwise greater than or equal to) the $j$th row, the $i$th row can be deleted without changing the value of $\{\boldsymbol{A}\}^1$.

The value of $\{\boldsymbol{A}\}^1$ remains the same when

**(S2)** a column of zeros is deleted,

**(S3)** the columns are permuted, $\qquad\qquad\qquad\qquad\qquad\qquad (6.16)$

**(S4)** the rows are permuted.

The same properties apply to $\{\boldsymbol{A}\}^2$ except that **S1** must be reversed. In other words,

**(S1′)** if the $i$th row of $A$ is dominated by (is componentwise less than or equal to) the $j$th row, the $i$th row can be deleted without changing the value of $\{A\}^2$.

These properties are all straightforward and we shall not prove them in detail. However, some comments are in order. The matrix $A$ is just a shorthand for writing a set of inequalities, with each row corresponding to a different inequality. Property **S1** simply states that redundant inequalities (rows) can be omitted. If the $k$th column of $A$ is zero, this means the spacing $G_k$ is not actually involved in any of the inequalities. Thus, we can drop $G_k$ from our inequalities and renumber the spacings $G_j$ for $j > k$ (justified by exchangeability) to fill in the resulting gap. This is the content of property **S2**. Noting that each column of $A$ is associated with a particular spacing, property **S3** follows immediately from the exchangeability of the spacings. Property **S4** is a consequence of the fact that each of the implied inequalities in (6.13) or (6.14) involves the same value $w$. If we had defined $\{A\}^1 = P(Y_i > w_i$ for all i) where the values of $w_i$ are not all equal, then property **S4** would no longer hold.

## 6.4   The Recursion for Exponential Variates

The joint distribution of the spacings is closely related to that of independent exponential random variables. Because of this, the recursion (6.12) [and the special case (6.15)] also holds for exponential random variables. This section is devoted to a brief discussion of this fact and its consequences. This section can be omitted with little loss of continuity.

Let $Z_1, Z_2, \ldots, Z_{N+1}$ be i.i.d. exponential random variables with mean 1. Define $Z = (Z_1, Z_2, \ldots, Z_{N+1})'$. Then we have

**Corollary 6.4.1** *The recursion (6.12) remains true when $G$ is replaced by $Z$.*

We give a brief argument for this corollary at the end of this section. Similarly, the recursion (6.15) and simplification properties (6.16) continue to hold if we replace $G$ by $Z$ in the definitions (6.13) and (6.14). Also, most of the formulas in Section 6.2 have obvious analogs for exponential random variables. In particular, if we replace the definition of $R$ in (6.10) by

$$R(j, \lambda) = \frac{w^j}{j!} e^{-\lambda w}, \tag{6.17}$$

then (6.11) continues to hold so long as we make the obvious notational change, replacing $G(\Delta_i)$ by the analogous quantity $Z(\Delta_i)$.

Because (6.11) and (6.15) also hold for exponential random variables, the results we obtain by our methods [which are expressions like (6.28), (6.31), and

(6.43) given later] have two valid interpretations: as results about spacings: and as results about exponential random variables. Alternatively, since the interarrival times in a Poisson process are independent exponential random variables, our results may be interpreted in terms of Poisson processes.

PROOF OF COROLLARY. Define $T = \sum_{i=1}^{N+1} Z_i$. It is well known that

$$T \text{ and } \boldsymbol{Z}/T \text{ are independent and } \boldsymbol{Z}/T \stackrel{d}{=} \boldsymbol{G}.$$

This implies that

$$\boldsymbol{Z} \stackrel{d}{=} T\boldsymbol{G}, \tag{6.18}$$

where $T$ is independent of $\boldsymbol{G}$ and has a gamma distribution with parameters $N+1$ and 1. Let $\boldsymbol{c}$, $\boldsymbol{\xi}$, and $B$ be as in (6.12), and let $g$ be the probability density function of the gamma random variable $T$ in (6.18). By conditioning on the value of $T$, we find

$$
\begin{aligned}
P(\boldsymbol{\Gamma}\boldsymbol{Z} \in B) &= \int_0^\infty P(t\boldsymbol{\Gamma}\boldsymbol{G} \in B)g(t)dt \\
&= \sum_{i=1}^{N+1} c_i \int_0^\infty P(t\boldsymbol{\Gamma}_{i,\xi}\boldsymbol{G} \in B)g(t)dt \\
&= \sum_{i=1}^{N+1} c_i P(\boldsymbol{\Gamma}_{i,\xi}\boldsymbol{Z} \in B).
\end{aligned}
$$

Here, we have used

$$P(t\boldsymbol{\Gamma}\boldsymbol{G} \in B) = \sum_{i=1}^{N+1} c_i P(t\boldsymbol{\Gamma}_{i,\xi}\boldsymbol{G} \in B)$$

which follows from the recursion (6.12) upon replacing $\boldsymbol{\Gamma}$ by $t\boldsymbol{\Gamma}$ and noting that $t\boldsymbol{\Gamma}_{i,\xi} = (t\boldsymbol{\Gamma})_{i,t\xi}$. $\blacksquare$

---

## 6.5   Binary Matrices With Two Rows

To illustrate the use of the basic recursion, we will consider the simple, but important case of a binary matrix with two rows. In particular, we will calculate the probability

$$P(G_1 + G_2 + G_3 + G_4 + G_5 + G_6 > w,\ G_5 + G_6 + G_7 + G_8 + G_9 > w). \tag{6.19}$$

This concrete example suffices to introduce all the necessary ideas. In the matrix notation defined in (6.13), this probability is $\{A\}_w^1$ where

$$A = \begin{pmatrix} 1\ 1\ 1\ 1\ 1\ 1\ 0\ 0\ 0 \\ 0\ 0\ 0\ 0\ 1\ 1\ 1\ 1\ 1 \end{pmatrix}. \tag{6.20}$$

Let $\gamma_i$ denote the $i$th column of $\boldsymbol{A}$ and $\boldsymbol{0}$ denote a column vector of zeros. The columns of $\boldsymbol{A}$ satisfy $\gamma_1 - \gamma_6 + \gamma_7 = \boldsymbol{0}$. Thus, we may apply the recursion (6.15) with $\boldsymbol{c} = (1, 0, 0, 0, 0, -1, 1, 0, 0)'$ and $\boldsymbol{\xi} = \boldsymbol{A}\boldsymbol{c} = \boldsymbol{0}$ to obtain

$$\{\boldsymbol{A}\} = \{\boldsymbol{A}_{1,0}\} - \{\boldsymbol{A}_{6,0}\} + \{\boldsymbol{A}_{7,0}\}$$

which can be written more explicitly as

$$\left\{ \begin{array}{l} 1\,1\,1\,1\,1\,1\,0\,0\,0 \\ 0\,0\,0\,0\,1\,1\,1\,1\,1 \end{array} \right\} =$$

$$\left\{ \begin{array}{l} 0\,1\,1\,1\,1\,1\,0\,0\,0 \\ 0\,0\,0\,0\,1\,1\,1\,1\,1 \end{array} \right\} - \left\{ \begin{array}{l} 1\,1\,1\,1\,1\,0\,0\,0\,0 \\ 0\,0\,0\,0\,1\,0\,1\,1\,1 \end{array} \right\} + \left\{ \begin{array}{l} 1\,1\,1\,1\,1\,1\,0\,0\,0 \\ 0\,0\,0\,0\,1\,1\,0\,1\,1 \end{array} \right\} .$$

Deleting the columns of zeros [using **S2** of (6.16)] leads to

$$= \left\{ \begin{array}{l} 1\,1\,1\,1\,1\,0\,0\,0 \\ 0\,0\,0\,1\,1\,1\,1\,1 \end{array} \right\} - \left\{ \begin{array}{l} 1\,1\,1\,1\,1\,0\,0\,0 \\ 0\,0\,0\,0\,1\,1\,1\,1 \end{array} \right\} + \left\{ \begin{array}{l} 1\,1\,1\,1\,1\,1\,0\,0 \\ 0\,0\,0\,0\,1\,1\,1\,1 \end{array} \right\} . \quad (6.21)$$

The matrices in these terms all have the same general form as $\boldsymbol{A}$ in (6.20). To capitalize on this fact, we shall introduce some notation. Define

$$Q(i, j, k) = P(G(\Delta_1) + G(\Delta_3) > w, G(\Delta_2) + G(\Delta_3) > w), \quad (6.22)$$

where $\Delta_1, \Delta_2, \Delta_3$ are disjoint sets and $|\Delta_1| = i, |\Delta_2| = j, |\Delta_3| = k$. The value of $k$ is the number of spacings common to both sums. This definition makes sense because it is clear from the exchangeability of the spacings that the probability depends on the sets $\Delta_1, \Delta_2, \Delta_3$ only through the values $i, j, k$.

With this notation, we have $\{\boldsymbol{A}\}_w^1 = Q(4, 3, 2)$. The result in (6.21) can be expressed as

$$Q(4, 3, 2) = Q(3, 3, 2) - Q(4, 3, 1) + Q(4, 2, 2) . \quad (6.23)$$

Essentially the same argument shows, more generally, that for any positive integers $i, j, k$ (satisfying $i + j + k \le N + 1$), we have

$$Q(i, j, k) = Q(i - 1, j, k) + Q(i, j - 1, k) - Q(i, j, k - 1) . \quad (6.24)$$

This general relation may now be used to break up each of the terms which appear in (6.23). Repeated application of (6.24) eventually leads to an expression for $Q(4, 3, 2)$ as a sum of "boundary" terms $Q(i, j, k)$ in which one of the values $i$, $j$, or $k$ is zero. For example, applying (6.24) to the middle term in (6.23) produces

$$Q(4, 3, 1) = Q(3, 3, 1) + Q(4, 2, 1) - Q(4, 3, 0)$$

which contains the boundary term $Q(4, 3, 0)$. From the definition (6.22), it is clear that

$$
\begin{aligned}
Q(i, 0, k) &= P(G(\Delta_3) > w), \\
Q(0, j, k) &= P(G(\Delta_3) > w), \text{ and} \\
Q(i, j, 0) &= P(G(\Delta_1) > w, G(\Delta_2) > w).
\end{aligned}
$$

Each of these expressions is easily evaluated using (6.11) yielding

$$
Q(i, 0, k) = Q(0, j, k) = \sum_{p=0}^{k-1} R(p, 1)
$$

and

$$
Q(i, j, 0) = \sum_{p=0}^{i-1} \sum_{q=0}^{j-1} \binom{p+q}{p} R(p+q, 2) . \tag{6.25}
$$

What we have outlined above is an entirely mechanical procedure for evaluating $Q(i, j, k)$ by repeated application of (6.24). The first step in the evaluation of $\{A\}_w^1 = Q(4, 3, 2)$ is given in (6.23). Pursuing this process to its conclusion leads to

$$
\begin{aligned}
\{A\}_w^1 = {}&-154R(0, 1) + 35R(1, 1) + 155R(0, 2) + 121R(1, 2) + 88R(2, 2) \\
&+ 57R(3, 2) + 30R(4, 2) + 10R(5, 2) .
\end{aligned}
$$

Note that this expression supplies an answer to (6.19) which is valid for all $w$ and for all $N \geq 8$.

---

## 6.6　The Distribution of the Scan Statistic

We now show how the recursion (6.15) can be used to compute the distribution of the scan statistic $S_w$. We do this by way of a particular example. Suppose there are $N = 6$ random points on the unit interval and we are interested in the probability $P(S_w < 3)$ that no interval of length $w$ contains more than 2 of these points. Then by (6.6), we have

$$
\begin{aligned}
P(S_w < 3)& \\
={}& P(X_{(3)} - X_{(1)} > w, X_{(4)} - X_{(2)} > w, X_{(5)} - X_{(3)} > w, X_{(6)} - X_{(4)} > w) \\
={}& P(G_2 + G_3 > w, G_3 + G_4 > w, G_4 + G_5 > w, G_5 + G_6 > w) \\
={}& P(G_1 + G_2 > w, G_2 + G_3 > w, G_3 + G_4 > w, G_4 + G_5 > w) .
\end{aligned}
$$

In the last step, we have made use of the exchangeability of the spacings to replace $G_i$ by $G_{i-1}$ everywhere. In the matrix notation of (6.13), we can now write $P(S_w < 3) = \{A\}_w^1$ where

$$A = \begin{pmatrix} 1\,1\,0\,0\,0 \\ 0\,1\,1\,0\,0 \\ 0\,0\,1\,1\,0 \\ 0\,0\,0\,1\,1 \end{pmatrix}. \tag{6.26}$$

Let $\gamma_i$ denote the $i$th column of $A$. Since $\gamma_1 - \gamma_2 + \gamma_3 - \gamma_4 + \gamma_5 = (0,0,0,0)' = \mathbf{0}$ and $1 - 1 + 1 - 1 + 1 = 1$, the recursion (6.15) with $c = (1,-1,1,-1,1)'$ and $\xi = \mathbf{0}$ implies that

$$\{A\} = \{A_{1,0}\} - \{A_{2,0}\} + \{A_{3,0}\} - \{A_{4,0}\} + \{A_{5,0}\}$$

$$= \left\{ \begin{smallmatrix} 0\,1\,0\,0\,0 \\ 0\,1\,1\,0\,0 \\ 0\,0\,1\,1\,0 \\ 0\,0\,0\,1\,1 \end{smallmatrix} \right\} - \left\{ \begin{smallmatrix} 1\,0\,0\,0\,0 \\ 0\,0\,1\,0\,0 \\ 0\,0\,1\,1\,0 \\ 0\,0\,0\,1\,1 \end{smallmatrix} \right\} + \left\{ \begin{smallmatrix} 1\,1\,0\,0\,0 \\ 0\,1\,0\,0\,0 \\ 0\,0\,0\,1\,0 \\ 0\,0\,0\,1\,1 \end{smallmatrix} \right\} - \left\{ \begin{smallmatrix} 1\,1\,0\,0\,0 \\ 0\,1\,1\,0\,0 \\ 0\,0\,1\,0\,0 \\ 0\,0\,0\,0\,1 \end{smallmatrix} \right\} + \left\{ \begin{smallmatrix} 1\,1\,0\,0\,0 \\ 0\,1\,1\,0\,0 \\ 0\,0\,1\,1\,0 \\ 0\,0\,0\,1\,0 \end{smallmatrix} \right\}.$$

By applying the simplification properties **S1** and **S2** of (6.16) to each term, we obtain

$$\{A\} = \left\{ \begin{smallmatrix} 1\,0\,0\,0 \\ 0\,1\,1\,0 \\ 0\,0\,1\,1 \end{smallmatrix} \right\} - \left\{ \begin{smallmatrix} 1\,0\,0\,0 \\ 0\,1\,0\,0 \\ 0\,0\,1\,1 \end{smallmatrix} \right\} + \left\{ \begin{smallmatrix} 1\,0 \\ 0\,1 \end{smallmatrix} \right\} - \left\{ \begin{smallmatrix} 1\,1\,0\,0 \\ 0\,0\,1\,0 \\ 0\,0\,0\,1 \end{smallmatrix} \right\} + \left\{ \begin{smallmatrix} 1\,1\,0\,0 \\ 0\,1\,1\,0 \\ 0\,0\,0\,1 \end{smallmatrix} \right\}.$$

Since we can freely permute the rows and columns in the above matrices [using properties **S3** and **S4** of (6.16)], we see that the first and fifth terms are equal. Also, the second and fourth terms are equal. Combining these equal terms leads to

$$\{A\} = 2 \left\{ \begin{smallmatrix} 1\,1\,0\,0 \\ 0\,1\,1\,0 \\ 0\,0\,0\,1 \end{smallmatrix} \right\} - 2 \left\{ \begin{smallmatrix} 1\,1\,0\,0 \\ 0\,0\,1\,0 \\ 0\,0\,0\,1 \end{smallmatrix} \right\} + \left\{ \begin{smallmatrix} 1\,0 \\ 0\,1 \end{smallmatrix} \right\}. \tag{6.27}$$

The second and third terms above can be evaluated directly using (6.11). Only the first term requires further work. The columns of the first term satisfy $\gamma_1 - \gamma_2 + \gamma_3 = (0,0,0)'$ so that using the recursion (6.15) with $c = (1,-1,1,0)'$ and $\xi = \mathbf{0}$ gives

$$\left\{ \begin{smallmatrix} 1\,1\,0\,0 \\ 0\,1\,1\,0 \\ 0\,0\,0\,1 \end{smallmatrix} \right\} = \left\{ \begin{smallmatrix} 0\,1\,0\,0 \\ 0\,1\,1\,0 \\ 0\,0\,0\,1 \end{smallmatrix} \right\} - \left\{ \begin{smallmatrix} 1\,0\,0\,0 \\ 0\,0\,1\,0 \\ 0\,0\,0\,1 \end{smallmatrix} \right\} + \left\{ \begin{smallmatrix} 1\,1\,0\,0 \\ 0\,1\,0\,0 \\ 0\,0\,0\,1 \end{smallmatrix} \right\}.$$

Using the simplification properties **S1** and **S2**, we obtain

$$\{A\} = \left\{ \begin{smallmatrix} 1\,0 \\ 0\,1 \end{smallmatrix} \right\} - \left\{ \begin{smallmatrix} 1\,0\,0 \\ 0\,1\,0 \\ 0\,0\,1 \end{smallmatrix} \right\} + \left\{ \begin{smallmatrix} 1\,0 \\ 0\,1 \end{smallmatrix} \right\} = 2 \left\{ \begin{smallmatrix} 1\,0 \\ 0\,1 \end{smallmatrix} \right\} - \left\{ \begin{smallmatrix} 1\,0\,0 \\ 0\,1\,0 \\ 0\,0\,1 \end{smallmatrix} \right\}.$$

Combining this with (6.27) gives

$$\{A\} = 5\left\{\begin{array}{cc} 1 & 0 \\ 0 & 1 \end{array}\right\} - 2\left\{\begin{array}{ccc} 1 & 0 & 0 \\ 0 & 1 & 0 \\ 0 & 0 & 1 \end{array}\right\} - 2\left\{\begin{array}{cccc} 1 & 1 & 0 & 0 \\ 0 & 0 & 1 & 0 \\ 0 & 0 & 0 & 1 \end{array}\right\}$$

$$= 5P(G_1 > w, G_2 > w) - 2P(G_1 > w, G_2 > w, G_3 > w)$$
$$- 2P(G_1 + G_2 > w, G_3 > w, G_4 > w).$$

Evaluating these terms using (6.11) leads to

$$\{A\}_w^1 = 5R(0,2) - 4R(0,3) - 2R(1,3) \tag{6.28}$$
$$= 5(1 - 2w)_+^N - 4(1 - 3w)_+^N - 2Nw(1 - 3w)_+^{N-1}.$$

Plugging in $N = 6$ now gives us $P(S_w < 3)$ for all $w \in (0,1)$. The expression
(6.28) also has a valid interpretation for $N > 6$; it gives the probability that
no three of the first six order statistics $X_{(1)}, \ldots, X_{(6)}$ lie in a window of length
$w$. Similarly, using (6.17), the expression has an interpretation in terms of the
Poisson process; it gives the probability that no three of the first six arrivals
(in a Poisson process with rate 1) lie in a window of length $w$.

The remarks above apply more generally. Our approach leads to answers
which are symbolic expressions (as opposed to numerical values). These expressions are valid for all window sizes $w$ and all sufficiently large values of $N$.
They also have a valid interpretation in terms of Poisson processes or sums of
exponential random variables.

We have so far tackled a very small problem; computing $P(S_w < 3)$ when
$N = 6$. The same approach works for much larger problems. We need notation
to describe matrices with patterns like that in (6.26). Let $I(p, k)$ denote a
$(k - p + 1) \times k$ matrix whose $(i, j)$th entry is 1 if $0 \leq j - i \leq p - 1$ and 0 otherwise.
This means that row $i$ contains a block of $p$ consecutive 1's which begins in
column $i$. Using this notation, the matrix in (6.26) is $I(2, 5)$. Repeating the
argument that led from (6.6) to (6.26), it is easy to see that, for any values of
$m$ and the sample size $N$, we have

$$P(S_w < m) = \{I(m - 1, N - 1)\}_w^1. \tag{6.29}$$

The evaluation of $\{I(p, k)\}$ for any values of $p$ and $k$ can be accomplished by
repeated application of the recursion (6.15) just as in the example above. For
small enough values of $p$ and $k$, this whole process can be carried out "by hand"
using the recursion in an ad hoc fashion. Most problems of interest are much
too large for this approach; they require a computer for their solution. In doing
a small problem by hand, we can rely on the human's ability to "see" that
a particular application of the recursion results in progress toward a solution.
For large problems, which may require using the recursion (6.15) hundreds or
thousands of times, we need to employ the recursion in a definite, systematic

fashion which is guaranteed to terminate in a solution. For the evaluation of $\{I(p,k)\}$, just such a systematic approach is supplied by the "marking" algorithm described in the next section.

As an example of a larger problem involving the scan statistic, we will find an expression for $P(S_w < 8)$ for a sample of size $N = 22$ points. By (6.29), this equals $\{I(7,21)\}$. In order to evaluate this, our software requires many thousands of applications of the recursion (6.15). We shall indicate only the first step in this process; we give below the matrix $I(7,21)$ along with the vector $c$ (given by the "marking" algorithm of the next section) used in the first application of (6.15):

$$\begin{pmatrix}
1 & 1 & 1 & 1 & 1 & 1 & 1 & 0 & 0 & 0 & 0 & 0 & 0 & 0 & 0 & 0 & 0 & 0 & 0 & 0 & 0 \\
0 & 1 & 1 & 1 & 1 & 1 & 1 & 1 & 0 & 0 & 0 & 0 & 0 & 0 & 0 & 0 & 0 & 0 & 0 & 0 & 0 \\
0 & 0 & 1 & 1 & 1 & 1 & 1 & 1 & 1 & 0 & 0 & 0 & 0 & 0 & 0 & 0 & 0 & 0 & 0 & 0 & 0 \\
0 & 0 & 0 & 1 & 1 & 1 & 1 & 1 & 1 & 1 & 0 & 0 & 0 & 0 & 0 & 0 & 0 & 0 & 0 & 0 & 0 \\
0 & 0 & 0 & 0 & 1 & 1 & 1 & 1 & 1 & 1 & 1 & 0 & 0 & 0 & 0 & 0 & 0 & 0 & 0 & 0 & 0 \\
0 & 0 & 0 & 0 & 0 & 1 & 1 & 1 & 1 & 1 & 1 & 1 & 0 & 0 & 0 & 0 & 0 & 0 & 0 & 0 & 0 \\
0 & 0 & 0 & 0 & 0 & 0 & 1 & 1 & 1 & 1 & 1 & 1 & 1 & 0 & 0 & 0 & 0 & 0 & 0 & 0 & 0 \\
0 & 0 & 0 & 0 & 0 & 0 & 0 & 1 & 1 & 1 & 1 & 1 & 1 & 1 & 0 & 0 & 0 & 0 & 0 & 0 & 0 \\
0 & 0 & 0 & 0 & 0 & 0 & 0 & 0 & 1 & 1 & 1 & 1 & 1 & 1 & 1 & 0 & 0 & 0 & 0 & 0 & 0 \\
0 & 0 & 0 & 0 & 0 & 0 & 0 & 0 & 0 & 1 & 1 & 1 & 1 & 1 & 1 & 1 & 0 & 0 & 0 & 0 & 0 \\
0 & 0 & 0 & 0 & 0 & 0 & 0 & 0 & 0 & 0 & 1 & 1 & 1 & 1 & 1 & 1 & 1 & 0 & 0 & 0 & 0 \\
0 & 0 & 0 & 0 & 0 & 0 & 0 & 0 & 0 & 0 & 0 & 1 & 1 & 1 & 1 & 1 & 1 & 1 & 0 & 0 & 0 \\
0 & 0 & 0 & 0 & 0 & 0 & 0 & 0 & 0 & 0 & 0 & 0 & 1 & 1 & 1 & 1 & 1 & 1 & 1 & 0 & 0 \\
0 & 0 & 0 & 0 & 0 & 0 & 0 & 0 & 0 & 0 & 0 & 0 & 0 & 1 & 1 & 1 & 1 & 1 & 1 & 1 & 0 \\
0 & 0 & 0 & 0 & 0 & 0 & 0 & 0 & 0 & 0 & 0 & 0 & 0 & 0 & 1 & 1 & 1 & 1 & 1 & 1 & 1
\end{pmatrix}. \qquad (6.30)$$

The vector $c$ is

$$\begin{pmatrix} 1 & 0 & 0 & 0 & 0 & 0 & -1 & 1 & 0 & 0 & 0 & 0 & 0 & -1 & 1 & 0 & 0 & 0 & 0 & 0 & 0 \end{pmatrix}.$$

In this case, we note that the vector $\boldsymbol{\xi} = A\boldsymbol{c}$ in (6.15) is not a vector of zeros (as it was in previous examples), but has a 1 as the last entry.

At the end of the entire process, we obtain the answer

$$\begin{aligned}
& 5903471R(0,3) + 250971R(1,3) - 334305R(2,3) - 43605R(3,3) \\
& +64719R(4,3) + 14331R(5,3) - 23425R(6,3) - 4932R(7,3) \\
& +21324R(8,3) + 22803R(9,3) + 32409R(10,3) + 110253R(11,3) \\
& +271293R(12,3) + 456885R(13,3) + 577005R(14,3) + 574860R(15,3) \\
& +453024R(16,3) + 262548R(17,3) + 87516R(18,3) - 5903470R(0,4) \\
& -6154438R(1,4) - 6071092R(2,4) - 5609800R(3,4) - 4791568R(4,4) \\
& -3716128R(5,4) - 2551872R(6,4) - 1494528R(7,4) \\
& -702464R(8,4) - 236544R(9,4) - 43008R(10,4) \, .
\end{aligned} \qquad (6.31)$$

Once we obtain the expression (6.31), we set $N = 22$ and then evaluate it numerically for the window sizes $w$ of interest to us. The expressions like (6.31) that we obtain by our methods are often quite large, but the individual terms are easily computed so that numerical computations using these expressions are fairly quick. Because of the very large integer coefficients (with both positive and negative values) that occur in these expressions, standard single or double precision calculations are generally not accurate enough to be useful. We usually do our final numerical calculations using a symbolic math package (such as Maple or Mathematica) which allows one to specify an arbitrarily large number of digits to be retained during the calculations. The process of constructing the expressions [such as (6.31)] by repeated use of the recursion (6.15) is carried out by a program written in the C programming language.

Using the definition of $R(\cdot, \cdot)$ in (6.10), we see by inspection that the expression (6.31) is a piecewise polynomial with two pieces; it is a distinct polynomial in $w$ within each of the disjoint intervals $(0, 1/4)$ and $(1/4, 1/3)$, and is identically zero for $w > 1/3$. Neff and Naus (1980) have presented these same piecewise polynomials in a different way. (They have also used totally different methods to compute the polynomials.) Our approach of writing the answer in terms of the $R$-functions has two advantages; the expressions we give are usually much more compact, and our expressions have interpretations for all sufficiently large $N$ and also for Poisson processes.

The amount of computation time and memory needed to obtain expressions for $\{I(p, k)\}$ increases rapidly with the value of $k$. With our current software, we can handle cases somewhat beyond those given in the tables of Neff and Naus (1980). It is difficult to summarize the exact capability of our current programs since the difficulty of computing $\{I(p, k)\}$ also varies with $p$, and the range of problems we can solve depends on machine-specific factors such as the amount of available RAM. Roughly speaking, we can handle most problems with $k \leq 20$ and some problems with $20 < k \leq 30$. But for $k > 30$, we can solve only a very narrow range of special cases. We plan to make various improvements to our programs, and hope to substantially extend the range of problems we can solve.

Because it is so time-consuming to construct expressions for $\{I(p, k)\}$ when $k$ is large, but the expressions themselves are easily evaluated, a sensible strategy is to store a library of such expressions for future use. Such a library would essentially be a computerized version of the tables of Neff and Naus (1980).

---

## 6.7   The Marking Algorithm

In this section, we give an algorithm for determining all probabilities of the form (6.4). When such probabilities are translated into the matrix notation of (6.13) or (6.14), we end up with binary matrices $A$ having a certain "descending" form; the 1's in each row form a contiguous block with these blocks moving to the right as one advances from row to row. Examples of such matrices are given in (6.20) and (6.26). A more formal definition follows. Suppose $A$ is an $r \times p$

binary matrix. For $i = 1, \ldots, r$, let $a_i$ and $b_i$ denote the position of the first and last 1 in row $i$. We say that $A = (A_{ij})$ has descending form when $A_{ij} = 1$ if and only if $a_i \leq j \leq b_i$ (that is, the 1's in each row are contiguous), and the values $a_i$ and $b_i$ satisfy $1 = a_1 < a_2 < \cdots < a_r$, $b_1 < b_2 < \cdots < b_r = p$ and $a_{i+1} \leq b_i + 1$ for $i = 1, \ldots, r - 1$. We note that, when translating a problem of the form (6.4) into matrix notation, it may be necessary to simplify and rearrange the resulting matrix using **S1–S4** of (6.16) in order to get it into descending form.

Suppose we have a matrix $A$ in descending form. If the blocks of 1's in the different rows are disjoint (that is, $a_{i+1} = b_i + 1$ for $i = 1, \ldots, r-1$), then we can immediately evaluate $\{A\}$ using (6.11). Our goal is to reduce any descending matrix $A$ to this special case (where the rows are disjoint) by repeated use of the recursion (6.15).

It suffices to deal with matrices in which consecutive rows always overlap (that is, $a_{i+1} \leq b_i$ for $i = 1, \ldots, r - 1$). We shall say that descending matrices with this property are "overlapping." It is clear that any descending matrix $A$ is either overlapping or can be written as a block diagonal matrix in which each of the blocks has this property. If we have an algorithm that works for overlapping matrices, we can simply apply this algorithm in turn to each of the blocks in $A$.

We now present an algorithm for applying the recursion (6.15) to an $r \times p$ overlapping matrix $A$. This algorithm constructs the vector $c = (c_k)$ needed in (6.15). First, we define some useful terms. We call two rows $i$ and $j$ (with $i < j$) "adjacent" if $b_i + 1 = a_j$. We "mark" a row (say row $i$) by setting $c_k$ to be $+1$ and $-1$ in the positions $k$ corresponding to the first and last nonzero entries in row $i$, that is, setting $c_{a_i} = +1$ and $c_{b_i} = -1$. We can now state the algorithm. Start with $c_k = 0$ for all $k$. Now search through rows $1, 2, \ldots, r$, and mark the sequence of adjacent rows that you observe (starting with row 1). Continue marking rows as long as possible. When you have finished with this, the vector $c$ has $\sum_k c_k = 0$, not 1 as required in (6.15). You must modify $c$ slightly (so that $\sum_k c_k = 1$) as follows. Let $m$ (where $1 \leq m \leq r$) denote the last row that was marked. If $m = r$, erase the last $-1$ in $c$, that is, set $c_{b_m} = 0$. If $m < r$, then add another $+1$ to $c$ immediately following the last $-1$ entry, that is, set $c_{b_m+1} = 1$. Now you are done. An equivalent, but shorter description of the marking algorithm is given in Figure 6.1 which is taken from Huffer and Lin (1997b). As an illustration, we give in (6.32) a typical matrix $A$ in descending form and the corresponding vector $c$ which is produced by the algorithm. We have underlined the rows which get marked by our procedure.

Now let us examine what happens when we apply the recursion (6.15) with the vector $c$ given by the marking algorithm. Consider $\xi = Ac$. By trying some cases, it is easy to see that either $\xi = 0$ or $\xi = e_r$ (where $e_r$ denotes a vector with a 1 in position $r$ and 0 everywhere else). The matrices in (6.20), (6.26), and (6.32) are cases with $\xi = 0$. The matrix in (6.30) illustrates the other situation with $\xi = e_r$. When $\xi = 0$, then all of the matrices $A_{i,\xi}$ have a

```
Initialize:
        c₁ := 1, cᵢ := 0  for i ≥ 2
        m := 1
Repeat:
        IF  m = r  THEN  STOP
                ELSE {
                        c_{b_m}   := -1
                        c_{b_m+1} := +1
                        IF ∃ j > m such that a_j = b_m + 1
                                THEN m := j
                                ELSE  STOP
                }
```

**Figure 6.1:** Procedure for obtaining the vector $c$ needed to apply (6.15) to a descending matrix with overlapping rows

$$A = \begin{pmatrix} 1 & 1 & 1 & 1 & 0 & 0 & 0 & 0 & 0 & 0 & 0 & 0 & 0 & 0 & 0 & 0 & 0 & 0 \\ 0 & 1 & 1 & 1 & 1 & 1 & 1 & 1 & 0 & 0 & 0 & 0 & 0 & 0 & 0 & 0 & 0 & 0 \\ 0 & 0 & 0 & 0 & 1 & 1 & 1 & 1 & 1 & 0 & 0 & 0 & 0 & 0 & 0 & 0 & 0 & 0 \\ 0 & 0 & 0 & 0 & 0 & 1 & 1 & 1 & 1 & 1 & 1 & 0 & 0 & 0 & 0 & 0 & 0 & 0 \\ 0 & 0 & 0 & 0 & 0 & 0 & 0 & 0 & 1 & 1 & 1 & 1 & 0 & 0 & 0 & 0 & 0 & 0 \\ 0 & 0 & 0 & 0 & 0 & 0 & 0 & 0 & 0 & 1 & 1 & 1 & 1 & 1 & 1 & 1 & 0 & 0 & 0 \\ 0 & 0 & 0 & 0 & 0 & 0 & 0 & 0 & 0 & 0 & 1 & 1 & 1 & 1 & 1 & 1 & 1 & 0 & 0 \\ 0 & 0 & 0 & 0 & 0 & 0 & 0 & 0 & 0 & 0 & 0 & 0 & 1 & 1 & 1 & 1 & 1 & 0 \\ 0 & 0 & 0 & 0 & 0 & 0 & 0 & 0 & 0 & 0 & 0 & 0 & 0 & 1 & 1 & 1 & 1 & 1 \end{pmatrix},$$

$$(6.32)$$

$$c = \begin{pmatrix} 1 & 0 & 0 & -1 & 1 & 0 & 0 & 0 & -1 & 1 & 0 & 0 & 0 & 0 & -1 & 1 & 0 & 0 \end{pmatrix}.$$

column of zeros which can be deleted using property **S2** of (6.16). After deleting this column of zeros, further simplifications may be possible using **S1** or **S1'** of (6.16). This leads to a list of matrices in descending form which are all simpler (in the sense of having fewer columns and fewer nonzero entries in each row) than their "parent" matrix $A$; see the examples in (6.21) and (6.27).

We give one small example to show what happens when $\xi = e_r$. Here is a matrix $A$ along with the corresponding vector $c$ given by the marking algorithm:

$$A = \begin{pmatrix} 1 & 1 & 1 & 1 & 0 & 0 & 0 \\ 0 & 0 & 1 & 1 & 1 & 0 & 0 \\ 0 & 0 & 0 & 0 & 1 & 1 & 1 \end{pmatrix},$$

$$c = \begin{pmatrix} 1 & 0 & 0 & -1 & 1 & 0 & 0 \end{pmatrix}.$$

This gives $\boldsymbol{\xi} = (0, 0, 1)'$. Application of the recursion (6.15) then leads to

$$
\begin{aligned}
\{A\}^1 &= \left\{ \begin{smallmatrix} 0 & 1 & 1 & 1 & 0 & 0 & 0 \\ 0 & 0 & 1 & 1 & 1 & 0 & 0 \\ 1 & 0 & 0 & 0 & 1 & 1 & 1 \end{smallmatrix} \right\} - \left\{ \begin{smallmatrix} 1 & 1 & 1 & 0 & 0 & 0 & 0 \\ 0 & 0 & 1 & 0 & 1 & 0 & 0 \\ 0 & 0 & 0 & 1 & 1 & 1 & 1 \end{smallmatrix} \right\} + \left\{ \begin{smallmatrix} 1 & 1 & 1 & 1 & 0 & 0 & 0 \\ 0 & 0 & 1 & 1 & 0 & 0 & 0 \\ 0 & 0 & 0 & 0 & 1 & 1 & 1 \end{smallmatrix} \right\} \\
&= \left\{ \begin{smallmatrix} 1 & 1 & 1 & 0 & 0 & 0 & 0 \\ 0 & 1 & 1 & 1 & 0 & 0 & 0 \\ 0 & 0 & 0 & 1 & 1 & 1 & 1 \end{smallmatrix} \right\} - \left\{ \begin{smallmatrix} 1 & 1 & 1 & 0 & 0 & 0 & 0 \\ 0 & 0 & 1 & 1 & 0 & 0 & 0 \\ 0 & 0 & 0 & 1 & 1 & 1 & 1 \end{smallmatrix} \right\} + \left\{ \begin{smallmatrix} 1 & 1 & 0 & 0 & 0 \\ 0 & 0 & 1 & 1 & 1 \end{smallmatrix} \right\}. \quad (6.33)
\end{aligned}
$$

The first and second terms in (6.33) have been put back into descending form by moving the column $(0, 0, 1)'$ into the last position. The third term has been simplified by using **S1** (of (6.16)) and then **S2**. All three matrices in (6.33) are simpler than $A$ in some sense. The third term has disjoint rows and can be evaluated immediately using (6.11). The second term has fewer nonzero entries than $A$ and also has less overlap between the first and second rows. The first term is perhaps not obviously simpler than $A$. This first term is the same as $A$ except that a 1 has been moved from the beginning of row one to the end of row $r(= 3)$. However, it is clear in general that repeating this maneuver (moving a 1 from row one to row $r$) must eventually force a simplification via **S1** or **S1**$'$ of (6.16).

We have seen that using the vector $c$ given by the marking algorithm always produces matrices which are in descending form and are simpler than the parent matrix. Repeated use of the marking algorithm eventually reduces any descending matrix $A$ to terms we can evaluate using (6.11). A more formal discussion of the marking algorithm is given by Huffer and Lin (1997b).

The marking algorithm determines only probabilities of the form (6.4). It cannot be used to evaluate $\{\,A\,\}$ for matrices which are not binary, or for binary matrices which cannot be put in descending form. We have developed methods which can solve some of these more general problems. These methods are described by Lin (1993).

---

## 6.8 The Moments of $C_w$

The results of the previous sections can be used to compute the moments of $C_w$, the number of $m : w$ clumps defined in (6.5). These moments will be used in the following chapter to construct bounds and approximations for the distribution of the scan statistic. In principle, our methods can be used to compute $E(C_w^k)$ for any positive integer $k$. However, the magnitude of the calculations increases rapidly with $k$, and we have mainly worked with $k \leq 4$. For $k = 2$, we have obtained a general formula valid for all clump sizes $m$ and window lengths $w$, and for all sufficiently large values of the sample size $N$. For $k = 3$ and $k = 4$, we have not found general formulas of this sort, but we have obtained expressions for the moments valid for particular values of the clump size $m$.

We shall sketch the derivation of the first and second moments. For convenience, we introduce the indicator random variables $Z_i = I(X_{(i+m-1)} - X_{(i)} < w)$ and let $g = N - m + 1$ so that $C_w = \sum_{i=1}^{g} Z_i$. Also, define $P_i = E(Z_i)$ and $P_{i,j} = E(Z_i Z_j)$. Then,

$$E(C_w) = \sum_{i=1}^{g} P_i$$

and

$$E(C_w^2) = E\left(\sum_{i,j} Z_i Z_j\right) = \sum_{i=1}^{g} P_i + 2 \sum_{1 \le i < j \le g} P_{i,j} \qquad (6.34)$$

since $P_{i,i} = P_i$ and $P_{i,j} = P_{j,i}$. The difference $X_{(i+m-1)} - X_{(i)}$ can be written as a sum of spacings. Define the set $\delta_i = \{i+1, i+2, \ldots, i+m-1\}$ so that $X_{(i+m-1)} - X_{(i)} = G(\delta_i)$ using the notation in (6.2). Since the spacings are exchangeable random variables, it is immediate that $P_i$ is the same for all $i$, and that $P_{i,j}$ depends only on the number of spacings in the overlap $\delta_i \cap \delta_j$ which in turn depends only on $|j - i|$. The set $\delta_i \cap \delta_j$ is empty when $|j - i| \ge m - 1$. Thus, for $i < j$, we see that $P_{i,j} = P_{1,j-i+1}$ for $j - i < m - 1$ and $P_{i,j} = P_{1,m}$ for $j - i \ge m - 1$. Combining identical terms in (6.34) leads to

$$E(C_w) = g P_1$$

and

$$E(C_w^2) = g P_1 + 2 \sum_{j=2}^{m-1} (g - j + 1)_+ P_{1,j} + 2 \binom{g - m + 2}{2} P_{1,m}. \qquad (6.35)$$

This formula is valid for all $m$ and $N$ so long as the binomial coefficient $\binom{a}{b}$ is defined to be zero when $a < b$.

Now note that

$$P_1 = 1 - P(G(\delta_1) > w)$$

and

$$
\begin{aligned}
P_{1,j} &= P(G(\delta_1) < w, G(\delta_j) < w) \\
&= 1 - P(G(\delta_1) > w) - P(G(\delta_j) > w) + P(G(\delta_1) > w, G(\delta_j) > w) \\
&= 1 - 2P(G(\delta_1) > w) + P(G(\delta_1) > w, G(\delta_j) > w).
\end{aligned}
$$

We obtain $P(G(\delta_1) > w)$ directly from (6.11). This gives us

$$E(C_w) = g\left(1 - \sum_{j=0}^{m-2} R(j,1)\right) = (N - m + 1)\left(1 - \sum_{j=0}^{m-2} \binom{N}{j} w^j (1-w)^{N-j}\right).$$

Recalling the definition of $Q$ in (6.22), we see that

$$P(G(\delta_1) > w, G(\delta_j) > w) = Q(j-1, j-1, m-j)$$

for all $j \leq m$. Thus, we may evaluate $P_{i,j}$ by repeated use of (6.24) and (6.25). All of the terms in (6.35) can be expressed in terms of the $R$-functions. After working out $E(C_w^2)$ explicitly for some small values of $m$, Lin (1993) was able to guess the general answer and prove it via induction. The principal tool in this induction argument was (6.24).

In order to write the general expression for $E(C_w^2)$, we must introduce some notation. For fixed values of $N$ and $w$, define

$$H^{(0)}(i) = \binom{N}{i} w^i (1-w)^{N-i}$$

and

$$F^{(0)}(i,j) = \binom{N}{i,j} w^{i+j}(1-2w)_+^{N-i-j} \,.$$

The quantities $H^{(0)}(i)$ and $F^{(0)}(i,j)$ are particular binomial and trinomial probabilities. Now for $p \geq 1$, define $H^{(p)}(i)$ and $F^{(p)}(i,j)$ by repeated cumulative summation as

$$H^{(p+1)}(i) = \sum_{j=0}^{i} H^{(p)}(j)$$

and

$$F^{(p+1)}(i,j) = \sum_{k=0}^{i} \sum_{\ell=0}^{j} F^{(p)}(k,\ell) \,.$$

In terms of this notation, we can write

$$E(C_w) = (N - m + 1)\left[1 - H^{(1)}(m-2)\right] \tag{6.36}$$

and

$$
\begin{aligned}
E(C_w^2) \;=\; & E(C_w) + (N - m + 1)(N - m)(1 - 2H^{(1)}(m-2)) \\
& + 4(N - m)H^{(3)}(m-3) - 12H^{(5)}(m-4) \\
& + (N - 2m + 3)(N - 2m + 2)F^{(1)}(m-2, m-2) \\
& - 2(N - 2m + 3)F^{(2)}(m-3, m-3) \\
& + 2F^{(3)}(m-4, m-4)
\end{aligned}
\tag{6.37}
$$

valid for $N \geq 2(m-1)$. A different expression for this formula has been given by Huffer and Lin (1997a).

Formulas for the first and second moment of $C_w$ were also given by Glaz and Naus (1983). Their formula for the second moment is difficult to use for large $N$. In contrast, for a fixed value of $m$, the computational effort and numerical accuracy of formula (6.37) changes very little with $N$. We have had no difficulty using the formula with $N = 1,000$. However, the computational effort does grow with $m$, with the rate of growth proportional to $m^2$. We have successfully used formula (6.37) for values of $m$ as large as 140, but the calculations become progressively more time-consuming.

We now go on to consider the third and fourth moments of $C_w$. Our results here are much less satisfactory, but still useful. We extend our earlier notation [used in (6.34)] and define $P_{i,j,k} = E(Z_i Z_j Z_k)$. Then

$$E(C_w^3) = \sum_{i=1}^{g} \sum_{j=1}^{g} \sum_{k=1}^{g} P_{i,j,k}$$

$$= \sum_i P_i + 6 \sum_{i<j} P_{i,j} + 6 \sum_{i<j<k} P_{i,j,k} . \tag{6.38}$$

As in (6.34), the exchangeability of the spacings implies that many of the terms in (6.38) are equal. The value of $P_{i,j,k}$ depends only on the amount of overlap among the sets $\delta_i$, $\delta_j$, $\delta_k$, that is, it depends only on the cardinalities of the sets $\delta_i \cap \delta_j$, $\delta_i \cap \delta_k$, $\delta_j \cap \delta_k$, and $\delta_i \cap \delta_j \cap \delta_k$. Using this, it is easy to show that

$$P_{i,j,k} = P_{1,j-i+1,k-i+1} \quad \text{if } j-i < m-1 \text{ and } k-j < m-1,$$
$$P_{i,j,k} = P_{1,m,m+k-j} \quad \text{if } j-i \geq m-1 \text{ and } k-j < m-1,$$
$$P_{i,j,k} = P_{1,m,2m-1} \quad \text{if } j-i \geq m-1 \text{ and } k-j \geq m-1$$

and other similar facts. Thus, we can greatly simplify (6.38) by grouping the terms $P_{i,j,k}$ into classes of equal terms and counting the number of terms in each of these classes. We select one term $P_{i,j,k}$ from each class and let $\#(i,j,k)$ denote the number of terms which are equivalent to this given term. The selected terms (one from each class) will be referred to as the "distinct" terms. Similarly, we group together the terms $P_{i,j}$ into classes and let $\#(i,j)$ denote the number of terms equivalent to $P_{i,j}$. Then we can write (6.38) as

$$E(C_w^3) = gP_1 + 6 \sum_{\substack{\text{distinct} \\ (i,j)}} \#(i,j)P_{i,j} + 6 \sum_{\substack{\text{distinct} \\ (i,j,k)}} \#(i,j,k)P_{i,j,k} . \tag{6.39}$$

The values $\#(i,j,k)$ can be obtained by elementary combinatorial arguments. When $m = 4$, carrying this process to its completion leads to the formula

$$\begin{aligned}
E(C_w^3) = \ & b(3,1)P_1 + 6b(4,1)P_{1,2} + 6b(5,1)P_{1,3} + 6b(5,2)P_{1,4} \\
& + 6b(5,1)P_{1,2,3} + 6b(6,1)P_{1,2,4} + 6b(6,1)P_{1,3,4} \\
& + 6b(7,1)P_{1,3,5} + 12b(6,2)P_{1,2,5} + 12b(7,2)P_{1,3,6} + 6b(7,3)P_{1,4,7},
\end{aligned} \tag{6.40}$$

where the combinatorial factors $b(j,k)$ are defined by

$$b(j,k) = \begin{cases} \binom{N-j}{k} & \text{for } N \geq j+k, \\ \\ 0 & \text{for } N < j+k. \end{cases} \tag{6.41}$$

The expression (6.40) is valid for all $N$, but only for $m = 4$. The corresponding expressions for other values of $m$ have the same general form. See formulas (25) and (26) of Huffer and Lin (1997a) for further details.

We have now reduced the original sum (6.38) down to a much smaller number of terms. These terms can all be evaluated by our methods since each term can be represented in the matrix notation (6.14) using a matrix which has the descending form required by the marking algorithm in Section 6.7. For example, when $m = 4$, we have $P_{1,2,4} = \{A\}_w^2$ with

$$A = \begin{pmatrix} 1 & 1 & 1 & 0 & 0 & 0 \\ 0 & 1 & 1 & 1 & 0 & 0 \\ 0 & 0 & 0 & 1 & 1 & 1 \end{pmatrix}.$$

We evaluate each term in (6.39), obtaining an expression involving the $R$-functions. Then we collect together terms having the same $R$-function and obtain a final expression in the form

$$E(C_w^3) = B + \sum_{j,k} B_{j,k} R(j, k), \tag{6.42}$$

where $B$ and $B_{j,k}$ denote functions of $N$ which are sums of integer multiples of the binomial coefficients $b(\cdot, \cdot)$ defined in (6.41). For example, evaluation of (6.40) leads to

$$
\begin{aligned}
E(C_w^3) = \\
(&+6b(7,3) + 30b(7,2) + 61b(7,1) + 37b(7,0) + 19b(6,0) + 7b(5,0) \\
&+1b(4,0)) + (-18b(7,3) - 12b(7,2) + 29b(7,1) + 23b(7,0) - 1b(6,0) \\
&-13b(5,0) - 1b(4,0))R(0,1) + (-18b(7,3) - 60b(7,2) - 73b(7,1) \\
&-31b(7,0) - 7b(6,0) - 1b(5,0) - 1b(4,0))R(1,1) + (-18b(7,3) - 84b(7,2) \\
&-157b(7,1) - 91b(7,0) - 43b(6,0) - 13b(5,0) - 1b(4,0))R(2,1) \\
&+(+18b(7,3) - 66b(7,2) - 54b(7,1) - 60b(7,0) - 18b(6,0) + 6b(5,0))R(0,2) \\
&+(+36b(7,3) - 36b(7,2) - 78b(7,1) - 66b(7,0) - 30b(6,0))R(1,2) \\
&+(+72b(7,3) + 72b(7,2) + 18b(7,1) - 18b(7,0) - 18b(6,0))R(2,2) \\
&+(+108b(7,3) + 252b(7,2) + 174b(7,1) + 48b(7,0))R(3,2) \\
&+(+108b(7,3) + 324b(7,2) + 282b(7,1) + 72b(7,0))R(4,2) \\
&+(-6b(7,3) + 48b(7,2) - 36b(7,1))R(0,3) \\
&+(-18b(7,3) + 96b(7,2) - 36b(7,1))R(1,3) \\
&+(-54b(7,3) + 168b(7,2) - 24b(7,1))R(2,3) \\
&+(-144b(7,3) + 216b(7,2))R(3,3) + (-324b(7,3) + 144b(7,2))R(4,3) \\
&+(-540b(7,3))R(5,3) + (-540b(7,3))R(6,3).
\end{aligned}
\tag{6.43}
$$

This expression is valid for all $N$ and $w$, but only for clumps of size $m = 4$. We have computed similar expressions for other values of $m$. Unfortunately, these expressions are so complicated that we have been unable to see a general pattern. So, we have been unable to deduce a general expession for the third moment valid for all $m$ (as we could for the second moment).

The discussion above has been in terms of the third moment of $C_w$, but carries over to moments of any order. For any given values of $k$ and $m$, our approach would lead to an expression for $E(C_w^k)$ having the general form in (6.42). But, the amount of computational effort required to obtain these expressions increases with both $k$ and $m$. The expressions also become progressively longer as $k$ and $m$ increase. With our current software, we can compute expressions for the fourth moment only for clumps of size $m \leq 10$. We have stored a library of such expressions for use in computing the bounds and approximations described in the next chapter.

The remarks following Eq. (6.31) can be echoed here. We construct expressions like (6.43) using a C program, but do our numerical calculations using Maple so that we can achieve the desired precision. (Also, Maple supplies many features which are useful in computing our bounds and approximations.) Numerical computations using these expressions are fairly fast.

As we have noted before, the expressions we obtain also give expressions for the Poisson process. Thus, for example, (6.43) gives the third moment of the number of clumps of size 4 among the first $N$ arrivals of a Poisson process. The expression is valid for all $N$ and $w$.

---

# References


1. Feller, W. (1971). *An Introduction to Probability Theory and Its Applications, Volume II*, New York: John Wiley & Sons.

2. Glaz, J. and Naus, J. (1983). Multiple clusters on the line, *Communications in Statistics—Theory and Methods*, **12**, 1961–1986.

3. Huffer, F. W. (1982). The moments and distributions of some quantities arising from random arcs on the circle, *Ph.D. Dissertation*, Department of Statistics, Stanford University, Stanford, CA.

4. Huffer, F. (1988). Divided differences and the joint distribution of linear combinations of spacings, *Journal of Applied Probability*, **25**, 346–354.

5. Huffer, F. and Lin, C. T. (1995). Approximating the distribution of the scan statistic using moments of the number of clumps, *Technical Report*, Department of Statistics, Florida State University, Tallahassee, FL.

6. Huffer, F. W. and Lin, C. T. (1996). Computing the exact distribution

of the extremes of linear combinations of spacings, *Technical Report*, Department of Statistics, Florida State University, Tallahassee, FL.

7. Huffer, F. and Lin, C. T. (1997a). Approximating the distribution of the scan statistic using moments of the number of clumps, *Journal of the American Statistical Association*, **92**, 1466–1475.

8. Huffer, F. W. and Lin, C. T. (1997b). Computing the exact distribution of the extremes of sums of consecutive spacings, *Computational Statistics & Data Analysis*, **26**, 117–132.

9. Johnson, N. L., Kotz, S. and Balakrishnan, N. (1999). *Continuous Multivariate Distributions–1*, Second edition, New York: John Wiley & Sons.

10. Khatri, C. G. and Mitra, S. K. (1969). Some identities and approximations concerning positive and negative multinomial distributions, In *Multivariate Analysis – II* (Ed., P. R. Krishnaiah), pp. 241–260, New York: Academic Press.

11. Lin, C. T. (1993). The computation of probabilities which involve spacings, with applications to the scan statistic, *Ph.D. Dissertation*, Department of Statistics, Florida State University, Tallahassee, FL.

12. Micchelli, C. A. (1980). A constructive approach to Kergin interpolation in $\mathcal{R}^k$: multivariate B-splines and Lagrange interpolation, *The Rocky Mountain Journal of Mathematics*, **10**, 485–497.

13. Neff, N. D. and Naus, J. I. (1980). The distribution of the size of the maximum cluster of points on a line, *IMS Series of Selected Tables in Mathematical Statistics*, Volume **6**, Providence, RI: American Mathematical Society.

14. Wilks, S. S. (1962). *Mathematical Statistics*, New York: John Wiley & Sons.

# 7

# Using Moments to Approximate the Distribution of the Scan Statistic

**Fred W. Huffer and Chien-Tai Lin**

*Florida State University, Tallahassee, FL*
*Tamkang University, Taiwan*

**Abstract:** Let $C_w$ denote the number of $m:w$ clumps among $N$ random points uniformly distributed in the interval $(0,1]$. (We say that an $m:w$ clump exists when $m$ points fall within an interval of length $w$.) The previous chapter described how to compute the lower-order moments of $C_w$. In the present chapter, we discuss ways these moments can be used to obtain bounds and approximations for the distribution of the (continuous conditional) scan statistic $S_w$. We give upper and lower bounds based on the use of four moments. In some situations, these bounds improve considerably on the previously available bounds. We present an approximation based on a simple Markov chain model, and also give a variety of compound Poisson approximations. These approximations are compared with others in the literature. Finally, we present a compound Poisson approximation to the distribution of the number of clumps $C_w$.

**Keywords and phrases:** Compound Poisson approximation, Markov chain approximation, linear programming, probability bounds, method of moments, spacings, number of clumps

## 7.1 Introduction

Let $X_{(1)} < X_{(2)} < \cdots < X_{(N)}$ be the order statistics from a sample of size $N$ from the uniform distribution on the interval $(0,1]$. Let $Y_t(w)$ be the number of $X_{(i)}$'s contained in the scanning interval (window) $(t, t+w]$ and define the scan statistic $S_w$ by $S_w = \max_{0 < t < 1-w} Y_t(w)$. We say that $m$ $X_{(i)}$'s form an $m:w$ clump if these points are all contained in some interval of length $w$, and we define the number of $m : w$ clumps $C_w$ to be

$$C_w = \sum_{i=1}^{N-m+1} I(X_{(i+m-1)} - X_{(i)} < w) \, .$$

In this chapter, we use the moments of the number of clumps $C_w$ to obtain bounds and approximations for the tail probabilities of the scan statistic. These tail probabilities are denoted by $P(m; N, w) = P(S_w \geq m)$. When no confusion can result we use the abbreviation $p = P(m; N, w)$. The chapter is organized into Sections 7.2 and 7.3 which discuss bounds for $p$ and approximations for $p$, respectively. The results in this chapter are largely taken from Huffer and Lin (1995, 1997).

The moments of $C_w$ are computed by the methods described in Chapter 6 (see Section 6.8). However, this chapter can be read entirely independently of the earlier one. Define $\mu_k = E(C_w^k)$. The amount of computational effort needed to obtain $\mu_k$ increases with $k$ and with the clump size $m$. For clump sizes $m \leq 10$, our current software allows us to compute $\mu_k$ for $k \leq 4$. Most of the approximations and bounds in this chapter require four moments and thus are currently restricted to $m \leq 10$. In particular, the bounds UB and LB discussed in Section 7.2 and the approximations LP4 and CPG4 discussed in Section 7.3 all require four moments. The approximations MC2 and CPG2 discussed in Section 7.3 use only the first two moments so that we can compute these approximations for much larger values of $m$.

---

## 7.2  Bounds

Throughout this section, we shall suppose that $m$, $N$, and $w$ are fixed and that we have computed numerical values for the first four moments $\mu_1$, $\mu_2$, $\mu_3$, $\mu_4$.

Our bounds and approximations for $p$ are all based on the simple observation [see previous chapter Eq. (6.6)] that $P(S_w \geq m) = P(C_w \geq 1)$. The largest possible value of $C_w$ is $g = N - m + 1$. Since the random variable $C_w$ takes values in the set $\mathcal{B} = \{0, 1, \ldots, g\}$, it is immediate that

$$\text{LB} \leq P(C_w \geq 1) \leq \text{UB},$$

where the upper bound UB is the maximum value of $P(X \geq 1)$ attained by random variables $X$ which take values in $\mathcal{B}$ and have $E(X^k) = \mu_k$ for $k = 1, \ldots, 4$. The lower bound LB is the corresponding minimum. Restating this in terms of the probabilities $p_k = P(X = k)$, we see that we may compute UB and LB by solving the following linear programming problems:

$$UB \ = \ \max \ (p_1 + p_2 + \cdots + p_g) \tag{7.1}$$

$$\text{subject to} \quad \sum_{i=1}^{g} i^k p_i = \mu_k \quad \text{for } 1 \le k \le 4$$

$$\text{and} \qquad p_i \ge 0 \quad \text{for all } i\,;$$

$$LB \ = \ \min \text{ of same}\,. \tag{7.2}$$

The solution of these linear programming problems is routine. The necessary software is widely available; for example, it is available within Maple.

A few comments are in order on the form of these linear programming problems. In what follows, we rely heavily on the theoretical results in Prékopa (1988). These results are stated in terms of binomial moments, but are easily restated in terms of ordinary moments.

First, note that, since $p_0$ does not occur explicitly in the formulas for the moments, we do not need to include any constraints (such as $p_0 \ge 0$) on $p_0$ in our formulation. One might expect that we would have to include in our problems the constraint that $p_1 + p_2 + \cdots + p_g \le 1$, but this turns out not to be necessary. To be precise, suppose we amended the above problems to include the constraint $p_1 + \cdots + p_g \le 1$ and denoted the resulting values of the maximum and minimum $UB'$ and $LB'$, respectively. We would find that $LB' = LB$ (always) and that $UB' = \min(UB, 1)$. Thus, nothing useful is accomplished by imposing the constraint. See Section 2 of Prékopa (1988) for further details.

Second, the values of $g = N - m + 1$ occurring in applications can be quite large (perhaps several hundreds), so that, in principle, the linear programming problems in (7.1) and (7.2) might be very high-dimensional. This would make it difficult to solve these problems using standard software. Luckily, Theorems 9 and 10 in Prékopa (1988) allow us to safely replace the problems (7.1) and (7.2) by low-dimensional problems. These theorems characterize the form of the optimal solutions to (7.1) and (7.2). Let $\boldsymbol{p} = (p_1, p_2, \ldots, p_g)$ be any "feasible" solution of (7.1) or (7.2), that is, any vector of values which satisfies the constraints $\sum_i i^k p_i = \mu_k$ for $1 \le k \le 4$ and $p_i \ge 0$ for all $i$. A consequence of Theorems 9 and 10 of Prékopa (1988) is the following:

If $\{i : p_i > 0\} = \{1, j, j+1, g\}$ with $1 < j$ and $j + 1 < g$,

then $\boldsymbol{p}$ maximizes $\sum_i p_i$. $\tag{7.3}$

If $\{i : p_i > 0\} = \{j, j+1, k, k+1\}$ with $1 \le j,\ j+1 < k,\ k+1 \le g$,

then $\boldsymbol{p}$ minimizes $\sum_i p_i$. $\tag{7.4}$

Slightly more general conditions are obtained by replacing "$=$" with "$\subset$" in both (7.3) and (7.4). However, when applying these conditions in practice, the sets in question are typically equal.

These results allow us to replace the problems (7.1) and (7.2) by "reduced" versions in which most of the values $p_i$ are assumed to be zero *a priori*. Let $h$

be some suitably small integer. The "reduced" problems are

$$\max \ \sum_{i=1}^{h} p_i + p_g \tag{7.5}$$

$$\text{subject to} \quad \sum_{i=1}^{h} i^k p_i + g^k p_g = \mu_k \quad \text{for } 1 \leq k \leq 4$$

$$\text{and} \qquad p_i \geq 0 \quad \text{for } i \in \{1, 2, \ldots, h, g\},$$

and the corresponding

$$\min \ \sum_{i=1}^{h} p_i \tag{7.6}$$

$$\text{subject to} \quad \sum_{i=1}^{h} i^k p_i = \mu_k \quad \text{for } 1 \leq k \leq 4$$

$$\text{and} \qquad p_i \geq 0 \quad \text{for } i \in \{1, 2, \ldots, h\}.$$

Any solution of (7.5) is automatically a feasible solution of (7.1). Thus, if it satisfies the condition (7.3), it will be the desired solution of the original problem (7.1). Similarly, any solution of (7.6) is automatically a feasible solution of (7.2), so that, if it satisfies condition (7.4), it will be a solution of the original problem (7.2).

The value of $h$ should be chosen small enough so that the resulting reduced problems are easily solved. If $h$ is chosen too small, one or both of the reduced problems may fail to have a solution [or may have a solution that does not satisfy the required condition (7.3) or (7.4)]. If this happens, you must try again with a larger value of $h$.

### Numerical example

We now illustrate the computation of UB and LB in one particular case. Take $w = .0025$, $N = 1,000$, and $m = 10$ so that $g = 991$. Using the approach described in Section 6.8, we find the first four moments of $C_w$ to be

$$\begin{aligned}
\mu_1 &= 1.1089315629, \\
\mu_2 &= 3.9886505165, \\
\mu_3 &= 20.3814923874, \\
\mu_4 &= 134.6837629145.
\end{aligned}$$

Solving the reduced problem (7.5) with $h = 10$ leads to the values

$$\begin{aligned}
p_1 &= .487685, \\
p_5 &= .045307, \\
p_6 &= .065785, \\
p_g &= .213808 \times 10^{-10}, \\
p_i &= 0 \quad \text{otherwise}.
\end{aligned} \tag{7.7}$$

This solution clearly satisfies condition (7.3) so that it also gives a solution for (7.1). Using these values, we obtain UB $= \sum_i p_i = .59878$. We note that the small value of $p_g$ we see in this example is fairly typical.

Solving the reduced problem (7.6) with $h = 10$ leads to the values

$$
\begin{aligned}
p_2 &= .260498, \\
p_3 &= .117033, \\
p_7 &= .000189, \\
p_8 &= .029439, \\
p_i &= 0 \quad \text{otherwise}.
\end{aligned}
\tag{7.8}
$$

This solution satisfies condition (7.4) so that it also gives a solution for (7.2). Using these values, we obtain LB $= \sum_i p_i = .40716$.

### 7.2.1  The dual problem

We have described how to obtain the bounds UB and LB via linear programming. There is an entirely different way to derive these bounds which is more familiar to statisticians because it is the same approach used to derive the well-known Markov and Chebyshev inequalities. In general, if we wish to find bounds for $P(X \in A)$, we can look for "tractable" functions $\phi$ and $\psi$ satisfying $\phi(x) \le I_A(x) \le \psi(x)$ for all $x$, and then use the bounds given by $E(\phi(X)) \le P(X \in A) \le E(\psi(X))$. Here, $I_A(x)$ denotes the indicator function of the set $A$. If we have classes of tractable functions $\phi$ and $\psi$, the tightest such bounds are given by $\max_\phi E(\phi(X)) \le P(X \in A) \le \min_\psi E(\psi(X))$.

In our situation, given knowledge of the moments $\mu_1$ to $\mu_4$, the tractable functions are clearly the polynomials of order four; for $\phi(y) = a_0 + \sum_{i=1}^4 a_i y^i$, we have $E(\phi(X)) = a_0 + \sum_{i=1}^4 a_i \mu_i$. Taking $A = \{1, 2, \ldots, g\}$ in the general approach above leads to the following bounds:

$$
\text{UB} = \min \ (a_0 + a_1\mu_1 + a_2\mu_2 + a_3\mu_3 + a_4\mu_4)
\tag{7.9}
$$

$$
\text{over all polynomials} \ \ \psi(y) = \sum_{i=0}^4 a_i y^i
$$

$$
\text{satisfying} \ \ \psi(0) \ge 0
$$

$$
\text{and} \ \ \psi(i) \ge 1 \ \ \text{for} \ \ i = 1, 2, \ldots, g;
$$

$$
\text{LB} = \max \ (a_0 + a_1\mu_1 + a_2\mu_2 + a_3\mu_3 + a_4\mu_4)
\tag{7.10}
$$

$$
\text{over all polynomials} \ \ \phi(y) = \sum_{i=0}^4 a_i y^i
$$

$$
\text{satisfying} \ \ \phi(0) \le 0
$$

$$
\text{and} \ \ \phi(i) \le 1 \ \ \text{for} \ \ i = 1, 2, \ldots, g.
$$

These bounds can be computed by linear programming. In fact, the optimization problems (7.9) and (7.10) above are the "duals" of the earlier problems (7.1) and (7.2), and the values UB and LB obtained by the two approaches are identical. See equations (13) and (14) of Prékopa (1988). We note that the polynomials $\psi(y)$ and $\phi(y)$ which achieve the bounds always have $a_0 = 0$, so that the coefficient $a_0$ and the constraints $\psi(0) \geq 0$ and $\phi(0) \leq 0$ can be dropped from (7.9) and (7.10).

Expressing the bounds UB and LB in the above form has one advantage: it permits us to calculate the bounds easily without explicit use of linear programming. That is because we can identify, for both (7.9) and (7.10), small classes of polynomials which are guaranteed to achieve the bounds. These polynomials are now defined. For integer $j$ satisfying $1 < j$ and $j + 1 < g$, define

$$\psi_j(y) = 1 - \frac{(y-1)(y-g)(y-j)(y-j-1)}{gj(j+1)}.$$

For integer pairs $(j, k)$ satisfying $1 \leq j$, $j + 1 < k$, and $k + 1 \leq g$, define

$$\phi_{j,k}(y) = 1 - \frac{(y-j)(y-j-1)(y-k)(y-k-1)}{j(j+1)k(k+1)}.$$

It is easy to show that these polynomials satisfy the constraints in (7.9) and (7.10). Consider the polynomial $\phi_{j,k}(y)$. It is immediate that $\phi_{j,k}(0) = 0$ and $\phi_{j,k}(j) = \phi_{j,k}(j+1) = \phi_{j,k}(k) = \phi_{j,k}(k+1) = 1$. Since a polynomial of order four can have at most four crossings of any given level, we know that $\phi_{j,k}(y) \neq 1$ for $y \notin \{j, j+1, k, k+1\}$. Now the fact that $\phi_{j,k}(0) < 1$ forces $\phi_{j,k}(y) < 1$ for $y \in (-\infty, j)$, $\phi_{j,k}(y) > 1$ for $y \in (j, j+1)$, $\phi_{j,k}(y) < 1$ for $y \in (j+1, k)$, $\phi_{j,k}(y) > 1$ for $y \in (k, k+1)$, and $\phi_{j,k}(y) < 1$ for $y \in (k+1, \infty)$. Since the intervals $(j, j+1)$ and $(k, k+1)$ contain no integers, we conclude that $\phi_{j,k}(i) \leq 1$ for $i = 1, 2, \ldots, g$ as desired. The argument for $\psi_j(y)$ is similar. We note that $\psi_j(y) = 1$ only when $y \in \{1, j, j+1, g\}$, and $\phi_{j,k}(y) = 1$ only when $y \in \{j, j+1, k, k+1\}$; these are precisely the sets occurring in (7.3) and (7.4).

Using Theorems 9 and 10 of Prékopa (1988), we can show that

$$\text{LB} = \max_{j,k} E(\phi_{j,k}(C_w)) \quad \text{and} \quad \text{UB} = \min_j E(\psi_j(C_w)).$$

This allows us to compute UB and LB by a simple systematic search. We choose some suitable upper limit $h$, evaluate

$$\max_{j,k \leq h} E(\phi_{j,k}(C_w)) \quad \text{and} \quad \min_{j \leq h} E(\psi_j(C_w)),$$

and take these as our bounds. The values of $j$ and $k$ which attain UB and LB are usually fairly small; setting $h = 10$ is generally adequate. We note that, if we should err and set $h$ too small, the bounds obtained are still valid, but they are no longer the tightest possible such bounds.

## 7.2.2 Performance of bounds

In this section, we examine the performance of the bounds UB and LB. Glaz (1989, 1992) gave extensive tables comparing various bounds and approximations for the distribution of the scan statistic. We shall compare our upper bound UB to the $m$th order bound obtained by taking $L = m$ in equation (2.5) of Glaz (1992); we refer to this bound as GUB in what follows. This bound was the best of the upper bounds that Glaz studied. It is based on a Bonferroni-type inequality given in Theorem 2.1 of Glaz (1989).

Since UB and GUB are based on entirely different principles, there is no *a priori* reason to expect one bound to be superior to the other. In fact, for the clump sizes $m \leq 10$ that we can currently work with, they are very close competitors. In Table 7.1, we present some numerical results comparing UB and GUB for sample sizes $N$ of 100 and 1,000. This table also gives approximate values of $p = P(m; N, w)$ obtained from simulations with 1,000,000 trials. From the table, we see that neither bound is uniformly superior to the other. UB tends to be superior to GUB for smaller values of $m$ and larger values of $p$. However, the region of superiority changes with the value of $N$ and it is difficult to give a good rule of thumb for which bound to use. Both bounds are usually quite good for small values of $p$. Glaz (1989) also studied the performance of one lower bound, given in his equation (2.12). He refers to this as the Kwerel lower bound (which we abbreviate as KLB) since it is based on an inequality due to Kwerel (1975). The bound KLB uses only the first two moments of $C_w$. We know *a priori* that LB is uniformly superior to KLB. That is because KLB is the solution of a linear programming problem similar to (7.2), but using only the first two moments. Our Table 7.1 lists the values of LB and KLB. In some situations, LB improves considerably upon KLB.

When $p$ is small, the bounds UB and LB are usually fairly tight.

## 7.2.3 Improving the bounds

The bounds UB and LB are the best possible bounds for $P(C_w \geq 1)$ which use *only* the first four moments of $C_w$ and no other information about the distribution of $C_w$. The distributions $p_i$ which actually achieve the values of UB and LB [such as (7.7) and (7.8)] may bear little resemblance to the true distribution $P(C_w = i)$. In particular, linear programming always produces solutions with $|\{i : p_i > 0\}| \leq 4$ which is clearly highly artificial.

In simulations, the distribution of $C_w$ seems to be reasonably well behaved. It is apparently unimodal, and when $E(C_w)$ is small (say, less than 0.5), the mode is at zero. If we could prove these empirical observations, we could use this knowledge to improve our bounds. For example, if we knew the distribution was unimodal at zero, we could modify the linear programming problems (7.1) and (7.2) to include the constraints $p_0 \geq p_1 \geq \cdots \geq p_g$. (It would now be

**Table 7.1:** Bounds for $p = P(m; N, w)$

|         | $m$ | $w$ | $p$ | KLB | LB | UB | GUB |
|---------|-----|-----|-----|-----|-----|-----|-----|
| N=100   | 4  | .0009  | .01029 | .01019 | .01021 | .01021 | .01025 |
|         | 4  | .0015  | .04345 | .04292 | .04330 | .04334 | .04413 |
|         | 4  | .002   | .09425 | .09238 | .09433 | .09460 | .09842 |
|         | 4  | .003   | .25912 | .23688 | .25790 | .26424 | .29385 |
|         | 4  | .004   | .47379 | .36548 | .46537 | .50195 | .61576 |
|         | 6  | .005   | .01290 | .01231 | .01273 | .01282 | .01280 |
|         | 6  | .007   | .05386 | .04940 | .05358 | .05501 | .05495 |
|         | 6  | .009   | .14490 | .11859 | .14332 | .15184 | .15381 |
|         | 6  | .011   | .29367 | .22103 | .28287 | .31802 | .33362 |
|         | 6  | .013   | .48245 | .34373 | .43766 | .55193 | .61055 |
|         | 8  | .015   | .05079 | .03839 | .04972 | .05363 | .05161 |
|         | 8  | .018   | .12789 | .09453 | .12229 | .14011 | .13339 |
|         | 8  | .021   | .25671 | .18028 | .23330 | .29327 | .28218 |
|         | 8  | .025   | .49465 | .33132 | .42846 | .61723 | .61310 |
|         | 10 | .020   | .01163 | .00856 | .01123 | .01222 | .01155 |
|         | 10 | .025   | .05033 | .03662 | .04760 | .05577 | .05098 |
|         | 10 | .028   | .10013 | .07013 | .09155 | .11385 | .10297 |
|         | 10 | .033   | .24460 | .16236 | .21074 | .29797 | .26494 |
|         | 10 | .040   | .55968 | .35628 | .46235 | .78054 | .70036 |
| N=1,000 | 4  | .00005 | .01917 | .01925 | .01927 | .01927 | .01946 |
|         | 4  | .00007 | .05025 | .05057 | .05076 | .05077 | .05207 |
|         | 4  | .0001  | .13633 | .13419 | .13615 | .13637 | .14622 |
|         | 4  | .0002  | .64604 | .50666 | .62806 | .69831 | $> 1$  |
|         | 6  | .0003  | .01404 | .01378 | .01397 | .01399 | .01406 |
|         | 6  | .0004  | .05134 | .04976 | .05135 | .05168 | .05272 |
|         | 6  | .0005  | .13396 | .12381 | .13297 | .13606 | .14305 |
|         | 6  | .0006  | .27199 | .22532 | .26914 | .28499 | .31642 |
|         | 6  | .0007  | .45839 | .34785 | .44065 | .49936 | .60771 |
|         | 8  | .0008  | .01626 | .01531 | .01627 | .01660 | .01646 |
|         | 8  | .0010  | .06081 | .05330 | .06026 | .06302 | .06258 |
|         | 8  | .0011  | .10362 | .08536 | .10202 | .10816 | .10885 |
|         | 8  | .0013  | .24459 | .18268 | .23582 | .26621 | .27903 |
|         | 8  | .0015  | .45770 | .31912 | .41034 | .53285 | .60433 |
|         | 10 | .0015  | .01923 | .01608 | .01905 | .02017 | .01948 |
|         | 10 | .0019  | .10027 | .07377 | .09609 | .10872 | .10480 |
|         | 10 | .0022  | .24697 | .17311 | .22431 | .28244 | .28040 |
|         | 10 | .0025  | .47486 | .31449 | .40716 | .59878 | .63265 |

$p = P(m; N, w)$ was estimated from 1,000,000 simulations. KLB and GUB are from Glaz [1989, Eq. (2.12)], and Glaz [1992, Eq. (2.5)], respectively.

necessary to also explicitly include the constraints $p_0 \geq 0$ and $\sum_i p_i = 1$ in the problems.) The simplest way to do this is probably to restate the linear programming problems entirely in terms of the differences $u_i = p_i - p_{i+1}$ for $i = 0, 1, \ldots, g$. These modified problems are still quick and easy to solve using standard software.

Similarly, if we knew the distribution was unimodal with mode $k$ (known), we just modify the problems to include the constraints $p_0 \leq p_1 \leq \cdots \leq p_{k-1} \leq p_k \geq p_{k+1} \geq \cdots$. If the location of the mode is unknown, then maximizing (or minimizing) the desired probability over all unimodal distributions (with given moments) would require solving a series of linear programming problems, varying the value of the mode $k$. In general, any information about the distribution of $C_w$, which can be stated in terms of linear inequalities, can be incorporated into the linear programming approach.

### 7.2.4   Using more than four moments

We have discussed in detail the construction of bounds which use four moments. If a larger (or smaller) number of moments is available, it is easy to modify the discussion and compute the corresponding bounds. For example, let us suppose that five moments are available. To obtain bounds corresponding to UB and LB, we simply modify the linear programming problems (7.1) and (7.2) and the reduced problems (7.5) and (7.6) by adding an additional constraint for the fifth moment $\mu_5$. From Theorems 9 and 10 of Prékopa (1988), the conditions analogous to (7.3) and (7.4) are:

> If $\{i : p_i > 0\} = \{1, j, j+1, k, k+1\}$ with $1 < j$, $j+1 < k$, and $k+1 \leq g$,
> then $\boldsymbol{p}$ maximizes $\sum_i p_i$ ;
> If $\{i : p_i > 0\} = \{j, j+1, k, k+1, g\}$ with $1 \leq j$, $j+1 < k$ and $k+1 < g$,
> then $\boldsymbol{p}$ minimizes $\sum_i p_i$ .

Solutions of the reduced problems satisfying these conditions are guaranteed to be solutions of the original linear programming problems. In the "dual" approach to obtaining the bounds UB and LB, we use families of polynomials $\psi(x)$ and $\phi(x)$ of order five which satisfy $\psi(0) = \phi(0) = 0$, $\psi(y) = 1$ for $y \in \{1, j, j+1, k, k+1\}$, and $\phi(y) = 1$ for $y \in \{j, j+1, k, k+1, g\}$.

The bounds based on moments that we use are special cases of Bonferroni-type inequalities. A thorough treatment of these inequalities is given in the book by Galambos and Simonelli (1996). This book also includes discussion on the use of linear programming to compute probability bounds. We note that the bounds based on two and three moments can be given in closed form so that it is not necessary to use linear programming software (or systematic search) to compute these bounds. The best lower bound based on two moments is the Kwerel lower bound (KLB) mentioned earlier. The closed formulae for the best upper and lower bounds based on three moments may be found in Boros and Prékopa (1989) and Galambos and Simonelli (1996).

---

## 7.3  Approximations

In this section, we present a number of approximations to $p = P(m; N, w)$. These approximations use the moments of $C_w$ to approximate $p = P(C_w \geq 1)$. Let $g = N - m + 1$ and define the sequence of indicator random variables $Z_1, Z_2, \ldots, Z_g$ by $Z_i = I(X_{(i+m-1)} - X_{(i)} < w)$ so that $C_w = \sum_{i=1}^{g} Z_i$. Our approximations are based on finding simple stochastic models for $Z_1, Z_2, \ldots, Z_g$ and then using the "method of moments."

The stochastic model we choose should exhibit groups or clusters of 1's qualitatively similar to those in the sequence $Z_1, Z_2, \ldots, Z_g$. The random variables $Z_i$ and $Z_j$ are positively correlated when $|i - j| < m - 1$ since in this case the spacings involved in the definitions of $Z_i$ and $Z_j$ overlap. But for $|i - j| \geq m - 1$, we expect $Z_i$ and $Z_j$ to be very close to being independent (assuming $N$ is sufficiently large). Thus, the sequence $Z_1, \ldots, Z_g$ exhibits short-range dependence (in the form of positive correlation) and long-range independence. Any convenient stochastic model with these properties might lead to reasonable approximations.

### 7.3.1  The approximation MC2

This approach is best illustrated by an example. The simplest stochastic model exhibiting short-range dependence and long-range independence is a two-state Markov chain. If we assume the sequence $Z_1, \ldots, Z_g$ behaves roughly like a two-state Markov chain, we are led to a simple, but often surprisingly good approximation for $P(m; N, w)$.

Let $\tilde{Z}_1, \tilde{Z}_2, \ldots, \tilde{Z}_g$ be a Markov chain with two states (0 and 1) having the transition matrix

$$\begin{pmatrix} 1 - a & a \\ b & 1 - b \end{pmatrix}.$$

We shall suppose this chain is started from the stationary distribution which is $P(\tilde{Z}_1 = 1) = \pi$ and $P(\tilde{Z}_1 = 0) = 1 - \pi$ where $\pi = a/(a + b)$. Define $\tilde{C} = \sum_{i=1}^{g} \tilde{Z}_i$. There are simple closed formulas for $P(\tilde{C} \geq 1)$, $E(\tilde{C})$ and $E(\tilde{C}^2)$. Our approach consists of the following steps: (i) computing $\mu_1 = E(C_w)$ and $\mu_2 = E(C_w^2)$ using the methods of Section 6.8, (ii) solving the system of equations

$$\begin{aligned} E(\tilde{C}) &= \mu_1 \\ E(\tilde{C}^2) &= \mu_2 \end{aligned}$$

(7.11)

for $a$ and $b$, and then (iii) using these values of $a$ and $b$ to compute $P(\tilde{C} \geq 1)$ which we take as our approximation to $P(C_w \geq 1)$. This leads to an approximation we call MC2. Since MC2 uses only the first two moments of $C_w$ (which we can obtain using formulas (6.36) and (6.37) in the previous chapter), we can compute MC2 even for rather large values of $m$.

We now fill in some of the details. The necessary formulas are more cleanly stated in terms of the parameters $\pi$ and $s$ given by

$$\pi = \frac{a}{a+b} \quad \text{and} \quad s = \frac{1}{a+b}$$

instead of $a$ and $b$. For the two-state Markov chain, it is straightforward to compute the following:

$$P\{\tilde{C} \geq 1\} = 1 - (1 - \pi)\left(1 - \frac{\pi}{s}\right)^{g-1}, \tag{7.12}$$

$$E(\tilde{C}) = g\pi, \quad \text{and} \tag{7.13}$$

$$E(\tilde{C}^2) = g^2\pi^2 + g\pi(1 - \pi) + 2\pi(1 - \pi)(s - 1)(g - s(1 - \varepsilon)),$$

where $\varepsilon = \left(1 - \frac{1}{s}\right)^g$. Substituting these formulas for $E(\tilde{C})$ and $E(\tilde{C}^2)$ into (7.11), we find that $\pi = \mu_1/g$ and the system reduces to a single equation for $s$ which can be solved by a simple iterative scheme. If we drop $\varepsilon$ from this equation, it becomes a quadratic equation in $s$, so that we obtain a closed form solution for $s$ given by

$$s = \frac{1}{2}\left(g + 1 - \sqrt{(g - 1)^2 - 4c}\right),$$

where $c = \frac{\mu_2 - \mu_1^2 - g\pi(1-\pi)}{2\pi(1-\pi)}$. (The value of $\varepsilon$ is extremely small in most applications so that dropping it has very little effect on the answers.) Plugging these values for $\pi$ and $s$ into Eq. (7.12) gives us the value of MC2.

We have extensively studied the performance of MC2 and report some numerical results in Tables 7.2 and 7.3. Glaz (1989) surveyed various approximations for $P(m; N, w)$. He recommended three different approximations for general use. These are the approximations given by Naus [1982, Eq. (6.1)], Wallenstein and Neff [1987, Eq. (1)], and Glaz [1989, Eq. (3.3)]. Later, Glaz (1992) improved upon his earlier approximation. As our standard of comparison, we shall use the $m$th order approximation obtained by taking $L = m$ in Glaz [1992, Eq. (2.8)]. This approximation is accurate in a wide range of circumstances, and is probably the best of the currently available approximations for the clump sizes $m \leq 10$ of primary interest to us. In our tables, this approximation is listed under the heading GLAZ.

We see from Tables 7.2 and 7.3 that MC2 is fairly reliable; there is no region where it does really badly. It is generally very accurate for small clump sizes $m$, but tends to underestimate $p$ for larger $m$ with the downward bias increasing as $m$ increases. As a general rule, MC2 tends to do better than GLAZ for smaller $m$ and for larger values of $p$. But the value of $N$ also affects the comparison, and it is difficult to state a simple rule for deciding which approximation to use. For $m \leq 10$, it appears that MC2 is better (on the whole) than GLAZ when $N = 100$, but that GLAZ is better when $N = 1,000$. When $p$ is small, MC2 and GLAZ often produce answers which are extremely close, so that it matters very little which is used. For $m > 10$, GLAZ is better than MC2 except for values $p$ which are close to 1.

For smaller values of $N$, the approximation GLAZ can be rapidly computed using equation (2.10) of Glaz (1992). However, for larger values of $N$, this approach breaks down because of the accumulation of round-off error. For large $N$ (such as $N = 1,000$ used in Table 7.2), GLAZ must be obtained by computing a series of numerical integrals so that the computation is much more time-consuming. On the other hand, the amount of computation required for MC2 increases with $m$, but is roughly constant in $N$. Thus, for large $N$, MC2 is faster to compute than GLAZ unless $m$ is quite large.

The relatively poor performance of MC2 for larger clump sizes $m$ is perhaps to be expected. When $m$ is large, the sequence $Z_1, \ldots, Z_g$ has a higher-order dependence which cannot be well approximated by the first-order dependence in the two-state Markov chain. It is intuitive that $P(Z_i = 1 \mid Z_{i-1} = 0, Z_{i-2} = 1)$ will be larger than (and perhaps substantially larger than) $P(Z_i = 1 \mid Z_{i-1} = 0)$ unless $m$ is small. Another problem with the two-state Markov chain is that it has only two parameters $a$ and $b$, so that we cannot easily incorporate higher order moments (when we are able to compute them) into our approximation. A natural way to tackle both of these problems is to try using a Markov chain with more states and more parameters. We have tried using a three-state Markov chain (with states labeled 1, 2, 3) with transition probability matrix given by

$$\begin{pmatrix} 1-a & a & 0 \\ c & 1-b-c & b \\ 0 & d & 1-d \end{pmatrix}.$$

Let $W_1, W_2, \ldots, W_g$ be a Markov chain with this transition matrix. We now define a process $\tilde{Z}_i$, $i = 1, \ldots, g$ by lumping together the states 1 and 2. More precisely, define $\tilde{Z}_i = I(W_i = 3)$ for $i = 1, \ldots, g$. The process $\{\tilde{Z}_i\}$ has two states 0 and 1, and can exhibit (with appropriate choice of the parameter values $a$, $b$, $c$, $d$) the kind of higher-order dependence we desire. Let $\tilde{C} = \sum_{i=1}^{g} \tilde{Z}_i$. We are able to derive formulas for $P(\tilde{C} \geq 1)$ and $E(\tilde{C}^k)$ for $k = 1, 2, 3, 4$. Unfortunately, these formulas are much more complicated than those in (7.12) and (7.13), and when we attempt to solve for the values $a, b, c, d$ which produce given values of $\mu_1, \mu_2, \mu_3, \mu_4$, we run into various difficulties (such as numerical inaccuracies and multiple roots). We may eventually overcome these difficulties, but in any case this example indicates the general sort of stochastic models we are seeking.

## 7.3.2 Compound Poisson approximations

We have worked with a number of approximations to $P(m; N, w)$ based on another sort of stochastic model. If the random variables $Z_i$'s were independent Bernoulli trials, then $C_w = \sum_i Z_i$ would have a binomial distribution with mean $\lambda = gE(Z_1)$ which [if $E(Z_1)$ is sufficiently small] could be well approximated by a Poisson($\lambda$) distribution. Because of the short-range dependence among the $Z_i$'s, there is a tendency for the 1's in the sequence $Z_1, \ldots, Z_g$ to occur in clusters. For this reason, a simple Poisson approximation to the distribution of $C_w$ often performs badly. However, the long-range independence among the $Z_i$'s suggests that the number of such clusters of 1's will have approximately a Poisson distribution so that $C_w$ itself will have approximately a *compound* Poisson distribution. This is an instance of the "Poisson clumping heuristic" treated in the book by Aldous (1989). This heuristic idea can be made precise in our situation: Roos (1993) and Glaz *et al.* (1994) have shown that, for any fixed value of $m$, if $N \to \infty$ and $w \to 0$ in such a way that $E(C_w)$ remains constant, then there is a sequence of compound Poisson distributions which converges to the distribution of $C_w$ at the rate $O(1/N)$. Much work has been done recently on the subject of Poisson and compound Poisson approximations to the distribution of $C_w$; see Dembo and Karlin (1992) and the previously cited work by Roos (1993) and Glaz *et al.* (1994).

### Review of compound Poisson distribution

Before proceeding, we will give a formal definition of the compound Poisson (CP) distribution and review some of its basic properties. Let $V_1, V_2, V_3, \ldots$ be a sequence of independent random variables with $V_i \sim$ Poisson($\lambda_i$) where $\boldsymbol{\lambda} = (\lambda_1, \lambda_2, \lambda_3, \ldots)$ satisfies $\lambda_i \geq 0$ for all $i$ and $\sum_i \lambda_i < \infty$. Then,

$$C^* = \sum_{i=1}^{\infty} iV_i \tag{7.14}$$

has a CP distribution denoted by $C^* \sim \text{CP}(\boldsymbol{\lambda})$. In our setting, we wish to approximate the distribution of $C_w$ by that of $C^*$; the random variable $V_i$ represents the number of clusters of size $i$, and $\lambda_i$ is the expected number of such clusters.

Compound Poisson distributions are very convenient for the sort of "method of moments" approximations we are using. That is because of the simplicity of the formulas (7.15) and (7.20) given below. Let $C^* \sim \text{CP}(\boldsymbol{\lambda})$. It is clear from (7.14) that

$$P\{C^* \geq 1\} = 1 - \exp\{-\textstyle\sum_{k=1}^{\infty} \lambda_k\}. \tag{7.15}$$

**Table 7.2:** Approximations for $p = P(m; N, w)$ with $m \leq 10$

|  | $m$ | $w$ | $p$ | MC2 | LP4 | CPG4 | GLAZ |
|---|---|---|---|---|---|---|---|
| N=100 | 4 | .0009 | .01029 | .01021 | .01021 | .01021 | .01021 |
|  | 4 | .0015 | .04345 | .04331 | .04331 | .04331 | .04324 |
|  | 4 | .002 | .09425 | .09439 | .09435 | .09436 | .09405 |
|  | 4 | .003 | .25912 | .25942 | .25909 | .25912 | .25702 |
|  | 4 | .004 | .47379 | .47494 | .47387 | .47392 | .46761 |
|  | 6 | .005 | .01290 | .01273 | .01273 | .01275 | .01273 |
|  | 6 | .007 | .05386 | .05381 | .05376 | .05386 | .05364 |
|  | 6 | .009 | .14490 | .14518 | .14472 | .14526 | .14379 |
|  | 6 | .011 | .29367 | .29422 | .29251 | .29410 | .28861 |
|  | 6 | .013 | .48245 | .48393 | .48056 | .48323 | .46987 |
|  | 8 | .015 | .05079 | .05055 | .05032 | .05087 | .05051 |
|  | 8 | .018 | .12789 | .12728 | .12592 | .12822 | .12618 |
|  | 8 | .021 | .25671 | .25642 | .25352 | .25864 | .25108 |
|  | 8 | .025 | .49465 | .49639 | .49100 | .50225 | .47687 |
|  | 10 | .020 | .01163 | .01141 | .01135 | .01154 | .01150 |
|  | 10 | .025 | .05033 | .04979 | .04939 | .05058 | .04997 |
|  | 10 | .028 | .10013 | .09917 | .09820 | .10108 | .09888 |
|  | 10 | .033 | .24460 | .24391 | .24061 | .25069 | .23881 |
|  | 10 | .040 | .55968 | .56164 | .55868 | na | .53309 |
| N=1,000 | 4 | .00005 | .01917 | .01927 | .01927 | .01927 | .01927 |
|  | 4 | .00007 | .05025 | .05076 | .05076 | .05076 | .05075 |
|  | 4 | .0001 | .13633 | .13618 | .13617 | .13617 | .13610 |
|  | 4 | .0002 | .64604 | .64654 | .64638 | .64640 | .64498 |
|  | 6 | .0003 | .01404 | .01396 | .01397 | .01397 | .01397 |
|  | 6 | .0004 | .05134 | .05135 | .05137 | .05139 | .05137 |
|  | 6 | .0005 | .13396 | .13342 | .13344 | .13357 | .13339 |
|  | 6 | .0006 | .27199 | .27202 | .27199 | .27235 | .27166 |
|  | 6 | .0007 | .45839 | .45787 | .45757 | .45837 | .45657 |
|  | 8 | .0008 | .01626 | .01628 | .01631 | .01634 | .01633 |
|  | 8 | .0010 | .06081 | .06049 | .06052 | .06077 | .06069 |
|  | 8 | .0011 | .10362 | .10289 | .10286 | .10342 | .10322 |
|  | 8 | .0013 | .24459 | .24356 | .24307 | .24501 | .24395 |
|  | 8 | .0015 | .45770 | .45579 | .45366 | .45851 | .45517 |
|  | 10 | .0015 | .01923 | .01914 | .01916 | .01931 | .01929 |
|  | 10 | .0019 | .10027 | .09874 | .09826 | .09996 | .09959 |
|  | 10 | .0022 | .24697 | .24355 | .24219 | .24706 | .24513 |
|  | 10 | .0025 | .47486 | .47037 | .46774 | .47732 | .47104 |

$p = P(m; N, w)$ was estimated from 1,000,000 simulations.

The approximation GLAZ is from Glaz [1992, Eq. (2.8)].

**Table 7.3:** Approximations to $p = P(m; N, w)$ for $m > 10$

|  | $m$ | $w$ | $p$ | CPG2 | MC2 | GLAZ |
|---|---|---|---|---|---|---|
| $N = 100$ | 12 | .03 | .01155 | .01134 | .01133 | .01149 |
|  | 12 | .05 | .35777 | .35903 | .35712 | .34451 |
|  | 14 | .10 | .99902 | .99983 | .99953 | .99129 |
|  | 16 | .10 | .86152 | .88372 | .87070 | .80795 |
|  | 18 | .10 | .41351 | .41394 | .40925 | .39193 |
|  | 20 | .10 | .11888 | .11460 | .11420 | .11601 |
|  | 22 | .10 | .02419 | .02307 | .02304 | .02401 |
|  | 30 | .20 | .27376 | .26785 | .26366 | .26482 |
|  | 32 | .20 | .09983 | .09463 | .09400 | .09811 |
|  | 35 | .20 | .01466 | .01375 | .01372 | .01463 |
|  | 35 | .30 | .94639 | .96874 | .94152 | .91355 |
|  | 38 | .30 | .58769 | .60007 | .57305 | .56627 |
|  | 40 | .25 | .02270 | .02135 | .02127 | .02277 |
|  | 40 | .35 | .92461 | .94973 | .91048 | .89966 |
|  | 42 | .35 | .70489 | .73005 | .68774 | .68688 |
|  | 43 | .30 | .08640 | .08230 | .08144 | .08615 |
|  | 45 | .30 | .02873 | .02731 | .02715 | .02904 |
|  | 45 | .35 | .31233 | .31274 | .30205 | .31167 |
| $N = 1,000$ | 12 | .005 | .99589 | .99603 | .99585 | .99365 |
|  | 14 | .005 | .57313 | .56259 | .56218 | .56493 |
|  | 16 | .005 | .09775 | .09471 | .09470 | .09768 |
|  | 20 | .01 | .86481 | .85475 | .85336 | .84671 |
|  | 25 | .01 | .03905 | .03650 | .03649 | .03898 |
|  | 30 | .02 | .99741 | .99760 | .99718 | .99215 |
|  | 35 | .02 | .39131 | .36202 | .36145 | .37956 |
|  | 45 | .03 | .70957 | .67745 | .67466 | .67620 |
|  | 50 | .03 | .11954 | .10423 | .10416 | .11687 |
|  | 60 | .04 | .26199 | .23133 | .23089 | .25333 |
|  | 65 | .04 | .03271 | .02768 | .02768 | .03237 |
|  | 80 | .05 | .00859 | .00722 | .00722 | .00866 |
|  | 115 | .1 | .98272 | .98593 | .98066 | .94429 |
|  | 120 | .1 | .75276 | .72100 | .71136 | .69105 |
|  | 125 | .1 | .35499 | .31013 | .30812 | .33419 |
|  | 130 | .1 | .11234 | .09239 | .09221 | .10947 |
|  | 140 | .1 | .00490 | .00387 | .00387 | .00489 |

Let $p_k^* = P(C^* = k)$ for $k = 0, 1, 2, \ldots$. Define the generating functions $\Lambda(x)$ and $P(x)$ by

$$\Lambda(x) = \sum_{i=1}^{\infty} \lambda_i x^i \quad \text{and} \quad P(x) = \sum_{i=0}^{\infty} p_i^* x^i . \tag{7.16}$$

We see that

$$p_0^* = \exp(-\textstyle\sum_i \lambda_i) = e^{-\Lambda(1)} . \tag{7.17}$$

Using elementary properties of generating functions and the fact that the probability generating function of the Poisson($\lambda$) distribution is $e^{\lambda(x-1)}$, it is easy to show that

$$P(x) = p_0^* e^{\Lambda(x)} . \tag{7.18}$$

Let $\xi_j$ denote the $j$th cumulant of $C_w$ and $\xi_j^*$ the $j$th cumulant of $C^*$. The first two cumulants $\xi_1$ and $\xi_2$ are just the mean and variance of $C_w$. There are standard relations for computing moments from cumulants and vice versa. To compute the cumulants from the moments, we can use the recursion

$$\xi_{j+1} = \mu_{j+1} - \sum_{k=1}^{j} \binom{j}{k} \mu_k \xi_{j+1-k} . \tag{7.19}$$

Working directly from (7.14) and using elementary properties of cumulants leads to

$$\xi_j^* = \sum_{k=1}^{\infty} k^j \lambda_k . \tag{7.20}$$

By comparing this formula to what one obtains by successive differentiation of the generating function $\Lambda(x)$, we find that

$$\xi_j^* = D^j \Lambda(x) \Big|_{x=1} \tag{7.21}$$

where $D$ is the operator $D = x \frac{\partial}{\partial x}$ .

Our main interest is in using $P(C^* \geq 1)$ given in (7.15) to approximate $P(C_w \geq 1)$. If we wish to approximate the entire distribution of $C_w$ by that of $C^*$, then we need a way to compute the values $p_j^*$ for $j \geq 1$. There is no simple closed form for these values, but they may be easily calculated using the recursion

$$j p_j^* = \sum_{k=1}^{j} (k \lambda_k) p_{j-k}^* . \tag{7.22}$$

Alternatively, the values $p_j^*$ may be obtained by computing the power series expansion of $P(x)$ given in (7.18). This is easily done within software packages such as Maple or Mathematica. The recursion (7.22) is not new, but is probably not well known, so we shall sketch a proof. Differentiation of (7.18) gives us $P'(x) = \Lambda'(x) P(x)$. Equating coefficients in the power series expansions of $P'(x)$ and the product $\Lambda'(x) P(x)$ then leads immediately to (7.22).

Formula (7.20) says that the cumulants $\xi_j^*$ are linear in the $\lambda_k$'s. Thus, given the cumulants of $C^*$, it is typically straightforward to compute the $\lambda_k$'s. Moreover, Eq. (7.21) guarantees that if $\Lambda(x)$ has a simple closed form, then the lower-order cumulants are also given by simple closed formulas.

## General remarks

Our approximations all proceed as follows. Suppose we have computed the first $k$ moments of $C_w$. We use (7.19) to compute the first $k$ cumulants $\xi_1, \xi_2, \ldots, \xi_k$. Then we look for a "plausible" sequence $\boldsymbol{\lambda} = (\lambda_1, \lambda_2, \lambda_3, \ldots,)$ which matches these cumulants, that is, such that the distribution of $C^* \sim \mathrm{CP}(\boldsymbol{\lambda})$ has

$$\xi_j^* = \xi_j \quad \text{for } j = 1, 2, \ldots, k. \tag{7.23}$$

We then use (7.15) to compute $P(C^* \geq 1)$ which we use as our approximation to $P(C_w \geq 1)$. If we wish to approximate the entire distribution of $C_w$, we use the values $p_j^* = P(C^* = j)$ computed with (7.22).

For given values $\xi_1, \xi_2, \ldots \xi_k$, there are (typically) infinitely many sequences $\boldsymbol{\lambda}$ which match these values. The asymptotic results of Roos (1993) suggest a particular choice for $\boldsymbol{\lambda}$, but the computations required to compute this choice are difficult (except for small values of the clump size $m$). Our approximations are based on somewhat ad hoc choices of $\boldsymbol{\lambda}$, motivated largely by analytical convenience and simulation studies.

## An approximation computed via linear programming

The simplest approximation is obtained by assuming that $\lambda_j = 0$ for $j > k$. This assumption can be partly justified by simulation studies. Given particular values for $m$, $w$, and $N$, we can simulate many realizations of the sequence $Z_1, Z_2, \ldots, Z_g$ and count the number of clusters of 1's of various sizes that we see in these sequences. (To do this, we need a precise definition of "cluster." There is some arbitrariness in this, but one definition is to say that two groups of 1's constitute separate clusters if they are separated by at least $m - 2$ 0's.) When $m$ is small, such simulation studies usually produce very few large clusters, so that assuming $\lambda_j = 0$ for sufficiently large $j$ seems reasonable. If we do assume that $\lambda_j = 0$ for $j > k$, then (7.23) is just a system of $k$ linear equations in $k$ unknowns. This system has a unique solution $(\lambda_1, \ldots, \lambda_k)$. This leads to an approximation for $P(C_w \geq 1)$ (denoted APk) which was studied by Lin (1993). This approximation is often very good for small values of $m$ and $p$, but it suffers from one major drawback: in many situations, the approximation is not well defined because solving the equations (7.23) leads to a solution in which one or more of the values $\lambda_i$ is negative. When this happens, it means that there does not exist a CP distribution with $\lambda_j = 0$ for $j > k$ having the given values for the first $k$ cumulants.

Let us say that a compound Poisson distribution $CP(\boldsymbol{\lambda})$ has order $q$ if $\lambda_j = 0$ for $j > q$. The easiest way to fix the problem with the APk approximation in the previous paragraph is to use a CP distribution of higher order. If we assume the distribution of $C_w$ can be well approximated by a CP distribution of order $q > k$, then (7.23) becomes a system of $k$ linear equations in $q$ unknowns. When $q$ is chosen large enough, there will usually be a solution of this system which satisfies $\lambda_i \geq 0$ for $i = 1, 2, \ldots, q$ and thus corresponds to a genuine CP distribution. Our problem is one of solving a system of linear equations subject to linear constraints, and so we can solve it by linear programming. Unfortunately, there will now (typically) be infinitely many solutions, and we must find some way to choose which of these solutions we will use to construct our approximation. One way to narrow down the set of solutions is to impose additional conditions on the values $\lambda_i$, which are suggested by simulation studies. Examining the frequency distribution of the cluster sizes (obtained by simulation as in the previous paragraph) suggests that this distribution has a mode at 1 and drops off in a convex (actually, roughly geometric) fashion. Thus, it seems reasonable to impose the additional constraints that the sequence $\lambda_1, \lambda_2, \lambda_3, \ldots$ is nonincreasing and convex. These are linear constraints, and so are easily incorporated into our linear programming problem. Imposing these constraints will still not determine a unique solution: If there exists a solution to the resulting linear programming problem, then (typically) there exists infinitely many solutions. We are forced to make a final choice from this solution set in a fairly arbitrary fashion.

Suppose we have computed the first four cumulants of $C_w$. The considerations of the previous paragraph suggest the following procedure for determining $\boldsymbol{\lambda}$. Choose a sufficiently large value of $q$, set $\lambda_i = 0$ for $i > q$, and then take $\lambda_1, \ldots, \lambda_q$ to be the solution of the following linear programming problem:

$$\text{minimize } \sum_{i=1}^{q} \lambda_i$$

$$\text{subject to the constraints}$$

$$\begin{aligned}
&\text{(a)} \quad \sum_{i=1}^{q} i^r \lambda_i = \xi_r \text{ for } r = 1, 2, 3, 4\,, \\
&\text{(b)} \quad \lambda_i \geq 0 \text{ for } i = 1, \ldots, q\,, \\
&\text{(c)} \quad \lambda_i - \lambda_{i+1} \geq 0 \text{ for } i = 1, \ldots, q\,, \\
&\text{(d)} \quad \lambda_i - 2\lambda_{i+1} + \lambda_{i+2} \geq 0 \text{ for } i = 1, \ldots, q\,.
\end{aligned} \qquad (7.24)$$

Condition (a) above is just the system (7.23) written out explicitly. The choice to minimize in (7.24) is fairly arbitrary; maximizing $\sum \lambda_i$ also leads to a legitimate solution to our problem. Both choices usually lead to good approximations, but in our experience "minimizing" usually produces the better approximation; it also leads to a solution that (if it exists) does not depend on the particular value of $q$, provided that $q$ is chosen to be sufficiently large. Note that the constraints (b), (c), (d) are nested, that is, (d) implies (c) implies (b), so that (b) and (c) are actually redundant. There is no guarantee that a solution will exist satisfying (d). If no solution exists, we drop (d) and try to solve the problem with the constraint (c). If there is still no solution, we drop (c) and look for a solution satisfying (b) alone. Occasionally, usually when $N$ is small, there is no solution satisfying (b) and this approach fails altogether. If we do obtain a solution $\lambda$, we use this to compute our approximation to $P(C_w \geq 1)$. The resulting approximation is given in Table 7.2 under the heading LP4. (We shall comment later on the performance of LP4.) If the number of cumulants available is $k \neq 4$, we make fairly obvious changes to the linear programming problem and compute a corresponding approximation LPk.

The linear programming problem (7.24) is most easily solved by restating it in terms of the second differences of the $\lambda_i$'s.

## The CPGk approximation

We now describe another general approach to constructing compound Poisson approximations. Again, we suppose we are given the values of the first $k$ cumulants $\xi_1, \ldots, \xi_k$, and wish to find a plausible sequence $\lambda$ which matches these cumulants. In this approach, we choose a parametric family of sequences $\{\lambda(\theta) : \theta \in \Theta\}$ depending on $k$ parameters $\theta = (\theta_1, \ldots, \theta_k)$. Then the system (7.23) becomes a system of $k$ equations in $k$ unknowns. If we can solve this system for $\theta$, we use the resulting sequence $\lambda(\theta)$ to construct our approximation. The APk approximation can be viewed as a simple example of this approach in which we take $\theta = (\lambda_1, \ldots, \lambda_k)$; this leads to a system of linear equations. In more complicated examples, this approach leads to systems of nonlinear equations which must be solved numerically by iterative techniques. If the family $\{\lambda(\theta) : \theta \in \Theta\}$ leads to generating functions $\Lambda(x)$ [see (7.16)] that have a simple closed form, then the relation (7.21) gives a convenient way to explicitly construct the system of equations (7.23).

We have experimented with a number of approximations of this type. We give details on only one of these. This approximation (designated CPGk) assumes the values $\lambda_i$ decay geometrically starting with $\lambda_{k-1}$, that is, $\lambda_i = \lambda_{k-1} r^{i-k+1}$ for $i \geq k - 1$. The $k$ parameters $\theta = (\lambda_1, \ldots, \lambda_{k-1}, r)$ must satisfy $\lambda_i \geq 0$ for all $i$ and $0 \leq r < 1$ in order to correspond to a legitimate CP distribution. This approximation is suggested by the simulation results mentioned earlier in which the frequency distribution of the cluster sizes drops off in a roughly geometric fashion. The assumptions for CPGk imply

$$\Lambda(x) = \sum_{i=1}^{k-2} \lambda_i x^i + \frac{\lambda_{k-1} x^{k-1}}{1 - rx}$$

so that (7.21) and (7.23) lead immediately to an explicit system of $k$ nonlinear equations for $\lambda_1, \ldots, \lambda_{k-1}, r$. In our work, we have used the Maple procedure *fsolve* to find numerical solutions of these equations. In some cases, the equations have no solution or the solution is not legitimate. In these cases, the

approximation CPGk is not defined. Table 7.2 lists values for the approximation CPG4 based on four moments. The "na" entry marks a case where CPG4
was not defined.

When $k = 2$, various simplifications occur and the approximation can be
written in closed form. In this case,

$$\Lambda(x) = \frac{\lambda_1 x}{1 - rx}$$

so that the system of equations we need to solve is

$$\xi_1 = \frac{\lambda_1}{(1-r)^2} \text{ and } \xi_2 = \frac{\lambda_1(1+r)}{(1-r)^3}. \tag{7.25}$$

Taking the ratio gives

$$\frac{\xi_2}{\xi_1} = \frac{1+r}{1-r}$$

which is easily solved to obtain

$$r = \frac{(\xi_2/\xi_1) - 1}{(\xi_2/\xi_1) + 1}.$$

Plugging this back into (7.25) gives us

$$\lambda_1 = \xi_1(1 - r)^2.$$

This solution for $\lambda_1$ and $r$ is legitimate if $0 \le r < 1$ which is true so long as
$\xi_2 \ge \xi_1$. Using these results and noting that $\Lambda(1) = \lambda_1/(1 - r)$, we obtain our
final approximation (from (7.15) or (7.17)) as

$$\text{CPG2} = 1 - \exp\left\{\frac{-2\xi_1}{1 + (\xi_2/\xi_1)}\right\}.$$

If we assume that $C_w$ has approximately a simple (*not* compound) Poisson
distribution, the natural approximation to use is $P(C_w \ge 1) \approx 1 - e^{-\xi_1}$. It
is clear that CPG2 always produces a value which is smaller (and sometimes
substantially smaller) than this simple Poisson approximation.

## Performance of approximations

The approximations CPG2 and MC2 both depend on only two moments. These
approximations tend to give similar answers; their agreement is particularly
close for small $m$. For this reason, we list only MC2 in Table 7.2. For small $m$,
MC2 usually does a little better than CPG2 when $N$ is small or $p$ is close to
1. The earlier remarks comparing the performance of MC2 and GLAZ can be
repeated for CPG2; it is a close competitor of GLAZ for small $m$. It is clear from
Table 7.3 that CPG2 is inferior to GLAZ for large $m$ (unless $p$ is fairly large), but

it may still be useful because it leads naturally [via (7.22)] to an approximation for the entire distribution of $C_w$; there is no similar approximation produced by GLAZ.

There is an intuitive reason for expecting reasonably close agreement between CPG2 and MC2; both are making similar assumptions about the clusters of 1's occurring in the sequence $Z_1, Z_2, \ldots, Z_g$. If the two-state Markov chain $\{\tilde{Z}_i\}$ (which underlies MC2) is in state 1 at time $t$, it will remain in state 1 for a length of time which has a geometric distribution. Similarly, CPG2 is based on the assumption that the number of 1's in a cluster has a geometric distribution.

We now examine the performance of the approximations LP4 and CPG4 which use four moments. Our study is currently limited to clump sizes $m \leq 10$. From information like that given in Table 7.2, we see that the approximations MC2, CPG4, LP4, and GLAZ usually give answers which are very close when $p$ is small (say $p < .1$). Since we must compare all of these with simulated estimates of $p$, it is often difficult to determine which of the approximations is best. For very small $m$ (say $m \leq 6$), both LP4 and CPG4 tend to be more accurate than MC2. But for larger values of $m$, the situation is less clear. For $m \geq 7$, we find that LP4 tends to do worse than both CPG4 and MC2. The relative performance of CPG4 depends on the sample size $N$. When $N = 1,000$, if we discard those cases where the answers are so close that comparisons are difficult or meaningless, we find in the remaining cases that CPG4 is usually better than both MC2 and GLAZ. For $N = 1,000$, CPG4 is probably the best overall approximation. But for the smaller sample size $N = 100$, there are many cases in which CPG4 does worse than the much simpler approximation MC2.

For the smaller sample size $N = 100$, the approximations LP4 and CPG4 sometimes fail to exist. This problem occurs primarily when $m \geq 7$ and $p$ is close to 1, and is more pronounced for CPG4 than for LP4. The problem becomes more severe as $m$ increases, that is, it occurs for smaller and smaller values of $p$. When $m = 10$, there are cases with $p \approx .46$ in which CPG4 does not exist.

The performance of the "four moment" approximations LP4 and CPG4 is somewhat disappointing. They are not uniformly better than the "two moment" approximations CPG2 and MC2, and when they are better, the improvement is not as dramatic as we had hoped for. We are continuing to look for better four moment approximations. The higher-order moments $\mu_3$ and $\mu_4$ of $C_w$ are quite sensitive to the exact behavior of the distribution in the right tail. In simulations, it is clear that the frequency distribution of cluster sizes departs more and more from the geometric assumption (which underlies the CPGk approximation) as $m$ increases. In order to effectively use the higher-order moments in our compound Poisson approximations, we need to replace the simple "geometric tail" assumption by something more accurate.

## Approximating the distribution of $C_w$

The compound Poisson approximations LPk and CPGk lead immediately [via (7.22)] to approximations for the entire distribution of $C_w$. An approximate distribution constructed in this way will have the same first $k$ moments as $C_w$. In Tables 7.4 and 7.5, we give examples using CPG2 and CPG4 to approximate $P(C_w \geq j)$ for $j = 2, 3, 4, 5$ for samples of size $N = 100$ and $N = 1,000$. We compare these with a compound Poisson approximation due to Glaz *et al.* [1994, Eq. (3.4)] which is listed as GCP in Tables 7.4 and 7.5. GCP was the most accurate of the approximations studied by Glaz *et al.* (1994); it was superior to a number of simple Poisson approximations.

The approximations GCP and CPGk are closely related. GCP approximates the distribution of $C_w$ by a particular compound Poisson distribution $CP(\boldsymbol{\lambda})$ in which $\lambda_2, \lambda_3, \ldots, \lambda_{m-1}$ decay geometrically, $\lambda_j = 0$ for $j \geq m$, and $\lambda_1$ is chosen so that the first moments match. Thus, both GCP and CPGk use the ideas of matching moments and geometric decay of the $\lambda_j$'s.

From Tables 7.4 and 7.5, we see that all of the approximations GCP, CPG2, and CPG4 tend to perform better at $N = 1,000$ than at $N = 100$. This is not at all surprising since the original heuristic motivation for using compound Poisson approximations was asymptotic in nature and the theoretical results supporting their use are asymptotic. Both GCP and CPG2 do well, but CPG2 is usually more accurate than GCP. The approximation CPG4 typically improves on CPG2 for $m \leq 6$, but often does worse than CPG2 for $m \geq 7$. Again, it appears that we need more precise knowledge of the distribution of $C_w$ in the right tail before we can safely use the higher moments in our approximations. For general use, we (tentatively) recommend CPG2. This recommendation is tentative because the performance of CPG2 deteriorates as $m$ increases; it is possible that GCP will prove to be superior to CPG2 for large $m$. More study is needed on this point.

**Table 7.4:** Approximations to Prob $= P(C_w \geq j)$ for $N = 100$

| $m$ | $w$ | $j$ | Prob | CPG2 | CPG4 | GCP |
|---|---|---|---|---|---|---|
| 4 | .002 | 2 | .01242 | .01237 | .01245 | .01276 |
| 4 | .002 | 3 | .00157 | .00156 | .00155 | .00162 |
| 4 | .002 | 4 | .00018 | .00019 | .00018 | .00014 |
| 4 | .002 | 5 | .00002 | .00002 | .00002 | .00001 |
| 4 | .005 | 2 | .37818 | .37736 | .37854 | .37677 |
| 4 | .005 | 3 | .18163 | .18041 | .18273 | .18694 |
| 4 | .005 | 4 | .07759 | .07776 | .07842 | .08236 |
| 4 | .005 | 5 | .02964 | .03097 | .03070 | .03328 |
| 7 | .013 | 2 | .04293 | .04335 | .04234 | .04318 |
| 7 | .013 | 3 | .01504 | .01497 | .01571 | .01474 |
| 7 | .013 | 4 | .00522 | .00514 | .00520 | .00496 |
| 7 | .013 | 5 | .00179 | .00176 | .00171 | .00162 |
| 7 | .023 | 2 | .60117 | .60575 | .57753 | .58861 |
| 7 | .023 | 3 | .43627 | .43373 | .44336 | .42919 |
| 7 | .023 | 4 | .30359 | .29742 | .31475 | .30242 |
| 7 | .023 | 5 | .20259 | .19694 | .20099 | .20633 |
| 10 | .03 | 2 | .07150 | .07332 | .06649 | .07321 |
| 10 | .03 | 3 | .03616 | .03641 | .03944 | .03500 |
| 10 | .03 | 4 | .01827 | .01804 | .01903 | .01667 |
| 10 | .03 | 5 | .00908 | .00892 | .00906 | .00789 |
| 10 | .037 | 2 | .25821 | .26175 | .22146 | .26228 |
| 10 | .037 | 3 | .16287 | .16240 | .17844 | .16289 |
| 10 | .037 | 4 | .10193 | .09979 | .11187 | .10004 |
| 10 | .037 | 5 | .06322 | .06080 | .06264 | .06076 |

Prob $= P(C_w \geq j)$ was estimated from 1,000,000 simulations.
GCP is from Glaz *et al.* [1994, Eq. (3.4)].

**Table 7.5:** Approximations to Prob $= P(C_w \geq j)$ for $N = 1,000$

| $m$ | $w$ | $j$ | Prob | CPG2 | CPG4 | GCP |
|---|---|---|---|---|---|---|
| 4 | .0002 | 2 | .31241 | .31182 | .31194 | .31213 |
| 4 | .0002 | 3 | .12371 | .12338 | .12350 | .12418 |
| 4 | .0002 | 4 | .04252 | .04233 | .04234 | .04263 |
| 4 | .0002 | 5 | .01300 | .01303 | .01300 | .01303 |
| 4 | .0003 | 2 | .83715 | .83779 | .83749 | .83528 |
| 4 | .0003 | 3 | .65875 | .65899 | .65897 | .65741 |
| 4 | .0003 | 4 | .46633 | .46601 | .46629 | .46619 |
| 4 | .0003 | 5 | .29897 | .29887 | .29923 | .30023 |
| 7 | .001 | 2 | .14241 | .14334 | .14166 | .14278 |
| 7 | .001 | 3 | .05578 | .05596 | .05629 | .05466 |
| 7 | .001 | 4 | .02123 | .02116 | .02134 | .02018 |
| 7 | .001 | 5 | .00778 | .00780 | .00787 | .00721 |
| 7 | .0015 | 2 | .83679 | .83725 | .83566 | .83931 |
| 7 | .0015 | 3 | .70010 | .70070 | .69957 | .70329 |
| 7 | .0015 | 4 | .55640 | .55641 | .55638 | .55896 |
| 7 | .0015 | 5 | .42256 | .42193 | .42259 | .42393 |
| 10 | .002 | 2 | .05379 | .05558 | .05306 | .05436 |
| 10 | .002 | 3 | .02194 | .02241 | .02261 | .02054 |
| 10 | .002 | 4 | .00914 | .00900 | .00922 | .00772 |
| 10 | .002 | 5 | .00376 | .00360 | .00374 | .00288 |
| 10 | .003 | 2 | .70744 | .70939 | .69958 | .72400 |
| 10 | .003 | 3 | .56337 | .56617 | .56260 | .57691 |
| 10 | .003 | 4 | .43596 | .43779 | .43884 | .44346 |
| 10 | .003 | 5 | .32912 | .32967 | .33095 | .33071 |

# References


1. Aldous, D. (1989). *Probability Approximations via the Poisson Clumping Heuristic*, New York: Springer-Verlag.

2. Boros, E. and Prékopa, A. (1989). Closed form two-sided bounds for probabilities that at least $r$ and exactly $r$ out of $n$ events occur, *Mathematics of Operations Research*, **14**, 317–342.

3. Dembo, A. and Karlin, S. (1992). Poisson approximation for r-scan processes, *Annals of Applied Probability*, **2**, 329–357.

4. Galambos, J. and Simonelli, I. (1996). *Bonferroni-type Inequalities with Applications*, New York: Springer-Verlag.

5. Glaz, J. (1989). Approximations and bounds for the distribution of the scan statistic, *Journal of the American Statistical Association*, **84**, 560–566.

6. Glaz, J. (1992). Approximations for tail probabilities and moments of the scan statistic, *Computational Statistics & Data Analysis*, **14**, 213–227.

7. Glaz, J., Naus, J., Roos, M. and Wallenstein, S. (1994). Poisson approximations for the distribution and moments of ordered m-spacings, *Journal of Applied Probability*, **31**, 271–281.

8. Huffer, F. and Lin, C. T. (1995). Approximating the distribution of the scan statistic using moments of the number of clumps, *Technical Report*, Department of Statistics, Florida State University, Tallahassee, FL.

9. Huffer, F. and Lin, C. T. (1997). Approximating the distribution of the scan statistic using moments of the number of clumps, *Journal of the American Statistical Association*, **92**, 1466–1475.

10. Kwerel, S. M. (1975). Most stringent bounds on aggregated probabilities of partially specified dependent probability systems, *Journal of the American Statistical Association*, **70**, 472–479.

11. Lin, C. T. (1993). The computation of probabilities which involve spacings, with applications to the scan statistic, *Ph.D. Dissertation*, Department of Statistics, Florida State University, Tallahassee, FL.

12. Naus, J. I. (1982). Approximations for distributions of scan statistics, *Journal of the American Statistical Association*, **77**, 177–183.

13. Prékopa, A. (1988). Boole-Bonferroni inequalities and linear programming, *Operations Research*, **36**, 145–162.

14. Roos, M. (1993). Compound Poisson approximations for the number of extreme spacings, *Advances in Applied Probability*, **25**, 847–874.

15. Wallenstein, S. and Neff, N. (1987). An approximation for the distribution of the scan statistic, *Statistics in Medicine*, **6**, 197–207.

# 8

# Applying Ballot Problem Results to Compute Probabilities Required for a Generalization of the Scan Statistic

**Sylvan Wallenstein**

*Mount Sinai School of Medicine, New York, NY*

**Abstract:** The scan statistic, the maximum number of events in a sliding window of width $w$, has previously been used to test the null hypothesis of uniformity against the pulse alternative that the density takes on two values – a high one on some interval $[b, b + w)$, $b$ unknown, and a low density elsewhere. We generalize the statistic to the case where the density under the null is an arbitrary step function, so that use of a single critical value would not be of interest. We find the probability that under an arbitrary step function density, the number of events in any interval $[t, t+w)$ is less than $g(t)$, where $g(t)$ is also a step function. The probabilities are derived based on a ballot counting problem of Karlin and McGregor concerning the amount of lead in a multi-candidate election.

**Keywords and phrases:** Scan statistic, ballot problem, Karlin–McGregor theorem

## 8.1 Introduction

The scan statistic, the maximum number of events in a sliding window of width $w$, is [Naus (1966b)] a generalized likelihood ratio test of the null hypothesis of uniformity against the pulse alternative, that the density takes on two values: a high one on some interval on length $w$, and a low density elsewhere. Naus (1966a) has used a corollary of Barton and Mallows (1965) to find the exact probability, that under a uniform distribution of events, the scan statistic exceeds a single critical value, when $w = 1/L$, $L$ an integer. The result was extended to rational $w$ by Wallenstein and Naus (1973), and to all $w$ by Hwang

(1977), and by Huntington and Naus (1975). A drawback in testing for disease clustering noted by Stroup, Williamson, and Herndon (1989) and others, is the apparent inability to adjust for temporal time trends, i.e., to test for a cluster superimposed on some given underlying linear or seasonal trend. To do so, while retaining the fixed window size $w$, we will have to let the critical value change as the postulated underlying probability of the event changes. Thus, we will have to solve a more complicated problem: What is the probability that for each $t$, $Y_t(w)$, the number of points in $[t, t+w)$, is less than a given function $g(t)$, given that the density of the events themselves is given by $f(t)$.

Without loss of generality, we assume throughout this chapter that the total time frame is $T = 1$. We will also assume that the scanning interval $w$ is of the form $1/L$, $L$ an integer, although this could perhaps be relaxed, to yield a slightly more complicated result. Formally, for $i = 1, \ldots, N$, let $X_i$ be a sequence of random variables independently distributed on $[0, 1)$, and let $Y_t(w)$ be the number of events in $[t, t+w)$. In this chapter, we find

$$\mathrm{Pr}_{f(t)}[Yt(w) < g(t), \text{ all } 0 \le t < 1 - w] \qquad (8.1)$$

when the density of the $X$'s is given by $f(t)$, and where $f(t)$ and $g(t)$ are step functions.

For the scan statistic, $g(t)$ is a constant, $k$, and $f(t)$ is usually the uniform distribution. Cressie (1977) and Wallenstein, Naus, and Glaz (1993) have calculated power of the scan statistic by allowing $f(t)$ to be a pulse (a step function with a single step of width $w$), but keeping the restriction that $g(t)$ is a constant.

In Section 8.2, we find the probability in (8.1) when the step functions are "as fine" as the scanning window, i.e., both step functions have equally sized steps of length $w$. In Section 8.3, we discuss the case when the density and critical values can change on intervals of width $w/2$, and in Section 8.4, we discuss arbitrary step functions.

---

## 8.2   Exact Distribution of a Statistic With Critical Values Changing on Each Interval of Length $w$

In this section, we find the probability in (8.1) when the functions $f$ and $g$ are constant over intervals of widths $w$, so that for $(i - 1)w \le t < iw$:

$$f(t) = \theta_i, \quad i = 1, \ldots, L; \qquad g(t) = k_i, \quad i = 1, \ldots, L - 1. \qquad (8.2)$$

Thus there are $L = 1/w$ cells, each of width $w$, on which the density is constant. For simplicity of notation, especially in the sequel, we let for $i = 1, \ldots, L,$

$\pi_i = w\theta_i$, and $n_i = Y_{(i-1)w}(w)$ (the number of events in each cell) so that $(n_1, n_2, \ldots, n_L) \sim \text{Mult}(N; \pi_1, \pi_2, \ldots, \pi_L)$ where $\Sigma\pi_i = 1$.

Under these conditions

$$\text{Pr}_{f(t)}[Yt(w) < g(t), \text{ all } 0 \leq t < 1 - w]$$

$$= \text{Pr}\left(\bigcap_{i=1,\ldots,L-1} \left\{ sup_{0 \leq s \leq w} Y_{(i-1)w+s}(w) < k_i \right\}\right).$$

**Theorem 8.2.1** *Given the density,* $f((i-1)w + s) = \theta_i$, $i = 1, \ldots, 1/w$, $0 < s < w$,

$$Pr\left(\bigcap_{i=1,\ldots,L-1} \left\{ sup_{0 \leq s \leq w} Y_{(i-1)w+s}(w) < k_i \right\}\right) = N!wN \sum_{v_1(N)} \det G \prod_{i=1}^{L} \theta_i^{n_i}$$

$$(8.3)$$

*here, $G$ is a square matrix with $L$ rows having elements*

$$g_{ij} = \begin{cases} 1 / \left[k_i + \sum_{u=i+1}^{j-1}(k_u - n_u)\right]! & for \quad i < j - 1, \\ 1/k_i! & for \quad i = j - 1, \\ 1/n_i! & for \quad i = j, \\ 1 / \left[n_i - \sum_{u=j}^{i-1}(n_u - k_u)\right]! & for \quad i > j, \end{cases} \qquad (8.4)$$

*where $1/x! = 0$ if $x < 0$, and $v_1(N)$ is the set of $\{n_1, \ldots, n_L\}$ such that $\Sigma n_i = N$, and $n_i < \min(k_i, k_{i-1})$, $i = 1, \ldots, L$; $k_0 = k_1$; $k_L \leq k_{L-1}$. (The notation suppresses the dependence of $v_1(N)$ on $k_1, k_2, \ldots, k_{L-1}$.) Note that when for all $i = 1, \ldots, L$, $k_i = k$, and $\theta_i = 1$, the result is identical to Naus (1966a).*

PROOF. Following Naus (1966a) and Cressie (1977), we prove this assertion using a corollary of Barton and Mallows (1965) to a theorem of Karlin and McGregor (1959). For clarity and for the purpose of further extensions, we elaborate on the proof which is presented tersely in previous versions. Suppose $N$ voters choose among $L$ candidates. Let $A_i(m)$ be the partial total of candidate $i$ after $m$ votes have been counted, so that $A_i(N)$ is the total number of votes for candidate $i$. Then, Barton and Mallows (1965) have shown that conditional on $A_i(N)$, and for $\gamma_1 > \gamma_2 > \cdots > \gamma_L$,

$$\text{Pr}\left[\bigcap_{i=1}^{L-1}\{A_i(m) + \gamma_i > A_{i+1}(m) + \gamma_{i+1}, \ m = 1, \ldots, N\}\right] = |C|, \qquad (8.5)$$

where $c_{ij} = A_i(N)!/(A_i(N) + \gamma_i - \gamma_j)!$. $\blacksquare$

We view the candidates in the election as the $L$ cells, and the $N$ votes as the $N$ events, so that $A_i(N) = n_i$. The data for the $j$th observation $X_j$,

$j = 1, \ldots, N$, is replaced by a pair $(c(X_j), d(X_j))$ where the cell number is $c(X_j) = \text{CEIL}[X_j/w] = \text{CEIL}[LX_j]$, and $d(X_j) = X_j - w[c(X_j) - 1]$ is the distance between $X_j$ and the left (lower) boundary of the cell. (CEIL($x$), is $x$ rounded upward, thus CEIL(2.3) = CEIL(3) = 3). The process then induces a total number of votes for candidate $i$ at position $x$, $0 \leq x \leq w$, given by

$$A_i'(x) = \sum_{j=1,N} I\{d(X_j) - x \leq 0\} I\{c(X_j) = i\}, \qquad (8.6)$$

where $I\{E\} = 1$ if $E$ is true, and $I\{E\} = 0$ if $E$ is false. The total number of votes at time $x$ is given by $M(x) = \Sigma_i A_i'(x)$.

The following table gives an example of such a setup when $L = 4$:

| $j$ | $x$ | $c(x)$ | $d(x)$ | $A_1'(d(x))$ | $A_2'(d(x))$ | $A_3'(d(x))$ | $A_4'(d(x))$ | $M(x)$ |
|---|---|---|---|---|---|---|---|---|
| 1 | .10 | 1 | .10 | 1 | 2 | 1 | 2 | 6 |
| 2 | .11 | 1 | .11 | 2 | 2 | 1 | 2 | 7 |
| 3 | .19 | 1 | .19 | 3 | 2 | 1 | 3 | 9 |
| 4 | .30 | 2 | .05 | 0 | 1 | 0 | 2 | 3 |
| 5 | .32 | 2 | .07 | 0 | 2 | 1 | 2 | 5 |
| 6 | .56 | 3 | .06 | 0 | 1 | 1 | 2 | 4 |
| 7 | .76 | 4 | .01 | 0 | 0 | 0 | 1 | 1 |
| 8 | .77 | 4 | .02 | 0 | 0 | 0 | 2 | 2 |
| 9 | .88 | 4 | .13 | 2 | 2 | 1 | 3 | 8 |

Thus,

$$Y_{(i-1)w+s}(w) = n_i - A_i'(s) + A_{i+1}'(s)$$

so that

$$\left\{ Y_{(i-1)w+s}(w) < k_i \right\} \rightarrow \left\{ A_{i+1}'(s) - A_i'(s) < k_i - n_i \right\}$$

and

$$\left\{ \sup_{0 \leq s \leq w} Y_{(i-1)w+s}(w) < k_i \right\} \rightarrow \left\{ \sup_{0 \leq s \leq w} \{A_{i+1}'(s) - A_i'(s) < k_i - n_i\} \right\}. \qquad (8.7)$$

To place the problem in the format of a ballot problem, we sort by $m$ and let $A_i(m)$ be the value for $A_i'(d(Xj))$ when $M(X_j) = m$, as in the following panel:

| $m$ | $A_1(m)$ | $A_2(m)$ | $A_3(m)$ | $A_4(m)$ |
|---|---|---|---|---|
| 1 | 0 | 0 | 0 | 1 |
| 2 | 0 | 0 | 0 | 2 |
| 3 | 0 | 1 | 0 | 2 |
| 4 | 0 | 1 | 1 | 2 |
| 5 | 0 | 2 | 1 | 2 |
| 6 | 1 | 2 | 1 | 2 |
| 7 | 2 | 2 | 1 | 2 |
| 8 | 2 | 2 | 1 | 3 |
| 9 | 3 | 2 | 1 | 3 |

Thus setting

$$\gamma_i - \gamma_{i+1} = k_i - n_i,$$

we rewrite (8.7) as

$$\left\{ \sup_{0 \le s \le w} Y_{(i-1)w+s}(w) < k_i \right\}$$
$$= \{A_i(m) + \gamma_i > A_{i+1}(m) + \gamma_{i+1}, \ m = 1, \ldots, N\}. \tag{8.8}$$

(The expression on the right-hand side really only depends on the set of $m$'s where $A_i(m)$ or $A_{i+1}(m)$ change.) Applying the corollary of Barton and Mallows (1965), we have

$$\Pr\left( \bigcap_i \sup_{0 \le s \le w} Y_{(i-1)w+s}(w) < k_i \mid n_1, n_2, \ldots, n_L \right)$$
$$= \Pr\left[ \bigcap_i \{A_i(m) + \gamma_i > A_{i+1}(m) + \gamma_{i+1}, \ m = 1, \ldots, N\} \right] = \det C,$$
$$\tag{8.9}$$

where

$$c_{ij} = \begin{cases} n_i!/(n_i + \gamma_i - \gamma_j)! & \text{if } (n_i + \gamma_i - \gamma_j) > 0 \\ 0 & \text{otherwise.} \end{cases}$$

Unconditioning, by multiplying (8.9) by

$$\Pr(n_1, n_2, \ldots, n_L) = N! \prod_{i=1}^{L} \pi_i^{n_i}/n_i!,$$

yields (8.3) with $g_{ij} = c_{ij}/n_i!$. To complete the proof, note that $(n_i + \gamma_i + \gamma_j)$ is the expression in square brackets in the denominator of (8.4), that the constraint $\gamma_i > \gamma_{i+1}$ stated prior to (8.5) implies $n_i < k_i$, while setting $m = N$ in (8.5) implies $n_i + \gamma_i - \gamma_{i+1} > n_{i+1}$, so that $n_{i+1} < k_i$.

## 8.3  Finding the Probability for Step Function With $2L$ Steps

In this section, we find the probability in (8.1) when the functions $f$ and $g$ are constant over intervals of widths $w/2$, so that for $(i-1)w/2 \le t < iw/2$:

$$f(t) = \theta_i, \quad i = 1, \ldots, 2L; \qquad g(t) = k_i, \quad i = 1, \ldots, 2L-2. \qquad (8.10)$$

As above, we set $\pi_i = w\theta_i/2$, and relabel the $2L$ cell occupancy numbers by $n_i = Y_{(i-1)w/2}(w/2)$, so that under (8.10), $(n_1, n_2, \ldots, n_{2L}) \sim \mathrm{Mult}(N; \pi_1, \pi_2, \ldots, \pi_{2L})$, where $\Sigma\pi_i = 1$. For $i = 1, 2, \ldots, 2L-2$, let

$$F_i = \left\{ \sup_{0 \le s < w/2} Y_{(i-1)w/2+s}(w) < k_i \right\}.$$

Under (8.10),

$$\mathrm{Pr}_{f(t)}[Y_t(w) < g(t), \text{ all } 0 \le t < 1-w] = \mathrm{Pr}\left( \bigcap_{i=1,\ldots,2L-2} F_i \right). \qquad (8.11)$$

To find this probability, we will consider two simultaneous arrangements of points, one in "odd numbered intervals" and the other in "even numbered intervals." The odd numbered intervals consist of the disjoint subintervals $[iw, iw + w/2)$, $i = 0, \ldots, L-1$, while the even numbered intervals consist of the disjoint subintervals $[iw + w/2, (i+1)w)$, $i = 0, \ldots, L-1$. Let $E_1 = \bigcap_{i=1,\ldots,L-1} F_{2i-1}$ and $E_2 = \bigcap_{i=1,\ldots,L-1} F_{2i}$. For given cell occupancy numbers $n_1, \ldots, n_{2L}$, the occurrence of $E_1$ depends on the distribution of points in odd numbered intervals, and the occurrence of $E_2$ depends on the distribution of points in even numbered intervals. Since these arrangements are conditionally independent,

$$\mathrm{Pr}\left( \bigcap_{i=1,\ldots,2L-2} F_i \mid \{n_i\} \right)$$
$$= \mathrm{Pr}(E_1 \cap E_2 \mid \{n_i\}) = \mathrm{Pr}(E_1 \mid \{n_i\})\mathrm{Pr}(E_2 \mid \{n_i\}). \qquad (8.12)$$

As above, for each $X_j$, define $c(X_j) = \mathrm{CEIL}[2X_j/w]$, and $d(X_j) = X_j - (w/2)[c(X_j) - 1]$, and define $A'_i(d(x))$ and $M(x)$ as in (8.6) except now $i$ goes from 1 to $2L$, and $x$ from 0 to $w/2$. Then

$$Y_{(i-1)w/2+s}(w) = n_i - A'_i(s) + n_{i+1} + A'_{i+2(s)},$$

so that

$$F_i = \left\{ \sup_{0 \le s < w/2} Y_{(i-1)w/2+s}(w) < k_i \right\}$$
$$\rightarrow \left\{ \sup_{0 \le s < w/2} A'_{i+2}(s) - A'_i(s) \right\} < k_i - n_{i+1} - n_i.$$

Then for $i \leq L - 2$, mapping from $A'(x)$ to $A(m)$ yields

$$\Pr(F_i | \{n_i\}) = \Pr\left\{ A_{i+2}(m) - A_i(m) < k_i - n_{i+1} - n_i,\ m = 1, \ldots, N \right\}.$$

Taking the intersection of events yields

$$\Pr\left(E_2 | \{n_i\}\right) = \Pr(F2 \cdot F4, \ldots, \cdot F_{L-2})$$
$$= \Pr\left[ \bigcap_{i=1}^{L-2} \left\{ A_{2i+2}(m) - A_{2i}(m) < k_{2i} - n_{2i+1} - n_{2i},\ m = 1, \ldots, N \right\} \right].$$

This event is basically the same as that in (8.8), and restricted to the difference in votes for "even numbered" candidates, with $\gamma_i - \gamma_{i+1}$ replaced by

$$\gamma_i^{(2)} - \gamma_{i+1}^{(2)} = k_{2i} - n_{2i+1} - n_{2i}.$$

Applying the corollary of Barton and Mallows (1965) then yields

$$\Pr(E_2 | n_1, n_2, \ldots, n_{2L}\}) = |C(2)|,$$

where $c_{ij}^{(2)} = n_{2i}/[n_{2i} + \gamma_i^{(2)} - \gamma_j^{(2)}]!$. Rewriting this difference in terms of the cell occupancy number and critical values yields that $C^{(2)}$ is the same as $C$, with $n_i$ replaced by $n_{2i}$, and $k_i$ by $k_{2i} - n_{2i+1}$, so that

$$c_{ij}^{(2)} = \begin{cases} n_{2i}! \left/ \left[ k_{2i} - n_{2i+1} + \sum_{u=i+1}^{j-1}(k_{2u} - n_{2u+1} - n_{2u}) \right]! \right. & \text{for } i < j - 1, \\ n_{2i}!/(k_{2i} - n_{2i+1})! & \text{for } i = j - 1, \\ n_{2i}!/n_{2i}! & \text{for } i = j, \\ n_{2i}! \left/ \left[ n_{2i} - \sum_{u=j}^{i-1}(n_{2u} - (k_{2u} - n_{2u+1})) \right]! \right. & \text{for } i > j. \end{cases}$$

A similar argument yields

$$\Pr(E_1 | \{n_1, n_2, \ldots, n_{2L}\}) = |C(1)|,$$

where

$$c_{ij}^{(1)} = n_{2i-1}/[n_{2i-1} + \gamma_i^{(1)} - \gamma_j^{(1)}]!,$$
$$\gamma_i^{(1)} - \gamma_{i+1}^{(1)} = k_{2i-1} - n_{2i} - n_{2i-1}.$$

Thus substituting into (8.12) yields

$$\Pr\left( \bigcap_{i=1}^{2L-2} F_i \mid \{n_i\} \right) = |C(1)|\,|C(2)|.$$

The unconditional probability is then

$$\Pr_{f(t)}[Y_t(w) < g(t),\ \text{all } 0 \leq t < 1 - w]$$
$$= \sum |C(1)|\,|C(2)|\Pr(n_1, n_2, \ldots, n_{2L}), \tag{8.13}$$

where the summation is over $v_2(N)$, the set of all $n_1, \ldots, n_{2L}$ such that $\sum n_i = N$, and $n_i + n_{i+1} < \min(k_i, k_{i-1})$, $i = 1, \ldots, 2L - 1$; $k_0 = k_1$; $k_{2L-1} = k_{2L-2}$. Substituting

$$\mathrm{Pr}(n_1, n_2, \ldots, n_{2L}) = N! \prod_{i=1}^{2L} \pi_i^{n_i} / n_i! \tag{8.14}$$

into (8.13) yields

$$\mathrm{Pr}\left(\bigcap_i \sup_{0 \leq s \leq w} Y_{(i-1)w+s}(w) < k_i\right) = (w/2)^N N! \sum_{v_2(N)} |G(1)|\, |G(2)| \prod_{i=1}^{2L} \theta_i^{n_i},$$

where $g_{ij}^{(1)} = c_{ij}^{(1)}/n_{2i-1}!$, $g_{ij}^{(2)} = c_{ij}^{(2)}/n_{2i}!$. Equivalently, $G^{(1)}$ and $G^{(2)}$ are $L \times L$ matrices with

$$g_{ij}^{(1)} = \begin{cases} 1 \Big/ \left[\sum_{u=i}^{j-1} k_{2u-1} - \sum_{u=2i}^{2j-2} n_u\right]! & \text{for } i < j \\ 1/n_{2i-1}! & \text{for } i = j \\ 1 \Big/ \left[\sum_{u=2j-1}^{2i-1} n_u - \sum_{u=j}^{i-1} k_{2u-1}\right]! & \text{for } i > j, \end{cases}$$

and

$$g_{ij}^{(2)} = \begin{cases} 1 \Big/ \left[\sum_{u=i}^{j-1} k_{2u} - \sum_{u=2i+1}^{2j-1} n_u\right]! & \text{for } i < j \\ 1/n_{2i}! & \text{for } i = j \\ 1 \Big/ \left[\sum_{u=2j}^{2i} n_u - \sum_{u=j}^{i-1} k_{2u}\right]! & \text{for } i > j, \end{cases}$$

and $1/x! = 0$, if $x < 0$.

---

## 8.4   General Result

Finally, we consider the more general case in which we evaluate the probability in (8.1) under arbitrarily fine step functions which are constant over intervals of length $w/r$, $r$ an integer. Formally, for $(i - 1)w/r \leq t < iw/r$,

$$f(t) = \theta_i, \quad i = 1, \ldots, Lr; \qquad g(t) = k_i, \quad i = 1, \ldots, r(L - 1). \tag{8.15}$$

By choosing $r$ large enough, any two step functions for $f(t)$ and $g(t)$ can be defined in this framework. However, the computations will become increasingly complex for large $r$. The concept of the argument is given in the previous section for the case $r = 2$, and also in Wallenstein and Naus (1973), although the latter paper concerns a different problem.

Set $\pi_i = w\theta_i/r$, and let $n_i = Y_{w(i-1)/r}(w/r)$, $i = 1, \ldots, rL$, so that under (8.15), $(n_1, n_2, \ldots, n_{rL}) \sim \mathrm{Mult}(N; \pi_1, \pi_2, \ldots, \pi_{rL})$. Then, for $i = 1, 2, \ldots, r(L-1)$, let

$$F_i = \left\{ \sup_{0 \le s < w/r} Y_{(i-1)w/r+s}(w) < ki \right\},$$

so that under (8.15),

$$\Pr_{f(t)}[Y_t(w) < g(t), \text{ all } 0 \le t < 1 - w] = \Pr\left( \bigcap_{i=1}^{Lr-r} F_i \right). \tag{8.16}$$

**Theorem 8.4.1** *Let $v_r(N)$ be the set of cell occupancy numbers, $n_1, n_2, \ldots, n_{rL}$, such that*

$$\Sigma n_i = N, \quad \sum_{u=i}^{i+r-1} n_u < \min(k_i, k_{i-1}), \quad i = 1, rL - r + 1,$$

*where $k_0 = k_1$, $k_{rL-r+1} = k_{rL-r}$. Then under (8.15),*

$$Pr\left( \bigcap_{i=1}^{Lr-r} F_i \right) = (w/r)^N N! \sum_{vr(N)} \left[ \prod m = 1^r |G(m)| \right] \left[ \prod_{i=1}^{rL} \theta_i^{n_i} \right], \tag{8.17}$$

*where $G(m)$ is a square matrix with $L$ rows and with elements*

$$G_{ij}^{(m)} = \begin{cases} 1 \Big/ \left[ \sum_{u=i-1}^{j-2} k_{ru+m} - \sum_{u=(i-1)r+1}^{(j-1)r-1} n_{u+m} \right]! & \text{for } i < j \\[2ex] 1 \Big/ \left[ \sum_{u=(j-1)r}^{(i-1)r} n_{u+m} - \sum_{u=j-1}^{i-2} k_{ru+m} \right]! & \text{for } i \ge j. \end{cases}$$

PROOF. As above for each $X_j$, $j = 1, \ldots, N$, we define $c(X_j) = \mathrm{CEIL}[rX_j/w]$, and $d(X_j) = X_j - (w/r)[c(X_j) - 1]$, and define $A'_i(x)$ and $M(x)$ as in (8.6), except now $i$ goes from 1 to $rL$ and $x$ from 0 to $w/r$. Then

$$Y_{(i-1)w/r+s}(w) = n_i - A'_i(s) + \sum_{j=i+1}^{i+r-1} n_j + A'_{i+r}(s),$$

so that

$$\Pr(F_i|\{n_i\}) = \Pr\left\{ \sup_{0 \le s < w/r} A'_{i+r}(s) - A'_i(s) < k_i - \sum_{j=i}^{i+r-1} n_j \right\}. \tag{8.18}$$

For $m = 1, \ldots, r$, let

$$I(m) = \{i \mid i = m(\mathrm{mod}\ r), i \le L - r\},$$

so that

$$E_m = \bigcap_{i=0}^{L-2} F_{m+ir} = \bigcap_{i \in I(m)} F_i$$

and

$$\bigcap_{m=1}^{r} E_m = \bigcap_{i=1}^{Lr-r} F_i.$$

Note that conditional on $n_1, n_2, \ldots, n_{rL}$, $E_m$ depends only on the position of points in cells $I(m)$, and the events $E_1, E_2, \ldots, E_r$ are conditionally independent. Thus

$$Pr\left(\bigcap_{i=1}^{Lr-r} F_i | \{n_i\}\right) = \Pr\left(\bigcap_{m=1}^{r} E_m | \{n_i\}\right) = \prod_{m=1}^{r} \Pr(E_m | \{n_i\}),$$

where by (8.18)

$$\Pr(E_m | \{n_i\})$$
$$= \Pr\left\{\bigcap_{i=0}^{L-2} \left[A_{ri+m}(v) - A_{r(i-1)+m}(v) < \gamma_i^{(m)} - \gamma_{i+1}^{(m)}, \; v = 1, \ldots, N\right]\right\},$$

with

$$\gamma_i^{(m)} - \gamma_{i+1}^{(m)} = \left[k_{r(i-1)+m} - \sum_{j=r(i-1)+m+1}^{ri+m-1} n_j\right] - n_{r(i-1)+m}$$
$$= k_{r(i-1)+m} - \sum_{j=r(i-1)+m}^{ri+m-1} n_j. \tag{8.19}$$

Apply the corollary of Barton and Mallows (1965) to find

$$\Pr(E_m | n_1, n_2, \ldots, n_{rL}) = |C(m)|,$$

where $C^{(m)}$ is identical to $C$ with $n_i$ replaced by $n_{r(i-1)+m}$ and with $k_i$ replaced by the expression in brackets in (8.19). Thus,

$$Pr\left(\bigcap_{i=1}^{Lr-r} F_i | \{n_j\}\right) = \prod_{m=1}^{r} |C^{(m)}|.$$

Unconditioning by multiplying by the probability of the cell occupancy numbers as in (8.14), proves the theorem with $g_{ij}^{(m)} = c_{ij}^{(m)} / n_{r(i-1)+m}!$. $\blacksquare$

# References


1. Barton, D. E. and Mallows, C. L. (1965). Some aspects of the random sequence, *Annals of Mathematical Statistics*, **36**, 236–260.

2. Cressie, N. (1977). On some properties of the scan statistic on the circle and the line, *Journal of Applied Probability*, **14**, 272–283.

3. Huntington, R. and Naus, J. I. (1975). A simpler expression for kth nearest neighbor coincidence probabilities, *Annals of Probability*, **3**, 894–896.

4. Hwang, F. K. (1977). A generalization of the Karlin–McGregor theorem on coincidence probabilities and an application to clustering, *Annals of Probability*, **5**, 814–817.

5. Karlin, S. and McGregor, G. (1959). Coincidence probabilities, *Pacific Journal of Mathematics*, **9**, 1141–1164.

6. Naus, J. (1966a). Some probabilities, expectations, and variances for the size of the largest clusters and smallest intervals, *Journal of the American Statistical Association*, **61**, 1191–1199.

7. Naus, J. (1966b). A power comparison of two tests of non-random clustering, *Technometrics*, **8**, 493–517.

8. Stroup, D. F., Williamson, G. D. and Herndon, J. L. (1989). Detection of aberrations in the occurrence of notifiable diseases surveillance data, *Statistics in Medicine*, **8**, 323–329.

9. Wallenstein, S. and Naus, J. (1973). Probabilities for the kth nearest neighbor problem on the line, *Annals of Probability*, **1**, 188–190.

10. Wallenstein, S., Naus, J. and Glaz, J. (1993). Power of the scan statistic for detection of clustering, *Statistics in Medicine*, **12**, 1829–1843.

# 9

# Scan Statistic and Multiple Scan Statistic

**Chien-Tai Lin**

*Tamkang University, Taiwan*

**Abstract:** The scan statistic and multiple scan statistic can be used in many areas of science. In this chapter, a survey of results on scan statistic for the continuous conditional case is presented. We discuss the exact distribution and asymptotic results, and study various approximations and bounds. Moreover, a general expression for the $k$th moment of the multiple scan statistic and a computational approach for its distribution are covered. Numerical results comparing various approximations in the literature are also presented.

**Keywords and phrases:** Bonferroni-type inequality, compound Poisson approximation, Markov chain approximation, multiple scan statistic, product-type approximation, scan statistic, spacings, Stirling numbers, symbolic computation

## 9.1 Introduction

Given $N$ points $X_1, \ldots, X_N$ randomly distributed on the interval $(0, 1]$, let $Y_t(w)$ be the total number of points that lie in the interval $(t, t + w]$. The (continuous conditional) scan statistic $S_w$ is defined as

$$S_w = \max_{0 \leq t \leq 1-w} Y_t(w).$$

It is commonly used to test for the presence of nonrandom clustering on an interval.

The field of scan statistic is a very fascinating area with wide variety of applications in various branches of sciences as already pointed out in Chapter 1 by Glaz and Balakrishnan (1999). The growth of literature in this area over the past forty years provides a testimonial to this fact. Naus (1965) was the first to carry out a detailed study of scan statistic. For a particular type of nonrandom clustering alternatives, the optimality properties of using the scan statistic to test the uniformity were well discussed by Cressie (1984) and Naus (1966). Under the assumption of uniformity, the developments of the exact distribution of scan statistic up to 1980 were thoroughly reviewed by Neff and Naus (1980). In the meantime, Cressie (1977, 1980) also established some of the asymptotic results.

When $N$ is large, $m$ is moderate, and $w$ is small, the computation of $P(S_w \geq m)$ becomes complicated and infeasible. Moreover, the applicability of the asymptotic results is limited. Hence, attention has since focused on the developments of the approximations and bounds to the distribution of the scan statistic. Recently, Huffer and Lin (1997b) proposed a computational approach to evaluate the exact distribution of scan statistic. The purpose of this chapter is to provide a review of published work on the scan statistic.

In addition to the scan statistic for applications in tests of hypotheses, we may also use for this purpose the multiple scan statistic, the number of clumps with clump size $m$ in one dimension as discussed by Dembo and Karlin (1992), Glaz and Naus (1983), Glaz *et al.* (1994), Huffer and Lin (1997a), and Roos (1993). This multiple scan statistic is defined as follows:

$$C_w = \sum_{i=1}^{N-m+1} I(X_{(i+m-1)} - X_{(i)} \leq w), \qquad (9.1)$$

where $X_{(1)} \leq X_{(2)} \leq \ldots \leq X_{(N)}$ are the $N$ ordered observations. Glaz and Naus (1983) looked at its mean and variance, and gave a product-type approximation for its distribution. Glaz *et al.* (1994), based on the results of Roos (1993), constructed a compound Poisson approximation for the distribution of $C_w$. Huffer and Lin (1997a) studied the lower-order moments of $C_w$ with order up to four. They also presented a compound Poisson approximation to the distribution of the multiple scan statistic. From tables presented in these articles, a criterion to evaluate the accuracy of various approximations for a wide range of applications is needed. We use the elementary facts in combinatorics and the exchangeability of spacings to obtain a general expression for $E(C_w^k)$. Based on these expressions and the algorithm proposed by Huffer and Lin (1997b), a simple procedure to compute the exact distribution of $C_w$ is established. For small $N$, the proposed method is readily computable and very accurate. The values should be of assistance in order to assess the performance of various approximations.

This chapter starts with a relatively long survey in the second section on aspects of the exact and asymptotic results, approximations and bounds for the distribution of the scan statistic. The following sections cover some results on the multiple scan statistic that have not appeared elsewhere. In Section 9.3, the moments of the multiple scan statistic are discussed. We then describe a procedure to compute the exact distribution of $C_w$ in Section 9.4. In Section 9.5, we perform an extensive numerical study of $P(C_w \geq 2)$ and compare it with existing approximations mentioned earlier. The results of a simulation study for large $N$ are also included in order to evaluate the accuracy of these approximations.

---

## 9.2 Methods of Evaluating the Tail Probabilities of $S_w$

We first discuss the developments on evaluating the exact tail probabilities of the scan statistic $P(S_w \geq m) = P(m; N, w)$ under the assumption of uniformity. Barton and David (1956), Darling (1953), and Parzen (1960) all studied the expression of $P(m; N, w)$ for $m = 2$. Naus (1965) used a combinatorial approach to express $P(m; N, w)$ as the sum of binomial probabilities for $m > N/2$.

Naus (1966) applied a result of Barton and Mallows (1965) and Karlin and McGregor (1959) to express $P(m; N, 1/L)$, where $L$ is an integer, as a sum of $L \times L$ determinants, in which the number of terms to sum is the number of all possible arrangements of partitioning $N$ into $L$ positive integers each less than $m$. Wallenstein and Naus (1974) combined the approaches underlying the previous work of Naus (1965, 1966) to calculate $P(m; N, 1/L)$ in terms of sums of determinants with smaller dimensions. They tabulated $P(m; N, 1/L)$ for various ranges of $N$ and $L$ when $N/3 < m \leq N/2$.

By using a different method of partitioning, the formula of $P(m; N, w)$ given by Naus (1966) was further extended by Huntington and Naus (1975). A modification was suggested by Neff and Naus (1980) in order to improve the computational effort, including reducing the summations of determinants. They provided piecewise polynomials for $P(m; N, w)$ and presented tables of $P(m; N, w)$ for $0 < w < 0.5$ and $3 \leq m < N \leq 25$.

The methods for evaluating $P(m; N, w)$ so far described are quite related to the scanning process and the calculation of sums of determinants. An alternative approach of obtaining $P(m; N, w)$ was given by Huffer and Lin (1997b). They developed a very general methodology for evaluating probabilities which involve linear combinations of spacings, and then applied this procedure to express $P(m; N, w)$ as a polynomial (in $w$) whose coefficients are computed exactly. The expressions of $P(m; N, w)$ could be stored and easily evaluated later. Their program has so far handled cases somewhat beyond those given in the tables of Neff and Naus (1980).

When $N$ gets larger, the exact computation of $P(m; N, w)$ becomes infeasible. Arguments about the asymptotic results for $P(m; N, w)$ can be found in Berman and Eagleson (1983), Cressie (1977, 1980), and Dembo and Karlin (1992). The performance of these results in many instances for $P(m; N, w)$, noticed especially by Glaz (1992b), tends to be not accurate enough for the specific use of the testing procedure based on the scan statistic. In view of this, recent research emphasis has settled more on finding computationally tractable approximations and bounds. Recently, Glaz (1993) presented a nice review of the developments on the approximations and bounds for $P(m; N, w)$ under the null hypothesis of uniformity. The related methods for evaluating $P(m; N, w)$ under the alternative hypothesis were discussed by Cressie (1977) and Wallenstein *et al.* (1993).

For the purpose of convenience in studying various approximations and bounds for $P(m; N, w)$, we will classify these results into three categories: (i) those that utilize the scanning process, (ii) those that are based on the order statistics representation of the scan statistic,

$$M_N^{(m)} = \min_{1 \leq i \leq N-m+1} \{X_{(i+m-1)} - X_{(i)}\},$$

and by the fact that

$$P(M_N^{(m)} \leq w) = P(S_w \geq m),$$

and (iii) those that are related to the multiple scan statistic $C_w$ for testing uniformity and by the relationship that

$$P(C_w \geq 1) = P(S_w \geq m).$$

Methods in the first category have been studied by Naus (1982) and Wallenstein and Neff (1987). Let

$$b(i; N, p) = \binom{N}{i} p^i (1-p)^{N-i} \tag{9.2}$$

denote the binomial probabilities for $0 \leq i \leq N$ and $0 \leq p \leq 1$. Employing a result of Naus (1966), we have an approximation for $Q(m; N, w) = 1 - P(m; N, w)$ as

$$Q(m; N, w) \approx V_1 (V_2/V_1)^{1/w-2},$$

where

$$V_1 = \sum_{i=0}^{2m-2} Q(m; i, 1/2) b(i; N, 2w),$$

$$V_2 = \sum_{i=0}^{3m-3} Q(m; i, 1/3) b(i; N, 3w).$$

Following the approach of Naus (1965), Wallenstein and Neff (1987) obtained an approximation that

$$P(m; N, w) \approx (m/w - N + 1) b(m; N, w) + 2 \sum_{i=m+1}^{N} b(i; N, w).$$

This expression is very similar to the exact formula of Naus (1965) for the case when $m > N/2$ and $w \leq 1/2$.

There are many results of the second type available in the literature. Berman and Eagleson (1985), Glaz (1989, 1992a), and Krauth (1988, 1991) have all discussed results of this form. Berman and Eagleson (1985), based on the second order Bonferroni-type inequality result due to Hunter (1976) and Worsley (1982), obtained an upper bound

$$P(m; N, w)$$
$$\leq \sum_{i=m-1}^{N} b(i; N, w) + (N - m) \sum_{i=m-1}^{N} (-1)^{i+m-1} b(i; N, w)$$
$$= (N - m + 1) \sum_{i=m-1}^{N} b(i; N, w) - (N - m) \sum_{i=m-1}^{N} (1 - (-1)^{i+m-1}) b(i; N, w),$$

$$(9.3)$$

where $b(i; N, w)$ is as defined in (9.2).

Glaz (1989) gave further results in this direction. Making use of Kwerel's (1975) lower bound result and the exchangeability of spacings, Glaz obtained the following lower bound for $P(m; N, w)$:

$$P(m; N, w) \geq \frac{2(N - m + 1)}{\ell} \sum_{i=m-1}^{N} b(i; N, w) - 2 \sum_{i=0}^{N-m-1} \frac{(N - m - i)\theta_{1i+2}}{\ell(\ell - 1)},$$

$$(9.4)$$

where $\ell$ is the integer part of

$$2 \sum_{i=0}^{N-m-1} \frac{(N - m - i)\theta_{1i+2}}{(N - m + 1) \sum_{j=m-1}^{N} b(j; N, w)} + 2$$

and

$$\theta_{1i+2} = \begin{cases} \sum_{j=m+i}^{N} b(j; N, w) \\ \quad + \sum_{j=0}^{N-m-i} (-1)^j b(m + i + j; N, w) \sum_{r=0}^{i} \sum_{s=0}^{i} \binom{r + s + j}{r, s}, \\ \qquad \qquad \qquad \qquad \text{for } i = 0, 1, \ldots, m - 2, \\[2ex] \sum_{j=m+i}^{N} b(j; N, w) \\ \quad + \sum_{j=0}^{N-m-i} (-1)^j b(m + i + j; N, w) \sum_{r=0}^{i} \sum_{s=0}^{\min(i, m+i-r-1)} \binom{r + s + j}{r, s} \\ \quad + \sum_{r=0}^{i-m+1} \sum_{s=m-1}^{i-r} \binom{N}{s, m + i - r - s - 1} w^{m+i-r-1}(1 - 2w)^{N-m-i+r+1}, \\ \qquad \qquad \qquad \qquad \text{for } i = m - 1, m, \ldots, N - m - 1. \end{cases}$$

In a follow-up paper, Glaz (1992a) extended the earlier work and proposed $m$th order product-type approximation and Bonferroni-type upper bound for $P(m; N, w)$. Let

$$Q_1^* = \sum_{i=m-1}^{N} b(i; N, w), \quad Q_2^* = \sum_{i=m-1}^{N} (-1)^{i+m-1} b(i; N, w),$$

and for $3 \leq L \leq m$

$$\begin{aligned}
Q_L^* &= b(m-1; N, w) - b(m; N, w) \\
&\quad + \sum_{i=L}^{N-m+1} (-1)^i \prod_{j=1}^{L-2} [1 - \frac{i(i-1)}{j(j+1)}] b(m+i-1; N, w).
\end{aligned}$$

Glaz presented the $m$th order product-type approximation for $P(m; N, w)$ as

$$P(m; N, w) \approx 1 - \left(1 - \sum_{i=1}^{m} Q_i^*\right) \left[\left(1 - \sum_{i=1}^{m} Q_i^*\right) \Big/ \left(1 - \sum_{i=1}^{m-1} Q_i^*\right)\right]^{N-2m+1}. \tag{9.5}$$

Glaz also used the $m$th order Bonferroni-type inequality introduced in Hoover (1990) and generalized the inequality in (9.3). This inequality is given by

$$P(m; N, w) \leq \sum_{i=1}^{m-1} Q_i^* + (N - 2m + 2)Q_m^*. \tag{9.6}$$

Other results that can be placed in this category are those of Krauth (1988, 1991). Krauth (1988) discussed a third order Bonferroni-type upper bound for $P(m; N, w)$. Later, Krauth (1991) presented a lower bound for $P(m; N, w)$ by Galambos' (1975) method of indicators, or by a linear programming approach similar to the result of Prékopa (1988).

Results in the third category deal with finding Poisson and compound Poisson approximations to the distribution of multiple scan statistic $C_w$. Glaz *et al.* (1994) considered several Poisson approximations and a compound Poisson approximation for $P(m; N, w)$. Numerical comparisons showed that the compound Poisson approximation gave more reasonable values. Let

$$\eta = \frac{1 - \sum_{i=0}^{m-1} (1 - (-1)^i) b(m - i - 1; N, w) + (-1)^m (1 - 2w)^N}{\sum_{i=m-1}^{N} b(i; N, w)},$$

and $b(i; N, w)$ be as defined in (9.2). The compound Poisson approximation is

$$P(m; N, w) \approx 1 - \exp\left(-\sum_{i=1}^{m-1} \lambda_i\right) \sum_{\beta_1 + 2\beta_2 + \cdots + (m-1)\beta_{m-1} = 0} \prod_{i=1}^{m-1} \frac{\lambda_i^{\beta_i}}{\beta_i!}, \tag{9.7}$$

where $\beta_i$'s are nonnegative integers,

$$\lambda_1 = (N - m + 1) \sum_{j=m-1}^{N} b(j; N, w) - \sum_{i=2}^{m-1} i\lambda_i,$$

and

$$\lambda_i = (N - m + 1)(1 - \eta)^2 \eta^{i-1} \sum_{j=m-1}^{N} b(j; N, w), \ i = 2, \ldots, m - 1.$$

Huffer and Lin (1997a), based on the theoretical results of Prékopa (1988) and the values of the first four moments $E(C_w^k)$ ($k = 1, 2, 3, 4$), obtained the upper and lower bounds for $P(m; N, w)$. They also investigated several approximations for $P(m; N, w)$. Among these results, a Markov chain approximation (MC2) and a compound Poisson approximation (CPG2) were fairly reliable and also computable for large $m$. Both these approximations depend only on the first two moments. Let

$$\mathrm{Var}(C_w) = E(C_w^2) - (E(C_w))^2$$

and

$$\xi = \mathrm{Var}(C_w)/E(C_w). \tag{9.8}$$

The Markov chain approximation MC2 is as follows:

$$P(m; N, w) \approx 1 - (1 - \pi)(1 - \pi/s)^{N-m}, \tag{9.9}$$

where

$$\pi = E(C_w)/(N - m + 1) \tag{9.10}$$

and

$$s = \frac{1}{2}\left(N - m + 2 - \sqrt{(N - m)^2 + 2(N - m + 1) - \frac{2\mathrm{Var}(C_w)}{\pi(1 - \pi)}}\right). \tag{9.11}$$

In contrast, the compound Poisson approximation CPG2 for $P(m; N, w)$ is given by

$$P(m; N, w) \approx 1 - \exp\left(-\frac{2E(C_w)}{1 + \xi}\right). \tag{9.12}$$

Other types of approximations and bounds to $P(m; N, w)$ have been discussed by Gates and Westcott (1984), Knox and Lancashire (1982), and Loader (1991). Gates and Westcott (1984) derived a recursive formula for $Q(m; N, w)$. Let $Q_N \equiv Q(m; N, w)$. The recursive approximation $\tilde{Q}_N$ for $Q_N$ is

$$\tilde{Q}_N = \tilde{Q}_{N-1} + \sum_{j=m}^{\min(N, 2m-1)} \binom{N-1}{j-1} \phi_j \tilde{Q}_{N-j}, \ N = m, m + 1, \ldots,$$

with

$$\tilde{Q}_N = Q_N = 1, \ N = 0, 1, \ldots, m - 1,$$

and

$$\phi_j = (-1)^{j-m+1} j \binom{j}{m} w^{j-1} + \left[ (-1)^{j-m}(j-1)\binom{j}{m} - 2 \sum_{i=0}^{j-m-1} (-1)^i \binom{j}{i} \right] w^j,$$

where $j = m, m+1, \ldots, 2m - 1$ and $w \le 1/2$. The recursive formulae of upper and lower bounds for $Q(m; N, w)$ were also discussed and illustrated with $m = 3$.

Knox and Lancashire (1982), based on a modification of the simple disjoint test procedure, found an entirely pragmatic approximation to $P(m; N, w)$. Their expression is

$$P(m; N, w) \approx 1 - e^{\lambda \sqrt{w}/w^2} \sum_{i=0}^{N-1} \lambda^i/i!,$$

where

$$\lambda = \begin{cases} Nw & \text{for a crude approximation,} \\ \dfrac{N - e/\ln N}{1/w + 1 + \sqrt{2}/(2 + \ln w)} & \text{for a more refined approximation.} \end{cases}$$

Another approximation for $P(m; N, w)$ was suggested by Loader (1991). Let $\epsilon = Nw/m - 1$, $\lambda_1 = (1 - w(1 + \epsilon))/(1 - w)$, and $\lambda_2 = 1 + \epsilon$. In this setting, Loader used results of boundary-crossing probabilities for random walks and Poisson processes to obtain a large-deviation approximation as

$$P(m; N, w) \approx N\epsilon b(m; N, w) + \sum_{i=m}^{N} b(i; N, w) + \sum_{i=0}^{m-1} (\lambda_1/\lambda_2)^{2(m-i)} b(i; N, w),$$

where $b(i; N, w)$ is as defined in (9.2).

Recommendations for using these approximations and bounds for $P(m; N, w)$ have been discussed by Glaz (1989, 1993) and Huffer and Lin (1997a). In general, the approximations in (9.5), (9.7), (9.9), and (9.12) may give us reasonable values. The lower bound given by Huffer and Lin (1997a) is superior to Glaz's (1989) lower bound in (9.4) when $m \le 10$; however, the situation is more complicated for the upper bound. It is recommended that we calculate by both methods of Glaz (1992a) and Huffer and Lin (1997a) first and then choose the smaller value as the upper bound. For $m > 10$, the current program of Huffer and Lin (1997a) is not durable to compute the higher-order moments of $C_w$ and thus the bounds for $P(m; N, w)$. Hence, the approaches of Glaz (1989, 1992a) in (9.4) and (9.6) can offer us the best bounds to $P(m; N, w)$. More details about the performance of approximations and bounds in the literature can be found in Chapter 7 by Huffer and Lin (1999).

## 9.3 The Moments of $C_w$

For $2 \leq m \leq N$, $0 < w < 0.5$, and $1 \leq i \leq N - m + 1$, define

$$A_i = \{X_{(m+i-1)} - X_{(i)} < w\}. \tag{9.13}$$

It follows from the definition of $C_w$ in (9.1) that the $k$th order moment of $C_w$ can be written as

$$E(C_w^k) = \sum_{1 \leq h_1, \cdots, h_k \leq N-m+1} P\left(\bigcap_{i=1}^{k} A_{h_i}\right).$$

In turn, the expression yields

$$E(C_w^k) = \sum_{i=1}^{k} d(k, i) T_i,$$

where

$$T_i = \sum_{1 \leq h_1 < h_2 < \cdots < h_i \leq N-m+1} P\left(\bigcap_{j=1}^{i} A_{h_j}\right) \quad \text{for } 1 \leq i \leq k, \tag{9.14}$$

and $d(k, i)$, the number of ways to partition a set of $k$ indices into exactly $i$ parts without taking order into account, is equal to the value of the Stirling number of the second kind times $i!$, i.e., $\left\{ {k \atop i} \right\} i!$. The values of $d(k, i)$ can be obtained directly from the identity of Stirling numbers of the second kind: for $a, b > 0$,

$$\left\{ {a \atop b} \right\} = b \left\{ {a-1 \atop b} \right\} + \left\{ {a-1 \atop b-1} \right\},$$

and the special values

$$\left\{ {a \atop 0} \right\} = 0, \quad \left\{ {a \atop 1} \right\} = 1, \quad \text{and} \quad \left\{ {a \atop a} \right\} = 1.$$

Thus, to evaluate $E(C_w^k)$, we just need to obtain the values of $T_i, i \leq k$.

The remainder of this section is devoted to obtaining expressions for $T_i, i \leq k$, in terms of summations of quantities involving $P_i = P(A_i)$ and $P_{i,j} = P(A_i \cap A_j)$, etc., as these expressions will be used later in Section 9.4.

For integers $a$ and $b$, let $\binom{a}{b}$ denote the usual binomial coefficient which we take to be zero when $a < b$. Using the exchangeability of the spacings and elementary combinatorics, we can show that the sum in (9.14) is equal to the following much smaller sum:

$$T_i = \sum_{\boldsymbol{h} \in \mathcal{C}_i} \binom{N - m + 1 - h_i + r(\boldsymbol{h})}{r(\boldsymbol{h})} P\left(\bigcap_{j=1}^{i} A_{h_j}\right), \qquad (9.15)$$

where $\mathcal{C}_i$ is the set of $i$-tuples of integers $\boldsymbol{h} = (h_1, \ldots, h_i)$ satisfying $1 = h_1 < h_2 < \cdots < h_i \leq N - m + 1$ and $h_j - h_{j-1} \leq m - 1$ for $j = 2, \ldots, i$, and $r(\boldsymbol{h}) = 1 + |\{j : h_j - h_{j-1} = m - 1\}|$. (For any set $A$, we use $|A|$ to denote the cardinality of $A$.) We think of the $i$-tuple $\boldsymbol{h}$ as consisting of $r(\boldsymbol{h})$ blocks with boundaries at those values of the index $j$ for which $h_j - h_{j-1} = m - 1$.

The above sum can be further reduced in size by noting that the terms corresponding to $i$-tuples $\boldsymbol{h}$ with similar block structure are indeed identical. This can be precisely stated as follows. For any $i$-tuple $\boldsymbol{h} \in \mathcal{C}_i$, let $\delta(\boldsymbol{h})$ denote the vector of consecutive differences $(h_2 - h_1, h_3 - h_2, \ldots, h_i - h_{i-1})$. If $r(\boldsymbol{h}) = r$, then $\delta(\boldsymbol{h}) = (d_1, m - 1, d_2, m - 1, \ldots, m - 1, d_r)$, where $d_1, d_2, \ldots, d_r$ are blocks of integers of varying lengths (denoted $|d_1|, |d_2|, \ldots, |d_r|$), some of which may be empty (i.e., 0-tuples). Define $P_{\boldsymbol{h}} = P(\cap_{j=1}^{i} A_{h_j})$. Suppose the $i$-tuples $\boldsymbol{h}$ and $\boldsymbol{h}'$ satisfy $r(\boldsymbol{h}) = r(\boldsymbol{h}') = r$. Let $d_1, d_2, \ldots, d_r$ be the blocks in $\delta(\boldsymbol{h})$, and $d_1', d_2', \ldots, d_r'$ be the blocks in $\delta(\boldsymbol{h}')$. We say that $\boldsymbol{h}$ and $\boldsymbol{h}'$ are equivalent if $d_1', d_2', \ldots, d_r'$ is a permutation of $d_1, d_2, \ldots, d_r$. It follows immediately from the exchangeability of the spacings that $P_{\boldsymbol{h}} = P_{\boldsymbol{h}'}$ whenever $\boldsymbol{h}$ and $\boldsymbol{h}'$ are equivalent. In particular, every $i$-tuple in $\mathcal{C}_i$ is equivalent to at least one $i$-tuple $\boldsymbol{h}$ for which $|d_1| \geq |d_2| \geq \cdots \geq |d_r|$. Thus, we may restrict the sum in (9.15) to this smaller class of $i$-tuples, provided we compensate by including the appropriate multiplicative factor. This is done below.

Let $\mathcal{C}_i^*$ consist of those $i$-tuples $\boldsymbol{h} \in \mathcal{C}_i$ whose blocks $d_1, d_2, \ldots, d_{r(\boldsymbol{h})}$ (defined above) satisfy $|d_1| \geq |d_2| \geq \cdots \geq |d_{r(\boldsymbol{h})}|$. For each $\boldsymbol{h} \in \mathcal{C}_i^*$, define $n(\boldsymbol{h})$ to be the number of distinct rearrangements of the integers $|d_1|, |d_2|, \ldots, |d_{r(\boldsymbol{h})}|$. Then, we have

$$T_i = \sum_{\boldsymbol{h} \in \mathcal{C}_i^*} \binom{N - m + 1 - h_i + r(\boldsymbol{h})}{r(\boldsymbol{h})} n(\boldsymbol{h}) P_{\boldsymbol{h}}. \qquad (9.16)$$

For illustrative purposes and for further use in Section 9.4, we now present in detail what formula (9.16) becomes in the case of $T_6$. The integer factor preceding each of the sums below is the value of $n(\boldsymbol{h})$ for that group of terms:

$T_6$

$$= \sum_{s_1=1}^{m-2} \sum_{s_2=s_1+1}^{s_1+m-2} \sum_{s_3=s_2+1}^{s_2+m-2} \sum_{s_4=s_3+1}^{s_3+m-2} \sum_{s_5=s_4+1}^{s_4+m-2} (N - m + 1 - s_5) P_{1,s_1+1,s_2+1,s_3+1,s_4+1,s_5+1}$$

$$+ 2 \sum_{s_1=1}^{m-2} \sum_{s_2=s_1+1}^{s_1+m-2} \sum_{s_3=s_2+1}^{s_2+m-2} \sum_{s_4=s_3+1}^{s_3+m-2} \binom{N - 2m + 3 - s_4}{2} P_{1,s_1+1,s_2+1,s_3+1,s_4+1,s_4+m}$$

$$+2\sum_{s_1=1}^{m-2}\sum_{s_2=s_1+1}^{s_1+m-2}\sum_{s_3=s_2+1}^{s_2+m-2}\sum_{s_4=s_3+1}^{s_3+m-2}\binom{N-2m+3-s_4}{2}P_{1,s_1+1,s_2+1,s_3+1,s_3+m,s_4+m}$$

$$+\sum_{s_1=1}^{m-2}\sum_{s_2=s_1+1}^{s_1+m-2}\sum_{s_3=s_2+1}^{s_2+m-2}\sum_{s_4=s_3+1}^{s_3+m-2}\binom{N-2m+3-s_4}{2}P_{1,s_1+1,s_2+1,s_2+m,s_3+m,s_4+m}$$

$$+3\sum_{s_1=1}^{m-2}\sum_{s_2=s_1+1}^{s_1+m-2}\sum_{s_3=s_2+1}^{s_2+m-2}\binom{N-3m+5-s_3}{3}P_{1,s_1+1,s_2+1,s_3+1,s_3+m,s_3+2m-1}$$

$$+6\sum_{s_1=1}^{m-2}\sum_{s_2=s_1+1}^{s_1+m-2}\sum_{s_3=s_2+1}^{s_2+m-2}\binom{N-3m+5-s_3}{3}P_{1,s_1+1,s_2+1,s_2+m,s_3+m,s_3+2m-1}$$

$$+\sum_{s_1=1}^{m-2}\sum_{s_2=s_1+1}^{s_1+m-2}\sum_{s_3=s_2+1}^{s_2+m-2}\binom{N-3m+5-s_3}{3}P_{1,s_1+1,s_1+m,s_2+m,s_2+2m-1,s_3+2m-1}$$

$$+4\sum_{s_1=1}^{m-2}\sum_{s_2=s_1+1}^{s_1+m-2}\binom{N-4m+7-s_2}{4}P_{1,s_1+1,s_2+1,s_2+m,s_2+2m-1,s_2+3m-2}$$

$$+6\sum_{s_1=1}^{m-2}\sum_{s_2=s_1+1}^{s_1+m-2}\binom{N-4m+7-s_2}{4}P_{1,s_1+1,s_1+m,s_2+m,s_2+2m-1,s_2+3m-2}$$

$$+5\sum_{s_1=1}^{m-2}\binom{N-5m+9-s_1}{5}P_{1,s_1+1,s_1+m,s_1+2m-1,s_1+3m-2,s_1+4m-3}$$

$$+\binom{N-6m+11}{6}P_{1,m,2m-1,3m-2,4m-3,5m-4}\,.$$

Given specific values of $N, m$ and $w$, $T_6$ can thus be obtained by passing on the expression to the program developed by Huffer and Lin (1997b). In a similar way, we can obtain all the values of $T_i$, $1 \le i \le k$, and thus the value of $E(C_w^k)$.

---

## 9.4 Method for Evaluating the Probabilities of $C_w$

Utilizing the expressions of $T_i$'s presented in Section 9.3, $P(C_w = j)$, the probability that exactly $j$ of the events $A_i$ [defined in (9.13)] occur, is

$$\sum_{i=j}^{N-m+1}(-1)^{i-j}\binom{i}{j}T_i\,. \tag{9.17}$$

For small values of $N$, we can apply the algorithm proposed by Huffer and Lin (1997b) to the expressions presented in Section 9.3 to obtain the exact polynomials and thus the accurate values of the distribution of $C_w$. Let us restrict our attention to the case when $N = 10$ and $m = 5$, and have

$$T_6 = P_{1,2,3,4,5,6},$$

$$T_5 = 2P_{1,2,3,4,5} + P_{1,2,3,4,6} + P_{1,2,3,5,6} + P_{1,2,4,5,6} + P_{1,3,4,5,6},$$

$$T_4 = 3P_{1,2,3,4} + 2P_{1,2,3,5} + P_{1,2,3,6} + 2P_{1,2,4,5} + P_{1,2,4,6} + P_{1,2,5,6}$$
$$+2P_{1,3,4,5} + P_{1,3,4,6} + P_{1,3,5,6} + P_{1,4,5,6},$$

$$T_3 = 4P_{1,2,3} + 3P_{1,2,4} + 2P_{1,2,5} + 3P_{1,3,4} + 2P_{1,3,5} + P_{1,3,6}$$
$$+2P_{1,4,5} + P_{1,4,6} + 2P_{1,2,6},$$

$$T_2 = 5P_{1,2} + 4P_{1,3} + 3P_{1,4} + 3P_{1,5},$$

$$T_1 = 6P_1 .$$

Translating these $T_i$'s into matrix notation as defined by Huffer and Lin (1997b) and then passing in to their program separately, we obtain

$$
\begin{aligned}
T_6 =\ & 1 - 2R(0,1) - 2R(1,1) + 4R(2,1) - 6R(3,1) + 15R(0,2) - 12R(1,2) \\
& +12R(2,2) - 20R(3,2) + 8R(4,2) - 14R(5,2) + 24R(6,2) - 14R(0,3), \\
T_5 =\ & 6 - 30R(0,1) + 6R(1,1) + 10R(2,1) - 30R(3,1) + 14R(0,2) \\
& +12R(1,2) - 6R(2,2) - 56R(3,2) - 32R(5,2) + 110R(6,2) \\
& +10R(0,3) + 10R(1,3), \\
T_4 =\ & 15 - 42R(0,1) + 24R(1,1) - 60R(3,1) + 39R(0,2) - 12R(1,2) \\
& -70R(2,2) - 90R(3,2) - 42R(4,2) + 10R(5,2) + 198R(6,2) \\
& -12R(0,3) - 12R(1,3) - 8R(2,3), \\
T_3 =\ & 20 - 8R(0,1) + 16R(1,1) - 20R(2,1) - 60R(3,1) - 32R(0,2) \\
& -48R(1,2) - 80R(2,2) - 80R(3,2) - 20R(4,2) + 88R(5,2) \\
& +172R(6,2) + 20R(0,3) + 20R(1,3) + 16R(2,3) + 8R(3,3), \\
T_2 =\ & 15 - 6R(1,1) - 20R(2,1) - 30R(3,1) - 15R(0,2) - 24R(1,2) \\
& -28R(2,2) - 12R(3,2) + 24R(4,2) + 60R(5,2) + 60R(6,2), \\
T_1 =\ & 6 - 6R(0,1) - 6R(1,1) - 6R(2,1) - 6R(3,1),
\end{aligned}
$$

where

$$R(\ell, \beta) = \begin{cases} \binom{N}{\ell} w^\ell (1 - \beta w)^{N-\ell} & \text{for } \beta w < 1, \\ 0 & \text{for } \beta w \geq 1. \end{cases}$$

and integers $\ell, \beta \geq 0$.

    Plugging in the values of $w$ and $j$ in (9.17), the exact value of $P(C_w = j)$ can be easily obtained. Table 9.1 compares the approximations that were studied by Glaz and Naus [1983, Eq. (2.5)(GN)] and Glaz *et al.* [1994, Eq. (3.4)(GCP)], with our results for various choices of $w$ to the distribution of $C_w$ when $N = 10$ and $m = 5$. Note that the exact results given by Neff and Naus (1980) are matching exactly with our results for $P(C_w = 0)$ here.

**Table 9.1:** Comparison of two approximations to $P(C_w = j)$ when $m = 5$ and $N = 10$

| $w$ | $j$ | $p$ | GN | GCP |
|---|---|---|---|---|
| 0.01 | 0 | .9387 | .9383 | .9422 |
| | 1 | .0484 | .0495 | .0433 |
| | 2 | .0106 | .0099 | .0109 |
| | 3 | .0020 | .0019 | .0028 |
| | 4 | .00028 | .00035 | .00071 |
| | 5 | .00003 | .00006 | .00005 |
| | 6 | .000001 | | .000001 |
| 0.15 | 0 | .7910 | .7871 | .8114 |
| | 1 | .1413 | .1500 | .1169 |
| | 2 | .0491 | .0451 | .0444 |
| | 3 | .0147 | .0129 | .0180 |
| | 4 | .0034 | .0036 | .0073 |
| | 5 | .0005 | .0010 | .0014 |
| | 6 | .00004 | | .00004 |
| 0.20 | 0 | .5670 | .5508 | .6202 |
| | 1 | .2398 | .2728 | .1951 |
| | 2 | .1206 | .1113 | .0919 |
| | 3 | .0509 | .0411 | .0500 |
| | 4 | .0172 | .0155 | .0274 |
| | 5 | .0039 | .0062 | .0093 |
| | 6 | .0005 | | .0004 |

$p = P(C_w = j)$ is evaluated from the proposed method.

GN is taken from Table III in Glaz and Naus (1983).

GCP is from Glaz *et al.* [1994, Eq. (3.4)].

From these results, we find that the approximations of Glaz and Naus (1983) and Glaz *et al.* (1994) perform poorly as the value of $w$ increases. A method that is more accurate is therefore needed.

## 9.5 Approximations for $P(C_w \geq 2)$

We now evaluate, for selected values of $N, m$, and $w$, $P(C_w \geq 2)$ derived in Section 9.4 and compare it with the approximations that were discussed by Glaz and Naus [1983, Eq. (2.5)], Glaz *et al.* [1994, Eq. (3.4)], and Huffer and Lin (1997a, CPG2 and MC2). In particular, the approximate values of $P(C_w \geq 2)$ obtained from CPG2 approximation and MC2 approximation are as follows:

CPG2 approximation for $P(C_w \geq 2)$ is

$$1 - (1 + \lambda_1)\exp\left(-\frac{2E(C_w)}{1+\xi}\right), \tag{9.18}$$

where $\lambda_1 = 4E(C_w)/(\xi+1)^2$ and $\xi$ is as defined in (9.8);

MC2 approximation for $P(C_w \geq 2)$ is

$$1 - (1-\pi)(1-\frac{\pi}{s})^{N-m}$$
$$- (1-\frac{\pi}{s})^{N-m-2}\left[\frac{2\pi(1-\pi)}{s}(1-\frac{\pi}{s}) + \frac{(N-m-1)\pi(1-\pi)^2}{s^2}\right],$$

where $\pi$ and $s$ are as defined in (9.10) and (9.11), respectively.

Employing the recursion

$$jP(C_w = j) = \sum_{i=1}^{j} i\lambda_i P(C_w = j - i), \ j \geq 1,$$

and substituting $j$ for 1, (9.18) can be obtained immediately. The MC2 approximation for $P(C_w \geq 2)$ is not included in Huffer and Lin (1997a), but its derivation can be easily obtained from the material given there.

From the numerical results presented in Table 9.2, we find that MC2 is as good or better than any other approximation for $P(C_w \geq 2)$. It is evident that the GN approximation is inaccurate for large values of $P(C_w \geq 2)$. The CPG2 approximation seems to be very reliable; it is very similar to MC2, but MC2 does a little better. When the values of $m$ and $N$ get larger, the performance of GCP approximation gets better, and it is the best approximation if $m = 5$ and $P(C_w \geq 2) > 0.1$.

We now turn to the problem of evaluating the approximations discussed above (except the GN approximation) for larger values of $N$. When $N$ gets larger, the exact value of $P(C_w \geq 2)$ is difficult to compute. Thus, to assess the accuracy of these approximations, we present results from a simulation study based on 1,000,000 trials.

**Table 9.2:** Comparison of four approximations to $P(C_w \geq 2)$

| $m$ | $N$ | $w$ | $p$ | GN | CPG2 | MC2 | GCP |
|---|---|---|---|---|---|---|---|
| 3 | 6 | .01 | .00012 | .0001 | .00011 | .00012 | .00016 |
| | | .03 | .00293 | .003 | .00281 | .00287 | .00424 |
| | | .05 | .01264 | .012 | .01192 | .01236 | .01797 |
| | | .07 | .03224 | .030 | .03014 | .03153 | .04373 |
| | | .09 | .06347 | .059 | .05946 | .06226 | .08096 |
| | 7 | .01 | .00027 | .0003 | .00027 | .00027 | .00036 |
| | | .03 | .00678 | .007 | .00648 | .00663 | .00923 |
| | | .05 | .02881 | .027 | .02718 | .02810 | .03764 |
| | | .07 | .07187 | .067 | .06781 | .07023 | .08751 |
| | | .09 | .13749 | .125 | .13111 | .13495 | .15469 |
| | 8 | .01 | .00054 | .0005 | .00053 | .00054 | .00070 |
| | | .03 | .01357 | .013 | .01296 | .01325 | .01763 |
| | | .05 | .05661 | .053 | .05364 | .05521 | .06876 |
| | | .07 | .13685 | .125 | .13080 | .13404 | .15208 |
| | 9 | .01 | .00099 | .001 | .00097 | .00098 | .00125 |
| | | .03 | .02457 | .023 | .02347 | .02398 | .03059 |
| | | .05 | .09971 | .093 | .09519 | .09743 | .11346 |
| | 10 | .01 | .00167 | .002 | .00164 | .00165 | .00207 |
| | | .03 | .04120 | .039 | .03942 | .04022 | .04930 |
| | | .05 | .16084 | .149 | .15499 | .15769 | .17274 |
| 4 | 8 | .05 | .00268 | .003 | .00255 | .00260 | .00332 |
| | | .07 | .00924 | .009 | .00865 | .00892 | .01136 |
| | | .09 | .02258 | .021 | .02084 | .02177 | .02722 |
| | | .11 | .04498 | .042 | .04116 | .04343 | .05247 |
| | | .13 | .07811 | .072 | .07133 | .07575 | .08729 |
| | 9 | .05 | .00565 | .005 | .00538 | .00548 | .00677 |
| | | .07 | .01898 | .018 | .01776 | .01832 | .02244 |
| | | .09 | .04523 | .042 | .04186 | .04366 | .05180 |
| | | .11 | .08773 | .081 | .08096 | .08497 | .09588 |
| | 10 | .05 | .01063 | .010 | .01010 | .01030 | .01244 |
| | | .07 | .03487 | .033 | .03262 | .03367 | .03984 |
| | | .09 | .08100 | .075 | .07531 | .07835 | .08838 |
| | 11 | .05 | .01840 | .018 | .01745 | .01782 | .02110 |
| | | .07 | .05894 | .055 | .05521 | .05698 | .06514 |
| | | .09 | .13311 | .123 | .12474 | .12919 | .13866 |

$p = P(C_w \geq 2)$ is evaluated from the proposed method.

GN is taken from Table III in Glaz and Naus (1983).

GCP is from Glaz *et al.* [1994, Eq. (3.4)].

**Table 9.2 (contd.):** Comparison of four approximations to $P(C_w \geq 2)$

| $m$ | $N$ | $w$ | $p$ | GN | CPG2 | MC2 | GCP |
|---|---|---|---|---|---|---|---|
| 5 | 10 | .07 | .00267 | .003 | .00254 | .00258 | .00305 |
|  |  | .09 | .00815 | .008 | .00767 | .00786 | .00925 |
|  |  | .11 | .01928 | .018 | .01794 | .01855 | .02157 |
|  |  | .13 | .03842 | .036 | .03544 | .03700 | .04201 |
|  |  | .15 | .06767 | .063 | .06213 | .06541 | .07173 |
|  |  | .17 | .10857 | .100 | .09972 | .10554 | .11075 |
|  | 11 | .07 | .00533 | .005 | .00509 | .00517 | .00596 |
|  |  | .09 | .01584 | .015 | .01493 | .01530 | .01753 |
|  |  | .11 | .03642 | .034 | .03399 | .03517 | .03952 |
|  |  | .13 | .07053 | .066 | .06546 | .06827 | .07415 |
|  |  | .15 | .12067 | .112 | .11199 | .11739 | .12178 |
|  | 12 | .07 | .00970 | .009 | .00926 | .00941 | .01067 |
|  |  | .09 | .02806 | .027 | .02647 | .02715 | .03042 |
|  |  | .11 | .06277 | .059 | .05875 | .06080 | .06614 |
|  |  | .13 | .11815 | .110 | .11041 | .11489 | .11943 |

From Table 9.3, it is evident that the most accurate approximation is given by the MC2 approximation. For larger value of $N$ and $P(C_w \geq 2)$ is smaller than 0.01, the GCP approximation is the best. The performance of the CPG2 approximation is still very similar to MC2, but MC2 does a better job in this study. In general, we would recommend the use of the MC2 approximation for $P(C_w \geq 2)$. However, a more extensive discussion is still needed in order to find a good approximation for the distribution of $C_w$.

**Table 9.3:** Comparison of three approximations to $P(C_w \geq 2)$

| $N$ | $m$ | $w$ | $p$ | CPG2 | MC2 | GCP |
|---|---|---|---|---|---|---|
| 100 | 4 | .01 | .98966 | .99097 | .99042 | .98294 |
| | 5 | .01 | .47064 | .46880 | .46924 | .46283 |
| | 6 | .01 | .07398 | .07449 | .07467 | .07517 |
| | 7 | .01 | .00953 | .00965 | .00965 | .00948 |
| | 15 | .10 | .94489 | .96314 | .95531 | .92159 |
| | 17 | .10 | .52467 | .54146 | .53732 | .53677 |
| | 19 | .10 | .16059 | .16522 | .16500 | .17764 |
| | 20 | .10 | .07665 | .07925 | .07922 | .08584 |
| | 22 | .10 | .01439 | .01473 | .01473 | .01564 |
| | 26 | .15 | .08336 | .08610 | .08594 | .09811 |
| | 28 | .20 | .49278 | .51357 | .50398 | .52953 |
| | 30 | .20 | .21149 | .21779 | .21599 | .25005 |
| | 32 | .20 | .07133 | .07323 | .07302 | .08663 |
| 1000 | 6 | .001 | .78877 | .78944 | .78930 | .78948 |
| | 7 | .001 | .14241 | .14333 | .14338 | .14278 |
| | 8 | .001 | .01528 | .01555 | .01555 | .01506 |
| | 9 | .001 | .00164 | .00167 | .00167 | .00160 |
| | 12 | .005 | .98617 | .98693 | .98656 | .98932 |
| | 13 | .005 | .78412 | .78582 | .78531 | .81161 |
| | 14 | .005 | .40288 | .40793 | .40790 | .43004 |
| | 15 | .005 | .15340 | .15779 | .15781 | .16368 |
| | 16 | .005 | .05018 | .05255 | .05255 | .05320 |
| | 17 | .005 | .01510 | .01601 | .01601 | .01591 |
| | 18 | .005 | .00426 | .00456 | .00456 | .00447 |
| | 19 | .005 | .00115 | .00122 | .00122 | .00119 |

$p = P(C_w \geq 2)$ was estimated from 1,000,000 simulations.

GCP is from Glaz *et al.* [1994, Eq. (3.4)].

# References

1. Barton, D. E. and David, F. N. (1956). Some notes on ordered random intervals, *Journal of the Royal Statistical Society, Series B*, **18**, 79–94.

2. Barton, D. E. and Mallows, C. L. (1965). Some aspects of the random sequence, *Annals of Mathematical Statistics*, **36**, 236–260.

3. Berman, M. and Eagleson, G. K. (1983). A Poisson limit theorem for incomplete symmetric statistics, *Journal of Applied Probability*, **20**, 47–60.

4. Berman, M. and Eagleson, G. K. (1985). A useful upper bound for the tail probabilities of the scan statistic when the sample size is large, *Journal of the American Statistical Association*, **80**, 886–889.

5. Cressie, N. (1977). On some properties of the scan statistic on the circle and the line, *Annals of Probability*, **14**, 272–283.

6. Cressie, N. (1980). The asymptotic distribution of the scan statistic under uniformity, *Annals of Probability*, **8**, 828–840.

7. Cressie, N. (1984). Using the scan statistic to test uniformity, *Colloquia Mathematica Societatis Járos Bolyai*, **45**, pp. 87–100, Debrecen, Hungary.

8. Darling, D. A. (1953). On a class of problems related to the random division of an interval, *Annals of Mathematical Statistics*, **24**, 239–253.

9. Dembo, A. and Karlin, S. (1992). Poisson approximation for r-scan processes, *Annals of Applied Probability*, **2**, 329–357.

10. Galambos, J. (1975). Methods for proving Bonferroni-type inequalities, *Journal of the London Mathematical Society*, **9**, 561–564.

11. Gates, D. J. and Westcott, M. (1984). On the distributions of scan statistics, *Journal of the American Statistical Association*, **79**, 423–429.

12. Glaz, J. (1989). Approximations and bounds for the distribution of the scan statistic, *Journal of the American Statistical Association*, **84**, 560–566.

13. Glaz, J. (1992a). Approximations for tail probabilities and moments of the scan statistic, *Computational Statistics & Data Analysis*, **14**, 213–227.

14. Glaz, J. (1992b). Extreme order statistics for a sequence of dependent random variables, In *Stochastic Inequalities* (Eds., M. Shaked and Y. L. Tong), pp. 100–115, IMS Lecture Notes – Monograph Series, Hayward, CA.

15. Glaz, J. (1993). Approximations for the tail probabilities and moments of the scan statistic, *Statistics in Medicine*, **12**, 1845–1852.

16. Glaz, J. and Balakrishnan, N. (1999). Introduction to scan statistics, *Chapter 1, of this volume.*

17. Glaz, J. and Naus, J. (1983). Multiple clusters on the line, *Communications in Statistics—Theory and Methods*, **12**, 1961–1986.

18. Glaz, J., Naus, J., Roos, M. and Wallenstein, S. (1994). Poisson approximations for the distribution and moments of ordered m-spacings, *Journal of Applied Probability*, **31**, 271–281.

19. Hoover, D. R. (1990). Subset complement addition upper bounds—an improved inclusion-exclusion method, *Journal of Statistical Planning and Inference*, **24**, 195–202.

20. Huffer, F. and Lin, C. T. (1997a). Approximating the distribution of the scan statistic using moments of the number of clumps, *Journal of the American Statistical Association*, **92**, 1466–1475.

21. Huffer, F. and Lin, C. T. (1997b). Computing the exact distribution of the extremes of sums of consecutive spacings, *Computational Statistics & Data Analysis*, **26**, 117–132.

22. Huffer, F. and Lin, C. T. (1999). Using moments to approximate the distribution of the scan statistic, *Chapter 7, of this volume.*

23. Hunter, D. (1976). An upper bound for the probability of a union, *Journal of Applied Probability*, **13**, 597–603.

24. Huntington, R. J. and Naus, J. I. (1975). A simpler expression for $k$th nearest neighbor coincidence probabilities, *Annals of Probability*, **3**, 894–896.

25. Karlin, S. and McGregor, J. (1959). Coincidence probabilities, *Pacific Journal of Mathematics*, **9**, 1141–1164.

26. Knox, E. G. and Lancashire, R. (1982). Detection of minimal epidemics, *Statistics in Medicine*, **1**, 183–189.

27. Krauth, J. (1988). An improved upper bound for the tail probability of the scan statistic for testing non-random clustering, In *Classification and Related Methods of Data Analysis: Proceedings of the First Conference of the International Federation of Classification Societies (IFCS), Technical University of Aachen, F.R.G., 29 June–1 July 1987* (Ed., H. H. Bock), pp. 237–244, New York: North-Holland.

28. Krauth, J. (1991). Lower bounds for the tail probabilities of the scan statistic, In *Classification, Data Analysis, and Knowledge Organization Models and Methods With Applications: Proceedings of the 14th Annual Conference of the Gesellschaft fur Klassification [sic] e.V., University of Marburg, March 12-14, 1990* (Eds., H. H. Bock and P. Ihm), pp. 61–67, New York: Springer-Verlag.

29. Kwerel, S. M. (1975). Most stringent bounds on aggregated probabilities of partially specified dependent probability systems, *Journal of the American Statistical Association*, **70**, 472–479.

30. Loader, C. R. (1991). Large-deviation approximations to the distribution of scan statistics, *Advances in Applied Probability*, **23**, 751–771.

31. Naus, J. I. (1965). The distribution of the size of the maximum cluster of points on a line, *Journal of the American Statistical Association*, **60**, 532–538.

32. Naus, J. I. (1966). Some probabilities, expectations and variances for the size of the largest clusters and smallest intervals, *Journal of the American Statistical Association*, **61**, 1191–1199.

33. Naus, J. I. (1982). Approximations for distributions of scan statistics, *Journal of the American Statistical Association*, **77**, 177–183.

34. Neff, N. D. and Naus, J. I. (1980). The distribution of the size of the maximum cluster of points on a line, *IMS Series of Selected Tables in Mathematical Statistics*, Volume **6**, Providence, RI: American Mathematical Society.

35. Parzen, E. (1960). *Modern Probability Theory and Its Applications*, New York: John Wiley & Sons.

36. Prékopa, A. (1988). Boole-Bonferroni inequalities and linear programming, *Operations Research*, **36**, 145–162.

37. Roos, M. (1993). Compound Poisson approximations for the number of extreme spacings, *Advances in Applied Probability*, **25**, 847–874.

38. Wallenstein, S. R. and Naus, J. I. (1974). Probabilities for the size of largest clusters and smallest intervals, *Journal of the American Statistical Association*, **69**, 690–697.

39. Wallenstein, S. and Neff, N. (1987). An approximation for the distribution of the scan statistic, *Statistics in Medicine*, **6**, 197–207.

40. Wallenstein, S., Naus, J. and Glaz, J. (1993). Power of the scan statistic for detection of clustering, *Statistics in Medicine*, **12**, 1829–1843.

41. Worsley, K. J. (1982). An improved Bonferroni inequality and applications, *Biometrika*, **69**, 297–302.

# 10

# On Poisson Approximation for Continuous Multiple Scan Statistics in Two Dimensions

**Marianne Månsson**

*Chalmers University of Technology, Göteborg, Sweden*

**Abstract:** In this chapter, Poisson approximation for multiple scan statistics in the continuous case is investigated. The setting is mainly two dimensional, but higher dimensions are also discussed. The scanning set can be any convex set, and in order to motivate the choice of parameters in the approximations, some geometrical arguments are given. The errors involved in the approximations are studied both by simulations and by giving bounds on the total variation distances by means of the Stein–Chen method. Furthermore, Poisson process approximations of some point processes, which occur in this context, are considered.

**Keywords and phrases:** Multiple scan statistics, Poisson approximation, Poisson process, Stein–Chen method

## 10.1   Introduction

Assume $\mathbf{A} \subset \mathbb{R}^d$, $d \geq 2$, is a $d$-dimensional rectangle in which $N$ points are independently and uniformly distributed, and let $W \subset \mathbf{A}$ be a convex set, small relative to $\mathbf{A}$. There are $\binom{N}{m}$ subsets consisting of $m \leq N$ points, referred to as *m-subsets* in the following, some of which are covered by some translate of the *scanning set* $W$. We define the *multiple scan statistic* in $d$ dimensions, $\xi(d, N, m, W)$, as the number of $m$-subsets which actually are covered by some translate of $W$. If the $m$-subsets are ordered in some fixed way, then the multiple scan statistic can be written as

$$\xi(d, N, m, W) = \sum_{i=1}^{\binom{N}{m}} I_i, \tag{10.1}$$

where

$$I_i = \begin{cases} 1 & \text{if there exists } x \in \mathbf{A} \text{ such that the } i\text{th } m\text{-subset is covered by } W(x), \\ 0 & \text{otherwise,} \end{cases}$$

and $W(x)$ denotes the translate of $W$ by $x \in \mathbb{R}^d$.

The purpose of this chapter is to investigate Poisson approximation for the multiple scan statistic as defined above, which means in the conditional continuous case. The setting will be primarily two dimensional, but also a short discussion of higher dimensions is included. Furthermore, to avoid problems with the boundaries of $\mathbf{A}$, the torus convention will be used.

The plan for this chapter is as follows. First, some notation and necessary theory concerning convex sets is introduced. In Section 10.3, references to previous works on scan statistics in the continuous conditional case are given. The probability of covering a given $m$-subset is derived in the following section. In Section 10.5, approximations are suggested and examined in terms of the total variation distance and by simulations. Some point processes, determined by positions and "sizes" of the $m$-subsets that are covered, are studied in Section 10.6. The chapter is concluded by a discussion on some generalizations and possible improvements of the results derived here.

---

## 10.2   Preliminaries

Let $\mathbb{R}^d$ denote the $d$-dimensional Euclidean space, with a fixed origin $O$, and orthogonal coordinate-axes. The *volume* of a (measurable) subset of $\mathbb{R}^d$ is its $d$-dimensional Lebesgue measure which we denote by $\mu$. We will mainly discuss $\mathbb{R}^2$ and then $\mu$ is the area.

For $B, C \subset \mathbb{R}^d$ and $c \in \mathbb{R}$, the *Minkowski sum* and *scalar multiple* are defined as

$$B + C = \{x + y : x \in B, y \in C\} \quad \text{and} \quad cB = \{cx : x \in B\},$$

respectively. If $c = -1$, we get $\check{B} = \{-x : x \in B\}$, which we call the *reflected set* of $B$. For $x \in \mathbb{R}^d$, $B + \{x\}$ is the *translate* of $B$ by $x$, which is denoted by $B(x)$. If $B = \check{B}(x)$ for some $x \in \mathbb{R}^d$, $B$ is said to be *centrally symmetric*. An alternative, and for us more useful, way of writing the Minkowski sum is

$$B + C = \{x : B \cap \check{C}(x) \neq \emptyset\}. \tag{10.2}$$

For the set $xB + yC$, where $x, y \in \mathbb{R}^+$ and $B, C \subset \mathbb{R}^2$ are nonempty convex sets, the area can be written as

$$\mu(xB + yC) = x^2\mu(B) + 2xy\nu(B, C) + y^2\mu(C), \tag{10.3}$$

where $\nu(B, C)$ is the *mixed area* of $B$ and $C$, which is actually defined by (10.3); see, for example, Bonnesen and Fenchel (1948, p. 40). It can be shown that if $C$ is a convex set in $\mathbb{R}^2$, then

$$\mu(C) \leq \nu(C, \check{C}) \leq 2\mu(C), \qquad (10.4)$$

where the lower bound is attained if and only if $C$ is centrally symmetric while the upper bound is attained if and only if $C$ is a triangle. For a proof of these facts, see Bonnesen and Fenchel (1948, p. 105), where equations corresponding to (10.3) and (10.4) in higher dimensions also can be found.

Of particular interest here is $\nu(W, \check{W})$, since this functional is involved in the expression for $E[\xi(2, N, m, W)]$, which is given in (10.14). To derive $\nu(W, \check{W})$ for an arbitrary convex set $W$ is not so easy. However, for centrally symmetric sets, such as discs and rectangles, and for triangles it is given in (10.4), and for arbitrary polygons a simple formula can be found in Eggleston (1958, p. 85).

We will assume that the rectangle in which the points are distributed, $\mathbf{A}$, is centered at the origin. The family of convex sets, $W \subset \mathbb{R}^2$, with the properties that $W + \check{W} \subset \mathbf{A}$ and that $O$ is an interior point of $W$, is denoted by $\mathcal{K}$.

Let $(\mathcal{X}, \mathcal{A})$ be any measurable space. The *total variation distance*, $d_{TV}$, between two probability measures $\mu$ and $\nu$ on $\mathcal{X}$ is defined to be

$$d_{TV}(\mu, \nu) = \sup_{A \in \mathcal{A}} |\mu(A) - \nu(A)|.$$

If the state space is discrete, then

$$d_{TV}(\mu, \nu) = \frac{1}{2} \sum_{i \in \mathcal{X}} |\mu\{i\} - \nu\{i\}|, \qquad (10.5)$$

and in this case convergence in total variation distance, $d_{TV}(\mathcal{L}(X_n), \mathcal{L}(X)) \to 0$, is equivalent to $\{X_n\}$ converging in distribution to $X$.

---

## 10.3  Historical Background

### 10.3.1  Multiple scan statistics

The multiple scan statistic in one dimension is defined as follows. Let $X_1, \ldots, X_N$ be independently and uniformly distributed in $\mathbf{A} = [0, T] \subset \mathbb{R}$, and let $X_{(1)}, \ldots, X_{(N)}$ denote the ordered sample. For a fixed $w$, $0 < w < T$, and $m$, $1 \leq m < N$, the multiple scan statistic is defined as

$$\xi(1, N, m, w) = \sum_{i=1}^{N-m} J_i, \qquad (10.6)$$

where

$$J_i = \begin{cases} 1 & \text{if } X_{(m+i)} - X_{(i)} \le w, \\ 0 & \text{otherwise,} \end{cases}$$

$i = 1, \ldots N - m$. By letting $\mathbf{A}$ be a circle instead of an interval, and replacing $N - m$ in the sum in (10.6) by $N$, we would get a torus version of this defintion. In contrary to the case of higher dimensions, where the torus convention will be used throughout, this convention will not be further discussed in one dimension.

In higher dimensions, we assume that $\mathbf{A} \subset \mathbb{R}^d$ is a $d$-dimensional rectangle in which $N$ points are independently and uniformly distributed. Since there is no natural total order relation in $\mathbb{R}^d$, $d \ge 2$, we cannot use the same definition of the multiple scan statistic as in one dimension. In this chapter, the $d$-dimensional multiple scan statistic $\xi(d, N, m, W)$ is defined as in (10.1), i.e., as the number of $m$-subsets which are covered by some translate of $W$, for a given convex set $W \subset \mathbf{A}$.

Note that $I_i = 1$ in the definition of the $d$-dimensional multiple scan statistic in (10.1), means that $m$ points are "close together," while $J_i = 1$ in the definition in the one-dimensional case in (10.6), means that $m + 1$ points are "close together." Hence, $m$ in the $d$-dimensional case, $d \ge 2$, corresponds to $m - 1$ in the one-dimensional case.

In one dimension, the *m-spacings* are defined by

$$Y_i^{(m)} = X_{(m+i)} - X_{(i)},$$

$i = 1, \ldots, N - m$. The relation between the ordered $m$-spacings, denoted by $Y_{(1)}^{(m)}, \ldots, Y_{(N-m)}^{(m)}$, and the multiple scan statistic is

$$P(\xi(1, N, m, w) \ge i) = P(Y_{(i)}^{(m)} \le w). \tag{10.7}$$

The lack of a total order relation in higher dimensions also prevents the definition of spacings to be directly generalized. To get a relation similar to (10.7), which involves the multiple scan statistic $\xi(d, N, m, W)$, we introduce

$$Z_i^{(m)} = T_i^d \mu(W), \tag{10.8}$$

where $T_i = \inf\{y \in \mathbb{R}^+ : \text{the } i\text{th } m\text{-subset} \in yW(x) \text{ for some } x \in \mathbf{A}\}$. We will call $Z_i^{(m)}$ the *size* of the $i$th $m$-subset. The relation between the ordered sizes $Z_{(i)}^{(m)}$, $i = 1, \ldots, \binom{N}{m}$, and the multiple scan statistic is

$$P(\xi(d, N, m, W) \ge i) = P(Z_{(i)}^{(m)} \le \mu(W)). \tag{10.9}$$

In one dimension, the multiple scan statistic has been studied by Glaz and Naus (1983) and Glaz *et al.* (1994). In the latter, several Poisson approximations have been proposed and investigated by means of simulations. In particular, a compound Poisson approximation, which was first derived by Roos (1993), has been studied, and the simulations show that this is the most accurate of the suggested approximations.

Barbour, Holst and Janson (1992) proposed a Poisson approximation for $m$-spacings in one dimension by means of the Stein–Chen method. By the relation (10.7), those results also hold for $\xi(1, N, m, w)$.

Other references in which the continuous multiple scan statistic has been investigated, however without using this notion, are Eggleton and Kormack (1944), Silberstein (1945), and Mack (1948, 1949). The setting in these references is mainly one dimensional, but they also cover the two-dimensional cases when the scanning set $W$ is a disc or a rectangle. Mack (1949) extended the reasoning in the two-dimensional case to arbitrary scanning sets. The cases of discs and squares are also studied by Aldous (1989).

In the special case of $m = 2$ and a circular scanning set $W$, the number of $m$-subsets that are covered equals the number of pairs of points with interpoint distance less than the diameter of $W$. Convergence of this number to a Poisson limit has been discussed by Silverman and Brown (1978, 1979).

Silverman and Brown (1978) have given the number of close pairs as an example of the classical $U$-statistic. Poisson approximation for $U$-statistics and sums of so-called dissociated variables are treated by Barbour and Eagleson (1984) and Barbour, Holst, and Janson (1992). In these two references, the Stein–Chen method is used to bound the total variation distance for Poisson approximations for the sums.

### 10.3.2 Scan statistics

The *scan statistic* $S_w$ in one dimension is defined as the maximal number of points which can be found in some translate of an interval of a given length $w < T$, when there are $N$ points uniformly and independently distributed in $[0, T]$. If we let $S_{t,t+w}$ be the number of the points that lie in the interval $(t, t + w]$, then

$$S_w = \max_{0 \leq t \leq T-w} S_{t,t+w}.$$

In this case, there is no problem to generalize the definition to higher dimensions: fix the scanning set $W$, and let $S_W$ be the maximal number of points which are covered by some translate of $W$.

In this chapter, the scan statistic will not be considered. However, its relation to multiple scan statistics and $m$-spacings should be noted:

$$P(S_w \geq m) = P(Y_{(1)}^{(m-1)} \leq w) = P(\xi(1, N, m - 1, w) \geq 1),$$

$$P(S_W \geq m) = P(Z_{(1)}^{(m)} \leq \mu(W)) = P(\xi(d, N, m, W) \geq 1), \qquad d \geq 2.$$

In one dimension, various approximations for the distribution of the scan statistic can be found; see, for example, Alm (1983), Barbour, Holst, and Janson (1992), Gates and Westcott (1985), Glaz (1989), Janson (1984), and Naus

(1982). In two dimensions, the case of a rectangular scanning set has been treated by Loader (1991). General convex scanning sets in two and higher dimensions have been investigated by Alm (1997). Kulldorff and Nagarwalla (1995) and Kulldorff (1997) have studied scan statistics in two dimensions from a somewhat different perspective, mainly in an epidemiological context.

## 10.4    The Probability of Covering a Given $m$-Subset

Assume $X_1 \ldots, X_m$ are independently and uniformly distributed points in the rectangle $\mathbf{A} \subset \mathbb{R}^d$ and let $W \in \mathcal{K}$. In this section, we will derive the probability

$$P(\exists x \in \mathbf{A} : X_1, \ldots, X_m \in W(x)) \tag{10.10}$$

by means of results in integral geometry, and give some historical background.

First, the case of two points will be considered in a two-dimensional setting, so as to give some understanding of the formula for the probability (10.10), given in Theorem 10.4.1. Given the position of $X_1$, the probability that $X_1, X_2 \in W(x)$ for some $x \in \mathbf{A}$ equals the quotient of the area of all possible positions for $X_2$ such that both points are covered by some translate of $W$, and $\mu(\mathbf{A})$:

$$P(\exists x \in \mathbf{A} : X_1, X_2 \in W(x) \mid X_1 = y) = \frac{\mu(\{z : y, z \in W(x) \text{ for some } x \in \mathbf{A}\})}{\mu(\mathbf{A})}.$$

By the independence and uniform distribution of $X_1$ and $X_2$, and the torus convention, it does not matter where the first point $X_1$ lies, and we can without loss of generality let $X_1 = O$.

**Example 10.4.1** If $W \in \mathcal{K}$ is a disc of radius $r$, the possible positions for the second point constitute a disc of radius $2r$. Then,

$$P(\exists x \in \mathbf{A} : X_1, X_2 \in W(x)) = 4\pi r^2 / \mu(\mathbf{A}) = 4\mu(W)/\mu(\mathbf{A}).$$

**Example 10.4.2** Let $W \in \mathcal{K}$ be a triangle. Figure 10.1 shows $W$ and the possible positions for the second point, given the position of $X_1$. As seen in the figure, this area is six times as large as that of the original triangle. Hence,

$$P(\exists x \in \mathbf{A} : X_1, X_2 \in W(x)) = 6\mu(W)/\mu(\mathbf{A}).$$

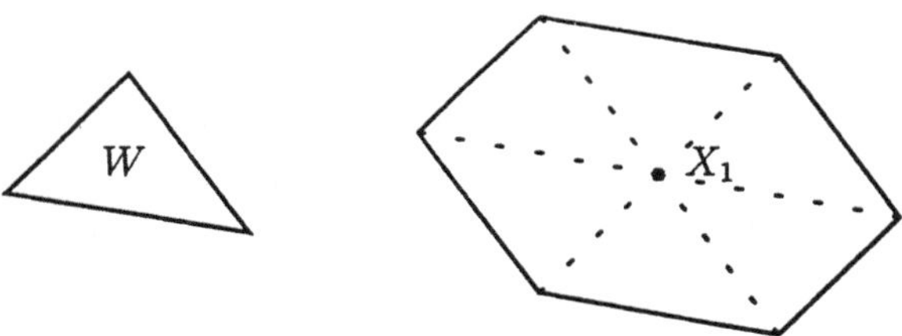


**Figure 10.1:** The possible area for the second point

By these examples (which will turn out to be extreme, see Corollary 10.4.1), we learn that it is not only the area of the sets which is of importance for the probability of covering the points.

To handle the case of a general $W \in \mathcal{K}$, we will view the problem in a different way. Recall that $\check{W} = \{-x : x \in W\}$ is the reflection of $W$ at the origin. First, note that for arbitrary points $x, y \in \mathbb{R}^d$ and $W \subset \mathbb{R}^d$,

$$x \in \check{W}(y) \Leftrightarrow x - y \in \check{W} \Leftrightarrow y - x \in W \Leftrightarrow y \in W(x).$$

From this equivalence, it follows that if $x_1, \ldots, x_m \in \mathbb{R}^d$, then

$$\exists x \in \mathbb{R}^d \text{ such that } x_1, \ldots, x_m \in W(x)$$

if and only if

$$\exists x \in \mathbb{R}^d \text{ such that } x \in \check{W}(x_i), \ i = 1, \ldots, m, \quad \text{i.e. } \cap_{i=1}^m \check{W}(x_i) \neq \emptyset.$$

This result implies that

$$P(\exists x \in \mathbf{A} : X_1, \ldots, X_m \in W(x)) = P(\cap_{i=1}^m W(X_i) \neq \emptyset)$$

when the torus convention is used. It is now easy to derive an expression for the probability in case of two points and a general convex set $W \in \mathcal{K}$. Since we may let $X_1 = O$, we are looking for the probability

$$P(W \cap W(X_2) \neq \emptyset) = P(X_2 \in \{x : W \cap W(x) \neq \emptyset\}).$$

Now $\{x : W \cap W(x) \neq \emptyset\}$ equals $W + \check{W}$ by (10.2), and thus

$$
\begin{aligned}
P(\exists x \in \mathbf{A} : X_1, X_2 \in W(x)) &= P(W \cap W(X_2) \neq \emptyset) \\
&= \mu(W + \check{W})/\mu(\mathbf{A}) \\
&= 2\left(\mu(W) + \nu(W, \check{W})\right)/\mu(\mathbf{A}),
\end{aligned}
$$

where the last equality follows from (10.3). What determines the probability is hence the area and the mixed area of the set, where the latter is dependent on the shape of the set. This carries over to the case of more than two points, as can be seen in Theorem 10.4.1.

In case of an arbitrary number of points, no proper proof of Theorem 10.4.1 will be given, but mainly references to previous results in integral geometry.

As already mentioned, a consequence of using the torus convention is that we may assume that the first point lies at the origin, so that the probability we aim at is

$$P(\exists x \in \mathbf{A} : O, X_2, \ldots, X_m \in W(x)) = P(W \cap W(X_2) \cap \ldots, \cap W(X_m) \neq \emptyset).$$

Because of the assumptions that the origin lies in the center of $\mathbf{A}$ and is an interior point of $W$, and of the restrictions on the sizes of $W$, $W \in \mathcal{K}$, all vectors $(x_2, \ldots, x_m)$ such that

$$W \cap W(x_2) \cap \ldots \cap W(x_m) \neq \emptyset$$

satisfy $x_i \in \mathbf{A}$, $i = 2, \ldots, m$, without using the torus convention. Hence, once the assumption that the first point lies at the origin is made, we may treat $\mathbf{A}$ as a "normal" $d$-dimensional rectangle.

Since $X_2, \ldots, X_m$ are independently and uniformly distributed in $\mathbf{A}$, the vector $(X_2, \ldots, X_m)$ is uniformly distributed in the product space $\mathbf{A}^{m-1}$. Let $\mu^{m-1}$ denote the $(m-1)$-fold product measure of $\mu$, the $d$-dimensional Lebesgue measure. Now $\mu^{m-1}(\mathbf{A}^{m-1}) = \mu(\mathbf{A})^{m-1}$, and

$$P(W \cap W(X_2) \cap \ldots \cap W(X_m) \neq \emptyset)$$
$$= \mu^{m-1}\{(x_2, \ldots, x_m) : x_i \in \mathbb{R}^d, W \cap W(x_2) \cap \ldots \cap W(x_m) \neq \emptyset\}/\mu(\mathbf{A})^{m-1}.$$

To see the connection with integral geometry, we write

$$\mu^{m-1}\{(x_2, \ldots, x_m) : x_i \in \mathbb{R}^d, W \cap W(x_2) \cap \ldots \cap W(x_m) \neq \emptyset\}$$
$$= \int_{\mathbb{R}^d} \cdots \int_{\mathbb{R}^d} V_0(W \cap W(x_2) \cap \ldots \cap W(x_m)) dx_2 \cdots dx_m, \quad (10.11)$$

where

$$V_0(C) = \begin{cases} 1 & \text{if } C \neq \emptyset, \\ 0 & \text{if } C = \emptyset. \end{cases}$$

The functional $V_0(C)$ is one of the so-called *intrinsic volumes* of $C$, $V_i(C)$, $i = 0, \ldots, d$, defined for compact, convex subsets of $\mathbb{R}^d$.

As early as 1937, explicit expressions for (10.11) were given for $m = 2, 3$ in two and three dimensions. Blaschke (1937) discussed both dimensions while Berwald and Varga (1937) handled three dimensions. In a probabilistic context, the planar case including an iterated version (i.e., for an arbitrary number of sets) was rediscovered by Miles (1974), and the two- and three-dimensional cases by Månsson (1996).

In an arbitrary dimension and for general $V_i$, the integral in (10.11) is handled in Weil (1990), where it is a special case of an even more general situation. Hence, the formula for $P(\cap_{i=1}^m W(X_i) \neq \emptyset)$ follows directly in an arbitrary dimension. However, in higher dimensions, the formulas involve complicated functionals for which explicit descriptions are known only in special cases. Since the setting in this chapter is mainly two dimensional, we present the probability only in this case here.

**Theorem 10.4.1** *Suppose* $W \in \mathcal{K}$, *and that* $X_i$, $i = 1, \ldots, m$, *are independently and uniformly distributed points in* $\mathbf{A} \subset \mathbb{R}^2$, *and* $m = 2, 3, \ldots$. *Then, using the torus convention,*

$$
\begin{aligned}
P(\exists x \in \mathbf{A} : X_1, \ldots, X_m \in W(x)) &= P(\cap_{i=1}^m W(X_i) \neq \emptyset) \\
&= \left( m + m(m-1) \frac{\nu(\breve{W}, W)}{\mu(W)} \right) \frac{\mu(W)^{m-1}}{\mu(\mathbf{A})^{m-1}}.
\end{aligned}
$$

The following corollary follows directly from Theorem 10.4.1 and (10.4).

**Corollary 10.4.1** *Under the assumptions of Theorem 10.4.1,*

$$
m^2 \frac{\mu(W)^{m-1}}{\mu(\mathbf{A})^{m-1}} \leq P(\exists x \in \mathbf{A} : X_1, \ldots, X_m \in W(x)) \leq (2m^2 - m) \frac{\mu(W)^{m-1}}{\mu(\mathbf{A})^{m-1}},
$$

*where there is equality on the left if and only if* $W$ *is centrally symmetric and on the right if and only if* $W$ *is a triangle.*

It should be noted that, in fact, no more effort is needed to handle $P(\cap_{i=1}^m W_i(X_i) \neq \emptyset)$, where the sets $W_i \in \mathcal{K}$ can be of different shape, than to derive the probability in Theorem 10.4.1. The result in this generalized case looks as follows:

**Theorem 10.4.2** *Suppose* $W_i \in \mathcal{K}$, $i = 1, \ldots, m$, *and that* $X_i$, $i = 1, \ldots, m$, *are independently and uniformly distributed points in* $\mathbf{A} \subset \mathbb{R}^2$, *and* $m = 2, 3, \ldots$. *Then, using the torus convention,*

$$
\begin{aligned}
&P(\cap_{i=1}^m W_i(X_i) \neq \emptyset) \\
&= \left( \sum_{\substack{i=1 \\ }}^m \prod_{\substack{j=1 \\ j \neq i}}^m \mu(W_j) + \sum_{\substack{i,j=1 \\ i \neq j}}^m \nu(W_i, \breve{W}_j) \prod_{\substack{l=1 \\ l \neq i,j}}^m \mu(W_l) \right) \frac{1}{\mu(\mathbf{A})^{m-1}}.
\end{aligned}
$$

When not all $W_i$'s are equal, we cannot reformulate the event $\cap_{i=1}^m W_i(X_i) \neq \emptyset$ in terms of covering points, and this case is hence not of any primary interest here.

The discussions in this chapter will be carried on in terms of covering points. But we shall bear in mind that when reading, for instance, "$m$ uniformly distributed points are covered by $W$," we equally well can read "$m$ uniformly translated copies of $\check{W}$ have a nonempty intersection."

---

## 10.5   Poisson Approximation

In this section, we will examine approximation of the multiple scan statistic in two dimensions, $\xi = \xi(2, N, m, W)$, by a Poisson variable with parameter $E[\xi]$. This will be done in terms of the total variation distance in Subsection 10.5.1, and by means of simulations in Subsection 10.5.2. Without loss of generality, we will henceforth assume that $\mu(\mathbf{A}) = 1$.

First, we need to find the expectation of $\xi$ which will be denoted by $\lambda$. This is easily done by results in the previous section. Recall from the introduction that $\xi$ is the number of $m$-subsets which are covered by some translate of $W$, which can be written as

$$\xi = \sum_{i=1}^{\binom{N}{m}} I_i, \tag{10.12}$$

where

$$I_i = \begin{cases} 1 & \text{if there exists } x \in \mathbf{A} \text{ such that the } i\text{th } m\text{-subset is covered by } W(x), \\ 0 & \text{otherwise.} \end{cases}$$

From Theorem 10.4.1, we know that

$$\begin{aligned} E[I_i] &= P(\exists x \in \mathbf{A} : X_1, \ldots, X_m \in W(x)) \\ &= \left( m + m(m-1)\frac{\nu(\check{W}, W)}{\mu(W)} \right) \mu(W)^{m-1}, \end{aligned} \tag{10.13}$$

$i = 1, \ldots, \binom{N}{m}$, and hence

$$\lambda = E[\xi] = \binom{N}{m} \left( m + m(m-1)\frac{\nu(\check{W}, W)}{\mu(W)} \right) \mu(W)^{m-1}. \tag{10.14}$$

The suggested approximation is thus

$$P(\xi = l) \approx e^{-\lambda}\lambda^l/l!, \tag{10.15}$$

where $\lambda$ is given in (10.14).

### 10.5.1 Total variation distance

We will now theoretically examine the approximation suggested in (10.15). This will be done in terms of total variation distance, which has been defined in (10.5).

$\xi$ is a sum of indicators, where those pairs of indicators which concern $m$-subsets with common points are dependent while those with no points in common are independent. In a situation such as this, the local approach of the Stein–Chen method is a suitable mean to get a bound on the total variation distance between the distribution of $\xi$ and a Poisson variable with parameter $\lambda$.

Before applying the Stein–Chen method to the problem at hand, it will be stated in general terms. Following the notation of Barbour, Holst, and Janson (1992), let $\Gamma$ be an arbitrary finite collection of indices and let

$$W = \sum_{\alpha \in \Gamma} I_\alpha \quad \text{and} \quad \lambda = E[W],$$

where $I_\alpha$, $\alpha \in \Gamma$, are, possibly dependent, indicator variables. For each $\alpha \in \Gamma$, let $\Gamma \backslash \{\alpha\}$ be divided into two subsets; one consisting of those $\beta \in \Gamma$ for which $I_\beta$ is weakly dependent on $I_\alpha$, and the other consisting of the indices of the indicators which are strongly dependent on $I_\alpha$. Denote these subsets by $\Gamma_\alpha^w$ and $\Gamma_\alpha^s$, respectively, and let

$$Z_\alpha = \sum_{\beta \in \Gamma_\alpha^s} I_\beta.$$

The local approach of the Stein–Chen method is suitable to use when there is a natural dependence structure which allows every pair of indicators to be classified in this way.

**Theorem 10.5.1** *[Theorem 1.A of Barbour, Holst, and Janson (1992)]. Let $\Gamma$ be an arbitrary finite collection of indices. With the above definitions, for any choice of the index sets $\Gamma_\alpha^s$, $\alpha \in \Gamma$,*

$$d_{TV}(\mathcal{L}(W), Po(\lambda)) \leq \sum_{\alpha \in \Gamma} (E[I_\alpha]^2 + E[I_\alpha]E[Z_\alpha] + E[I_\alpha Z_\alpha])\lambda^{-1}(1 - e^{-\lambda})$$

$$+ \sum_{\alpha \in \Gamma} \eta_\alpha \min(1, \lambda^{-1/2}), \tag{10.16}$$

*where $\eta_\alpha = E\left[|E[I_\alpha \mid (I_\beta, \beta \in \Gamma_\alpha^w)] - E[I_\alpha]|\right]$.*

In our case, $\Gamma = \{1, \ldots, \binom{N}{m}\}$, and a suitable choice of $\Gamma_i^s$, $i \in \Gamma$, is

$$\Gamma_i^s = \cup_{l=1}^{m-1} \Gamma_{i,l}^s,$$

where

$$\Gamma_{i,l}^s = \{j \neq i : \text{the } i\text{th and } j\text{th } m\text{-subsets have } l \text{ common points}\}$$

are disjoint. Then, $I_i$ and $I_j$ are independent if $j \in \Gamma_i^w$, and hence $\eta_i = 0$, so that the last sum in (10.16) is zero. Furthermore, note that the number of indices in $\Gamma_{i,l}^s$, i.e., the number of $m$-subsets with $l$ points in common with the $i$th $m$-subset, is $\binom{m}{l}\binom{N-m}{m-l}$. Now, for each $l = 1, \ldots, m-1$ choose any of the indicators pertaining to an $m$-subset that has $l$ particles in common with the 1st $m$-subset, and denote it by $I_1^l$. Månsson (1996) has shown that

$$
E[I_i]^2 + E[I_i]E[Z_i] = E[I_1]^2 \left(1 + \sum_{l=1}^{m-1} \binom{m}{1}\binom{N-m}{m-1}\right)
$$

$$
\leq E[I-1]^2 \binom{N}{m} m^2/N,
$$

and

$$
E[I_i Z_i] = \sum_{l=1}^{m-1} \binom{m}{l}\binom{N-m}{m-l} E[I_1]E[I_1^l \mid I_1 = 1]
$$

$$
\leq \sum_{l=1}^{m-1} \binom{m}{l}\binom{N-m}{m-l} E[I_1]\mu(\check{W} + W)^{m-l},
$$

for all $i = 1, \ldots, \binom{N}{m}$. Inserting these estimates in the bound in Theorem 10.5.1 yields the following result.

**Theorem 10.5.2** *Let $\xi$ and $\lambda$ be defined by (10.12) and (10.14), respectively. Then,*

$$
d_{TV}(\mathcal{L}(\xi), \mathrm{Po}(\lambda)) \leq \left\{\frac{\lambda m^2}{N} + \sum_{l=1}^{m-1} \binom{m}{l}\binom{N-m}{m-l}\mu(\check{W} + W)^{m-l}\right\}(1 - e^{-\lambda}).
$$

It should be noted that it is $E[I_i Z_i]$, $i = 1, \ldots, \binom{N}{m}$, and hence the sum in the bound in Theorem 10.5.2, which handles the dependence between the indicators. We will see later that this sum, or to be more precise, the term which concerns $m$-subsets with $m - 1$ common points ($l = m - 1$), is the critical part in the approximation.

We will now consider how the bound in Theorem 10.5.2 behaves as $N \to \infty$ for sequences of sets of decreasing area. Let $\{W_N\}_{N=1}^{\infty}$ be a sequence of sets in $\mathcal{K}$, $\{\xi_N\}_{N=1}^{\infty}$ the corresponding multiple scan statistics, and let $\lambda_N = E[\xi_N]$. To get a bound on the variation distance valid for all shapes of the set $W_N$ we use that

$$
\mu(\check{W}_N + W_N) \leq 6\mu(W_N)
$$

by (10.3) and (10.4). Furthermore, the expectation of $\xi_N$, given in (10.14), can be bounded by

$$
\lambda_N \leq \frac{N^m (2m^2 - m)\mu(W_N)^{m-1}}{m!}
$$

by Corollary 10.4.1. These bounds, together with $(1 - e^{-\lambda_N}) \leq \min(1, \lambda_N)$, inserted in Theorem 10.5.2, yield the following result.

**Theorem 10.5.3**

(i) *For any sequence of sets,* $\{W_N\}_{N=1}^{\infty}$, $W_N \in \mathcal{K}$,

$$d_{TV}(\mathcal{L}(\xi_N), Po(\lambda_N)) = O\left(\min\{1, N^m \mu(W_N)^{m-1}\} \sum_{l=1}^{m-1} (N\mu(W_N))^{m-l}\right).$$

(ii) *For a sequence of sets* $\{W_N\}_{N=1}^{\infty}$, $W_N \in \mathcal{K}$, *with* $\mu(W_N) = O(N^{-t})$, *where* $t > 1$ *is constant, the bound of the total variation distance tends to zero and is of the order*

$$d_{TV}(\mathcal{L}(\xi_N), Po(\lambda_N)) = \begin{cases} O(N^{1-t}) & \text{if } 1 < t \leq m/(m-1), \\ O(N^{m(1-t)+1}) & \text{if } t \geq m/(m-1). \end{cases}$$

**Remark 10.5.1** Note that it is the term in the sum in Theorem 10.5.2 for which $l = m - 1$ which is critical and determines the rates in the second part of the theorem above. This term concerns the dependence between two indicators connected to $m$-subsets with $m - 1$ common points. Hence it is, as expected, the tendency of clumping which is troublesome in the Poisson approximation.

**Remark 10.5.2** The bound on the total variation distance tends to zero for sequences $\{W_N\}_{N=1}^{\infty}$ such that $\mu(W_N)N \to 0$. By (10.14), this condition is equivalent to $\lambda_N N^{-1} \to 0$.

In Theorem 10.5.3 (ii), we can see that the rate of convergence is changed at the value $t = m/(m-1)$ if $\mu(W_N) = O(N^{-t})$. We will now consider how $\lambda_N$ behaves in the special case where $\mu(W_N) = cN^{-t}$, $c > 0$ and $t \in \mathbb{R}$, which further illustrates the special role of $t = m/(m-1)$. Then, the expected number of $m$-subsets which are covered by some translate of $W_N$ is

$$\lambda_N = \binom{N}{m}\left(m + m(m-1)\frac{\nu(W_N, \check{W}_N)}{\mu(W_N)}\right)c^{m-1}N^{-t(m-1)}. \tag{10.17}$$

The limit of this expectation depends on the value of $t$:

$$\lambda_N \to \begin{cases} \infty & \text{if } t < m/(m-1) \\ 0 & \text{if } t > m/(m-1) \end{cases} \quad \text{as } N \to \infty, \tag{10.18}$$

and if $t = m/(m-1)$

$$\frac{m^2 c^{m-1}}{m!} \leq \liminf_{n\to\infty} \lambda_N \leq \limsup_{n\to\infty} \lambda_N \leq \frac{(2m^2 - m)c^{m-1}}{m!},$$

where the bounds follow by Corollary 10.4.1. If $W_N$ are of the same shape for all $N$, then $\nu(W_N, \check{W}_N)/\mu(W_N) = a$ for some $1 \leq a \leq 2$, and

$$\lambda_N \to (m + m(m-1)a)c^{m-1}/m! \qquad \text{as } N \to \infty$$

if $t = m/(m-1)$.

Of particular interest is when $\lambda_N$ is held constant as $N \to \infty$. This means that $\mu(W_N) = O(N^{-m/(m-1)})$, and the following corollary follows immediately from Theorem 10.5.3 *(ii)*.

**Corollary 10.5.1** *Let $\{W_N\}_{N=1}^{\infty}$ be a sequence of sets such that $W_N \in \mathcal{K}$ and assume that $E[\xi_N] = \lambda$ for all $N$ where $0 < \lambda < \infty$. If $\mathcal{L}(Z) = \mathrm{Poisson}(\lambda)$, then*

$$\xi_N \xrightarrow{\mathcal{D}} Z \quad as \quad N \to \infty$$

*at rate $O(N^{-1/(m-1)})$.*

**Remark 10.5.3** This result can be compared with the Poisson approximation for $m$-spacings in one dimension which is carried out by means of the Stein–Chen method in Chapter 7 of Barbour, Holst, and Janson (1992). There, it is shown that the distribution of the number of "small" $m$-spacings converges to a Poisson distribution at the rate $O(N^{-1/m})$ as $N \to \infty$, when the expectation is held constant. Recall that in Section 10.1, it was argued for that $m$ in the context of $m$-spacings in one dimension corresponds to $m-1$ in our case. Hence, the bounds in these two cases are of the same order.

This subsection is concluded by an application of Theorem 10.5.3. Let $S_{W_N}$ be the maximal number of $N$ independent and uniformly distributed points on **A** which are covered by $W_N(x)$ for some $x \in \mathbf{A}$, i.e., the scan statistic. The following theorem shows that the asymptotic distribution of $S_{W_N}$ for sequences of sets decreasing at a certain rate, is either concentrated in one or in two values.

**Theorem 10.5.4** *Suppose that $m \in \{2, 3, \ldots\}$, $c > 0$, and let $\{W_N\}_1^{\infty}$ be a sequence of sets in $\mathcal{K}$ with $\mu(W_N) = cN^{-t}$. If $t = m/(m-1)$ and $\lambda_N \to \lambda$ as $N \to \infty$, then*

$$P(S_{W_N} = i) \to \begin{cases} e^{-\lambda} & i = m-1 \\ 1 - e^{-\lambda} & i = m \\ 0 & i \neq m-1, m \end{cases} \qquad as \ N \to \infty.$$

*If $(m+1)/m < t < m/(m-1)$, then*

$$P(S_{W_N} = i) \to \begin{cases} 1 & i = m \\ 0 & i \neq m \end{cases} \qquad as \ N \to \infty.$$

**Remark 10.5.4** Note that the condition $\lambda_N \to \lambda$ is equivalent to that $\nu(\check{W}_N, W_N)/\mu(W_N) \to a$ for some $1 \leq a \leq 2$. This condition is satisfied, for instance, if $W_N$ are of the same shape for all $N$.

A proof of this theorem has been given by Månsson (1996).

## 10.5.2  Simulations

In this subsection, the accuracy of the approximation which was proposed in (10.15) is evaluated by simulations. Here $\mathbf{A}$ is a square, and as earlier, we let $\mu(\mathbf{A}) = 1$.

Recall that the suggested approximation of $P(\xi \leq l)$ is

$$\sum_{i=0}^{l} e^{-\lambda}\lambda^i/i!, \tag{10.19}$$

where $\lambda$, given in (10.14), depends on $N$, $m$, $\mu(W)$ and on the functional $\nu(W, \check{W})$, which is determined by the shape of $W$. In Table 10.1, the empirical distribution of $P(\xi \leq l)$ is compared to (10.19) in the case where $W$ is a square of area $s^2$. In this case,

$$\lambda = \binom{N}{m} m^2 s^{2(m-1)}$$

by Corollary 10.4.1. In Table 10.2, the same comparison is made, but with an equilateral triangle $W$ with area $t^2/2$, which gives

$$\lambda = \binom{N}{m} (2m^2 - m)(t^2/2)^{m-1}$$

again by Corollary 10.4.1.

As can be seen from the tables, the approximations perform well in the examples chosen here when $m = 2$, but as $m$ increases the approximations rapidly become poor. The reason is that the covered $m$-subsets tend to occur in clumps, and the larger $m$ gets, the worse this problem gets. If, for instance, $m + k$ points are covered by $W(x)$, then $\binom{m+k}{m}$ $m$-subsets are covered by $W(x)$. This was also discussed in Remark 10.5.1, and in Corollary 10.5.1 it was shown that the total variation distance tends to zero at the rate $O(N^{-1/(m-1)})$ as $N \to \infty$. It can also be seen in the tables that when $m$ gets large, there is a tendency for the empirical distribution of $\xi$ to get more spread out than that of the approximating Poisson distribution. This also agrees with the conclusion that an approximating distribution allowing for clumps is more appropriate to use. A natural candidate is a suitably chosen compound Poisson distribution. This is further discussed in Section 10.7.

Marianne Månsson

**Table 10.1:** Comparison of Poisson approximations to $P(\xi \le l)$ with simulated values when $W$ is a square of area $s^2$

| $N$ | $m$ | $s$ | $l$ | empir. | (10.19) | $N$ | $m$ | $s$ | $l$ | empir. | (10.19) |
|---|---|---|---|---|---|---|---|---|---|---|---|
| 100 | 2 | .01 | 0 | .1364 | .1381 | 1000 | 2 | .002 | 0 | .0003 | .0003 |
| | | | 1 | .4115 | .4114 | | | | 5 | .1908 | .1920 |
| | | | 2 | .6847 | .6821 | | | | 10 | .8161 | .8167 |
| | | | 3 | .8626 | .8607 | | | | 15 | .9917 | .9918 |
| | | | 4 | .9492 | .9491 | | | | 20 | .9999 | .9999 |
| | | | 5 | .9837 | .9841 | 1000 | 2 | .005 | 30 | .0015 | 0016 |
| | | | 6 | .9953 | .9957 | | | | 40 | .0858 | .0872 |
| 100 | 2 | .05 | 30 | .0011 | .0020 | | | | 50 | .5415 | .5403 |
| | | | 40 | .0916 | .0974 | | | | 60 | .9285 | .9288 |
| | | | 50 | .5764 | .5657 | | | | 70 | .9970 | .9971 |
| | | | 60 | .9348 | .9374 | 1000 | 3 | .005 | 0 | .4131 | .3927 |
| | | | 70 | .9960 | .9976 | | | | 1 | .7668 | .7600 |
| 100 | 3 | .03 | 0 | .3680 | .3076 | | | | 2 | .9258 | .9313 |
| | | | 1 | .7050 | .6703 | | | | 3 | .9749 | .9848 |
| | | | 2 | .8758 | .8841 | | | | 4 | .9910 | .9972 |
| | | | 3 | .9376 | .9680 | | | | 5 | .9969 | .9996 |
| | | | 4 | .9685 | .9928 | 1000 | 3 | .01 | 5 | .0105 | .0029 |
| | | | 5 | .9852 | .9986 | | | | 10 | .1879 | .1207 |
| 100 | 3 | .05 | 0 | .0016 | .0001 | | | | 15 | .5842 | .5727 |
| | | | 5 | .2416 | .1100 | | | | 20 | .8652 | .9189 |
| | | | 10 | .6848 | .6946 | | | | 25 | .9678 | .9940 |
| | | | 15 | .9000 | .9760 | | | | 30 | .9934 | .9998 |
| | | | 20 | .9701 | .9995 | 1000 | 4 | .01 | 0 | .5776 | .5155 |
| 100 | 4 | .05 | 0 | .5300 | .3752 | | | | 1 | .8662 | .8570 |
| | | | 1 | .8034 | .7430 | | | | 2 | .9556 | .9702 |
| | | | 2 | .9076 | .9233 | | | | 3 | .9774 | .9952 |
| | | | 3 | .9400 | .9821 | | | | 4 | .9819 | .9994 |
| | | | 4 | .9491 | .9995 | | | | 5 | .9910 | .9999 |
| 100 | 4 | .08 | 5 | .1292 | .0010 | 1000 | 4 | .015 | 0 | .0067 | .0005 |
| | | | 10 | .3827 | .0634 | | | | 3 | .1831 | .0573 |
| | | | 15 | .6000 | .4231 | | | | 6 | .5059 | .3716 |
| | | | 20 | .7425 | .8418 | | | | 9 | .7376 | .7709 |
| | | | 25 | .8338 | .9822 | | | | 12 | .8695 | .9555 |
| | | | 30 | .8913 | .9991 | | | | 15 | .9333 | .9951 |

**Table 10.2:** Comparison of Poisson approximations to $P(\xi \leq l)$ with simulated values when $W$ is a triangle of area $t^2/2$

| $N$ | $m$ | $t$ | $l$ | empir. | (10.19) | $N$ | $m$ | $t$ | $l$ | empir. | (10.19) |
|---|---|---|---|---|---|---|---|---|---|---|---|
| 100 | 2 | .01 | 0 | .2241 | .2265 | 1000 | 2 | .005 | 20 | .0012 | .0013 |
| | | | 1 | .5637 | .5629 | | | | 30 | .1247 | .1256 |
| | | | 2 | .8143 | .8126 | | | | 40 | .6984 | .6973 |
| | | | 3 | .9364 | .9362 | | | | 50 | .9795 | .9798 |
| | | | 4 | .9815 | .9821 | | | | 60 | .9997 | .9997 |
| | | | 5 | .9953 | .9958 | 1000 | 3 | .01 | 0 | .0042 | .0020 |
| 100 | 3 | .05 | 0 | .0525 | .0226 | | | | 3 | .1766 | .1317 |
| | | | 2 | .3748 | .2705 | | | | 6 | .5873 | .5692 |
| | | | 4 | .6854 | .6698 | | | | 9 | .8668 | .8992 |
| | | | 6 | .8582 | .9101 | | | | 12 | .9666 | .9883 |
| | | | 8 | .9382 | .9843 | | | | 15 | .9924 | .9992 |
| | | | 10 | .9720 | .9981 | 1000 | 4 | .02 | 0 | .0023 | .0000 |
| 100 | 4 | .07 | 0 | .3736 | .1990 | | | | 5 | .2609 | .1000 |
| | | | 2 | .8117 | .7796 | | | | 10 | .6769 | .6726 |
| | | | 4 | .9077 | .9755 | | | | 15 | .8833 | .9721 |
| | | | 6 | .9588 | .9986 | | | | 20 | .9565 | .9994 |
| | | | 8 | .9786 | 1.0000 | | | | 25 | .9829 | 1.0000 |

## 10.6 Point Processes

The purpose of this section is to introduce three point processes determined by positions and sizes of the $m$-subsets which are covered by some $W(x)$, and consider approximation of these processes by Poisson processes. The accuracy of the approximations will be examined by means of bounds on the total variation distance between the processes. It is worth noting that the same bounds are valid also for the total variation distance between the distribution of any functional of the approximated process and of its corresponding Poisson process. For a more detailed exposition than the one given below, the reader is referred to Månsson (1997).

The point processes are defined as follows. Let the leftmost points (the lowest of these in case of ambiguity) in the $m$-subsets which actually are covered by some translate of $W$ constitute the points of the point process $\Xi_\mathbf{A}$ on $\mathbf{A}$. Recall from (10.8) that the size of the $i$th $m$-subset $Z_i^{(m)}$, is defined to be $T_i^2 \mu(W)$, where $T_i$ is the smallest real number such that $T_i W(x)$ covers the $i$th $m$-subset, for some $x \in \mathbf{A}$. If the normalized sizes $Z_i^{(m)}/\mu(W)$ are attached to the points of $\Xi_\mathbf{A}$, we get a point process on the space $\mathbf{A} \times [0,1]$, which we denote by $\Xi$. These sizes are identically, but not independently, distributed with distribution function

$$F(y) \;=\; P(\exists x \in \mathbf{A} : X_1,\ldots,X_m \in yW(x) \mid \exists x \in \mathbf{A} : X_1,\ldots,X_m \in W(x))$$

$$= \frac{\left(m + m(m-1)\nu(y\breve{W},yW)/\mu(yW)\right)\mu(yW)^{m-1}}{\left(m + m(m-1)\nu(\breve{W},W)/\mu(W)\right)\mu(W)^{m-1}}$$

$$= y^{2(m-1)}, \tag{10.20}$$

$0 \leq y \leq 1$, by Theorem 10.4.1 and since $\nu(y\breve{W},yW) = y^2\nu(\breve{W},W)$ and $\mu(yW) = y^2\mu(W)$. If we drop the positions and just consider the sizes, the result is a point process on $[0,1]$, which we denote by $\Xi_{[0,1]}$.

All these three processes can be written as

$$\sum_{i=1}^{\binom{N}{m}} I_i\delta_{Y_i},$$

where $\delta_y$ denotes the unit mass at $y$, and the state space for $\{Y_i\}$, denoted by $\mathcal{Y}$, is $\mathbf{A}$, $[0,1]$ or $\mathbf{A} \times [0,1]$. We will first consider $\Xi$, for which $\mathcal{Y} = \mathbf{A} \times [0,1]$, and we need to derive the measure on $\mathcal{Y}$ defined by

$$\lambda(A) = \sum_{i=1}^{\binom{N}{m}} \lambda_i(A),$$

where

$$\lambda_i(A) = P(I_i = 1, Y_i \in A),$$

$i = 1,\ldots,\binom{N}{m}$. The position of an $m$-subset which is covered by some translate of $W$, i.e., its leftmost point, is uniformly distributed in $\mathbf{A}$, since the $m$ points themselves are uniformly and independently distributed in $\mathbf{A}$, and the torus convention is used. The size of an $m$-subset which is covered takes its value in $[0,1]$ and has density function

$$2(m-1)y^{2m-3}$$

by (10.20). Furthermore, the size is independent of the position of the $m$-subset. With $p = E[I_1]$ and $\lambda$ as defined in (10.13) and (10.14), respectively, the measure $\lambda_i$ is thus given by

$$d\lambda_i(x,y) = p\frac{1}{\mu(\mathbf{A})}2(m-1)y^{2m-3}dx\,dy = p\,2(m-1)y^{2m-3}dx\,dy, \tag{10.21}$$

$i = 1,\ldots,\binom{N}{m}$, and we get

$$\begin{aligned} d\lambda(x,y) &= \sum_{i=1}^{\binom{N}{m}} d\lambda_i(x,y) = \binom{N}{m}p\,2(m-1)y^{2m-3}dx\,dy \\ &= \lambda 2(m-1)y^{2m-3}dx\,dy. \end{aligned} \tag{10.22}$$

In the theorem below, a bound on the total variation distance between $\mathcal{L}(\Xi)$ and a Poisson process with intensity $\lambda$ is given.

**Theorem 10.6.1** *Let* $\Xi = \sum_{i=1}^{\binom{N}{m}} I_i \delta_{Y_i}$ *be the point process on* $\mathbf{A} \times [0,1]$ *defined above and let* $\lambda$ *be given by (10.22). Then,*

$$d_{TV}(\mathcal{L}(\Xi), \text{Poisson}(\lambda)) \;\leq\; \frac{\lambda^2 m^2}{N} + \lambda \sum_{l=1}^{m-1} \binom{m}{l}\binom{N-m}{m-l} \mu(\check{W} + W)^{m-l}.$$

**Remark 10.6.1** The bound in the theorem above equals the bound on the distance between $\mathcal{L}(\xi)$ and a Poisson variable with parameter $\lambda$ given in Theorem 10.5.2 if $\lambda \leq 1$. If $\lambda > 1$, the bound is unfortunately not as good in the process case.

**Remark 10.6.2** Since $\Xi_{\mathbf{A}}$ and $\Xi_{[0,1]}$ and the corresponding Poisson processes are obtained as measurable mappings from $\Xi$ and $\text{Poisson}(\lambda)$, respectively, it follows that

$$d_{TV}(\mathcal{L}(\Xi_{\mathcal{Y}}), \text{Poisson}(\lambda_{\mathcal{Y}})) \leq d_{TV}(\mathcal{L}(\Xi), \text{Poisson}(\lambda)),$$

where $\mathcal{Y} = \mathbf{A}$ or $[0,1]$, and $\lambda_{\mathcal{Y}}$ is the measure corresponding to (10.22). Hence, the bound in Theorem 10.6.1 holds also when these processes are concerned.

A proof of Theorem 10.6.1 has been given by Månsson (1997).

Let $\{\Xi_N\}_{N=1}^{\infty}$ and $\{\lambda_N\}_{N=1}^{\infty}$ be the sequences of processes and measures, respectively, introduced above, which correspond to the sequence of sets $\{W_N\}_{N=1}^{\infty}$. As noted in Remark 10.6.1, the bound on $d_{TV}(\mathcal{L}(\Xi_N), \text{Poisson}(\lambda_N))$ in Theorem 10.6.1 equals the bound on $d_{TV}(\mathcal{L}(\xi_N), \text{Poisson}(\lambda_N))$ in Theorem 10.5.2 if $\lambda_N \leq 1$, and is not as good when $\lambda_N > 1$. To obtain convergence of $d_{TV}(\mathcal{L}(\Xi_N), \text{Poisson}(\lambda_N))$ to zero in the latter case, we can therefore not allow the areas of the sets to decrease as slowly as in the previous case of sequences of variables, and the counterpart of Theorem 10.5.3 reads as follows.

**Theorem 10.6.2** *Let* $\Xi_N$ *and* $\lambda_N$ *be as defined in Theorem 10.6.1.*

*(i) For any sequence of sets* $\{W_N\}_{N=1}^{\infty}$, $W_N \in \mathcal{K}$,

$$d_{TV}(\mathcal{L}(\Xi_N), \text{Poisson}(\lambda_N)) = O\Big(\sum_{l=1}^{m-1} N^{2m-l} \mu(W_N)^{2m-l-1}\Big).$$

*(ii) For a sequence of sets* $\{W_N\}_{N=1}^{\infty}$, $W_N \in \mathcal{K}$, *with* $\mu(W_N) = O(N^{-t})$, *where* $t > (m+1)/m$ *is constant, this bound tends to zero and is of the order*

$$d_{TV}(\mathcal{L}(\Xi_N), \text{Poisson}(\lambda_N)) \;=\; O(N^{m(1-t)+1}).$$

Note that the condition for convergence in the variable case is that $\mu(W_N)N \to 0$, which does not depend on $m$. Here, the condition is $\mu(W_N)N^{(m+1)/m} \to 0$; the smaller the $m$, the faster must the areas decrease.

## 10.7  Miscellaneous

### 10.7.1  Higher dimensions

A natural extension of this chapter is to generalize the results to higher dimensions. The approximations make use of the probability of covering a number of independently and uniformly distributed points with some translate of a convex set. In three dimensions, this probability can be found in Månsson (1996) and, as shown in Section 10.4, the probability in an arbitrary dimension can be obtained directly from results in Weil (1990). It should be straightforward, but tedious, to extend the approximation results to an arbitrary dimension.

### 10.7.2  The unconditional case

In this chapter, we have studied the conditional case, i.e., when $N$, the total number of points, is fixed. In the unconditional case, the number of points in $\mathbf{A}$ is Poisson distributed rather than being fixed. In that case $\lambda$, i.e., the expected number of covered $m$-subsets, becomes slightly different: in the definition of $\lambda$ given in (10.14), $\binom{N}{m}$ should be replaced by $\theta^m/m!$, if the expected total number of points in $\mathbf{A}$ is $\theta$.

To derive results in the unconditional case, corresponding to these presented here, should be easy, and it would be surprising if it asymptotically would be any different.

### 10.7.3  The torus convention

In this chapter, the torus convention has been used throughout. The reason for this is computational convenience: it is then possible to calculate the exact parameters and easier to derive bounds on the total variation distances. However, since this convention does not seem very natural in applications, it is desirable to find approximations also when the convention is not used. For simple sets such as rectangles, it should not be difficult to find the exact parameter, and a bound on the total variation distance can be found by adding a term to the bound derived here. This new term is of minor importance if the sets are small enough, in which case any approximation suggested here will be reasonable. This is further discussed by Månsson (1996, p. 55).

### 10.7.4  Compound Poisson approximation

Recall that the multiple scan statistic $\xi(d, N, m, W)$, as defined in (10.1), is a sum of $\binom{N}{m}$ indicators, where the $i$th indicator is 1 if the $i$th $m$-subset is covered by some translate of $W$. The indicators pertaining to $m$-subsets with common points are not independent, but have a positive dependence, and the $m$-subsets that are covered tend to occur in clumps. The more common the points and the larger the sets, the stronger is the dependence. And, consequently, the worst is the Poisson approximation suggested here, which can be seen in the simulations and by the bounds on the total variation distance. Then, it seems natural to approximate $\xi(d, N, m, W)$ by some distribution other than Poisson, and the point process determined by the positions of the covered $m$-subsets by some process in which clumps are more likely to occur than in the usual Poisson process. Natural candidates are the compound Poisson distribution and the compound Poisson process. However, it is not obvious how to carry out such approximations. One problem is how to define clumps in such a way that the approximations can be theoretically examined, for instance, by means of Stein's method. Another problem is to calculate the parameters in the approximating distribution.

Alm (1983) suggested a method to handle the problem with clumping for scan statistics in one dimension. Alm (1997) generalized the technique to two and three dimensions. The scanning set can be any convex set, and the accuracy of the suggested approximations is verified by means of simulations.

In Corollary 10.5.1, it is stated that if $\lambda$ is kept fixed as $N \to \infty$, then $\xi(2, N, m, W_N)$ converges to a Poisson variable with parameter $\lambda$ at the rate $O(N^{-1/(m-1)})$. As noted in Remark 10.5.3, it has been shown by Barbour, Holst, and Janson (1992) that the rate at which the distribution of the number of small $m$-spacings converges to a Poisson distribution is $O(N^{-1/m})$. Recall that $m - 1$ in our case corresponds to $m$ in the one-dimensional case so that the rates are equal. Roos (1993) has shown that a suitably chosen compound Poisson approximation for the number of small $m$-spacings yields rates of order $O(N^{-1})$ for all $m$. It might be possible that a compound Poisson approximation would give rates of this order also in the case of multiple scan statistics in two and higher dimensions.

---

# References

1. Aldous, D. (1989). *Probability Approximations via the Poisson Clumping Heuristic*, New York: Springer-Verlag.

2. Alm, S. E. (1983). On the distribution of the scan statistic of a Poisson process, In *Probability and Mathematical Statistics: Essays in honour of Carl-Gustav Esseen*, pp. 1–10, Department of Mathematics, Uppsala University, Sweden.

3. Alm, S. E. (1997). On the distribution of scan statistics of a two-dimensional Poisson process, *Advances in Applied Probability*, **29**, 1–18.

4. Barbour, A. D. and Eagleson, G. K. (1984). Poisson convergence for dissociated statistics. *Journal of the Royal Statististical Society, Series B*, **46**, 397–402.

5. Barbour, A. D., Holst, L. and Janson, S. (1992). *Poisson Approximation*, Oxford, England: Oxford University Press.

6. Berwald, W. and Varga, O. (1937). Integralgeometrie 24, Über die Schiebungen im Raum, *Mathematisch Zeitschrift*, **42**, 710–736.

7. Blaschke, W. (1937). Integralgeometrie 21, Über Schiebungen, *Mathematisch Zeitschrift*, **42**, 399–410.

8. Bonnesen, T. and Fenchel, W. (1948). *Theorie der Konvexen Körper*, New York: Chelsea.

9. Eggleston H. G. (1958). *Convexity*, Cambridge, England: Cambridge University Press.

10. Eggleton, P. and Kermack, W. O. (1944). A problem in the random distribution of particles, *Proceedings of the Royal Society of Edinburgh, Section A*, **62**, 103–115.

11. Gates, D. J. and Westcott, M. (1985). Accurate and asymptotic results for distributions of scan statistics, *Journal of Applied Probability*, **22**, 531–542.

12. Glaz, J. (1989). Approximations and bounds for the distribution of the scan statistic, *Journal of the American Statistical Association*, **84**, 560–566.

13. Glaz, J. and Naus J. (1983). Multiple clusters on the line, *Communications in Statistics—Theory and Methods*, **12**, 1961–1986.

14. Glaz, J., Naus J., Roos, M. and Wallenstein, S. (1994). Poisson approximations for the distribution and moments of ordered $m$-spacings, *Journal of Applied Probability*, **31A**, 271–281.

15. Janson, S. (1984). Bounds on the distributions of extremal values of a scanning process. *Stochastic Processes and Their Applications*, **18**, 313–328.

16. Kulldorff, M. (1997). A spatial scan statistic. *Communications in Statistics—Theory and Methods*, **26**, 1481–1496.

17. Kulldorff, M. and Nagarwalla, N. (1995). Spatial disease clusters: detection and inference, *Statistics in Medicine*, **14**, 799–810.

18. Loader, C. R. (1991). Large-deviation approximations to the distribution of scan statistics, *Advances in Applied Probability*, **23**, 751–771.

19. Mack, C. (1948). An exact formula for $Q_k(n)$, the probable number of $k$-aggregates in a random distribution of $n$ points, *Philosophical Magazine*, **39**, 778–790.

20. Mack, C. (1949). The expected number of aggregates in a random distribution of $n$ points, *Proceedings of the Cambridge Philosophical Society*, **46**, 285–292.

21. Månsson, M. (1996). On clustering of random points in the plane and in space, *Thesis*, ISBN 91-7197-290-0/ISSN 0346-718X, Department of Mathematics, Chalmers University of Technology, Göteborg, Sweden.

22. Månsson, M. (1997). Poisson approximation in connection with clustering of random points, *Preprint* NO 1997-32/ISSN 0347-2809, Department of Mathematics, Chalmers University of Technology, Göteborg, Sweden.

23. Miles, R. E. (1974). The fundamental formula of Blaschke in integral geometry and geometrical probability, and its iteration, for domains with fixed orientations, *Australian Journal of Statistics*, **16**, 111–118.

24. Naus, J. I. (1982). Approximations for distributions of scan statistics, *Journal of the American Statistical Association*, **77**, 177–183.

25. Roos, M. (1993). Compound Poisson approximations for the numbers of extreme spacings, *Advances in Applied Probability*, **25**, 847–874.

26. Silberstein, L. (1945). The probable number of aggregates in random distributions of points, *Philosophical Magazine*, **36**, 319–336.

27. Silverman, B. and Brown, T. (1978). Short distances, flat triangles and Poisson limits, *Journal of Applied Probability*, **15**, 815–825.

28. Silverman, B. and Brown, T. (1979). Rates of Poisson convergence for $U$-statistics, *Journal of Applied Probability*, **16**, 428–432.

29. Weil, W. (1990). Iterations of translative formulae and non-isotropic Poisson processes of particles, *Mathematisch Zeitschrift*, **205**, 531–549.

# PART IV
## APPLICATIONS

# 11

# A Start-Up Demonstration Test Using a Simple Scan-Based Statistic

**Markos V. Koutras and N. Balakrishnan**

*University of Athens, Athens, Greece*
*McMaster University, Hamilton, Ontario, Canada*

**Abstract:** Recently, start-up demonstration tests and various extensions of them (in order to accommodate dependence between the trials, to allow for corrective action to be taken once the equipment fails for the first time, etc.) have been discussed in the literature. In this chapter, we propose a start-up demonstration test using a simple scan-based statistic that would facilitate an early rejection of a potentially bad equipment. We then derive the probability generating function of the waiting time for rejection of an equipment and the mean and variance of the waiting time. We also present some recurrence relations satisfied by the probability mass function and the moments. In addition, we indicate how the moment estimator for the unknown probability of success ($p$) of the individual trials can be derived. The case of start-up demonstration testing when consecutive attempts are dependent in a Markovian fashion is discussed next. Finally, we present a numerical example to illustrate the test proposed and the usefulness of the results established in this chapter.

**Keywords and phrases:** Start-up demonstration testing, scan statistics, Bernoulli trials, probability generating function, method-of-moment estimation, Markov dependent trials

## 11.1  Introduction

A start-up demonstration test, as first discussed by Hahn and Gage (1985), involves successive attempted start-ups of an equipment with each attempt resulting in either a success or a failure and accepts the equipment if a pre-specified number ($c$) of consecutive successful start-ups occur on or before a pre-specified number of attempts. While Hahn and Gage (1985) explained probability calculations for such start-up demonstration tests, Viveros and Bala-krishnan (1993) discussed some inferential methods (for the unknown probability of a successful start-up, $p$) based on data obtained from start-up demonstration tests. Viveros and Balakrishnan (1993) and Balakrishnan, Balasubramanian, and Viveros (1995) studied the start-up demonstration testing problem after allowing the outcomes of successive attempted start-ups to be dependent in a Markovian fashion [instead of being independent as in the original formulation of Hahn and Gage (1985)]. Balakrishnan, Balasubramanian, and Viveros (1995) also considered the start-up demonstration testing under corrective action model which allows for a corrective action to be taken by the experimenter immediately after observing the first failure (i.e., after the first failed attempt). An extension of the start-up demonstration testing in which the outcomes of successive attempted start-ups have a higher-order Markov dependence has been discussed recently by Aki, Balakrishnan, and Mohanty (1996). In a similar vein, Balakrishnan, Mohanty, and Aki (1997) have discussed the start-up demonstration testing with corrective actions under the higher-order Markov dependence model. For a synthesis of all these developments and work on related waiting time problems, we refer the interested readers to the book by Balakrishnan and Koutras (1999).

In this chapter, we first propose a start-up demonstration test using a simple scan-based statistic, which is as follows. Let an equipment be subjected to successive attempts to start-up. Then:

(i) If the equipment fails to start in any of the first $r - 1$ attempts, an additional failure in any subsequent attempt will lead to the rejection of the equipment;

(ii) If the equipment starts successfully in the first $r - 1$ attempts, the equipment will be rejected only if two attempted start-ups lying less than $k$ places apart result in failures;

(iii) If neither (i) nor (ii) occurs in a pre-specified number of attempts to start (say, $N$), then the equipment is accepted.

Observe that in (i) we regard it to be very important for the equipment to succeed in the first $r - 1$ trials and that if this does not happen, then the experimenter becomes very strict and does not allow any other unsuccessful start-up till the end of the testing process. Observe also that in (ii) the equipment is being rejected using a simple scan-based statistic with a window of size $k$.

Let $X_i$ correspond to the outcome of the $i$th start-up attempt, taking on the value 1 if it is successful and the value 0 if it is a failure. Let the corresponding probabilities be $p$ and $q = 1 - p$, respectively. Let us denote the waiting time for the rejection of the equipment by $T$. Finally, let us denote the probability mass function of $T$, viz., $P(T = n)$, by $g(n)$ and the probability generating function of $T$, viz., $\sum_{n=0}^{\infty} g(n)z^n$, by $G(z)$.

In this chapter, we first derive a recurrence relation for the probability mass function $g(n)$ and also an explicit expression for the probability generating function $G(z)$. Then, from $G(z)$ we also derive an explicit expression for the mean and variance of the waiting time for rejection of the equipment and a recurrence relation for determining the higher-order moments. We then show that the mean waiting time is monotonically decreasing in $q$ and use this fact to suggest a method-of-moment estimator for the unknown parameter $p$. Next, we present some results for the problem when the underlying attempts for start-up on the equipment are dependent in a Markovian fashion. These results naturally generalize the corresponding results based on Bernoulli trials presented in the earlier sections. Finally, we present a numerical example to illustrate the start-up demonstration testing procedure proposed in this chapter and also some applications of the results presented here.

---

## 11.2   Probability Mass Function of $T$

With $g(n)$ denoting the probability mass function of the waiting time variable $T$, it is clear that

$$g(0) = g(1) = 0 \quad \text{and} \quad g(n) = (n-1)q^2 p^{n-2} \quad \text{for } 2 \leq n \leq r.$$

For $n > r$, we may write

$$
\begin{aligned}
g(n) \;=\; & P(T = n \text{ and } X_i = 0 \text{ for exactly one } i \in \{1, 2, \ldots, r-1\}) \\
& + P(T = n \text{ and } X_1 = X_2 = \cdots = X_{r-1} = 1) \\
=\; & P(X_n = 0 \text{ and } X_i = 0 \text{ for exactly one } i \in \{1, 2, \ldots, r-1\} \\
& \quad \text{and } X_i = 1 \text{ for } i = r, r+1, \ldots, n-1) \\
& + P(X_1 = \cdots = X_{r-1} = 1 \text{ and the waiting time for the first} \\
& \quad \text{appearance of two failures which lie at most } k \text{ places apart} \\
& \quad \text{in the sequence } \{X_r, X_{r+1}, \ldots\} \text{ is } n - r + 1).
\end{aligned}
$$

If we now use $f(n)$ to denote the probability mass function of the waiting time for the first appearance of two failures separated by at most $k-2$ successes in a sequence of Bernoulli trials with success probability $p = 1 - q$, then $g(n)$ may be expressed as

$$g(n) = (r-1)q^2 p^{n-2} + p^{r-1} f(n - r + 1), \qquad n > r.$$

Probability mass function $f(n)$ satisfies the recurrence relations [see Koutras (1996)]

$$
\begin{aligned}
f(n) &= (n-1)q^2 p^{n-2} = pf(n-1) + q^2 p^{n-2} \text{ for } 2 \leq n \leq k, \\
f(n) &= pf(n-1) + qp^{k-1} f(n-k) \text{ for } n > k
\end{aligned}
\tag{11.1}
$$

with initial conditions $f(0) = f(1) = 0$. Thus, after computing $f(n)$ recursively from (11.1), the probability mass function $g(n)$ of $T$ can be determined by

$$
\begin{aligned}
g(n) &= 0 \qquad \text{for } n = 0, 1 \\
&= (n-1)q^2 p^{n-2} \qquad \text{for } 2 \le n \le r \\
&= (r-1)q^2 p^{n-2} + p^{r-1} f(n-r+1) \qquad \text{for } n > r.
\end{aligned} \tag{11.2}
$$

---

## 11.3   Probability Generating Function of $T$

From (11.2), the probability generating function of the waiting time variable $T$ can be written as

$$
\begin{aligned}
G(z) = E(z^T) &= \sum_{n=0}^{\infty} g(n) z^n \\
&= \sum_{n=2}^{r} (n-1)q^2 p^{n-2} z^n + \sum_{n=r+1}^{\infty} (r-1)q^2 p^{n-2} z^n \\
&\quad + \sum_{n=r+1}^{\infty} p^{r-1} f(n-r+1) z^n.
\end{aligned} \tag{11.3}
$$

Noting that

$$
\sum_{n=r}^{\infty} (pz)^{n-2} = \frac{(pz)^{r-2}}{1-pz},
$$

$$
\begin{aligned}
\sum_{n=2}^{r-1} (n-1)(pz)^{n-2} &= \frac{1}{p} \sum_{n=1}^{r-1} \frac{d}{dz} (pz)^{n-1} \\
&= \frac{1 - (r-1)(pz)^{r-2} + (r-2)(pz)^{r-1}}{(1-pz)^2},
\end{aligned}
$$

and

$$
\begin{aligned}
\sum_{n=r+1}^{\infty} f(n-r+1) z^{n-r+1} &= \sum_{n=2}^{\infty} f(n) z^n \\
&= \frac{(qz)^2 \left\{ \frac{1-(pz)^{k-1}}{1-pz} \right\}}{1 - pz - qp^{k-1} z^k}
\end{aligned}
$$

[see Koutras (1996)], substituting all these expressions in (11.3) and simplifying, we obtain

$$
G(z) = \frac{(qz)^2}{(1-pz)^2} \left\{ 1 - (r-1)(pz)^{r-2} + (r-2)(pz)^{r-1} \right\}
$$

$$+ (r-1) \frac{(qz)^2 (pz)^{r-2}}{1-pz} + (pz)^{r-1} \frac{(qz)^2 \{1 - (pz)^{k-1}\}}{(1-pz)(1-pz-qp^{k-1}z^k)} \; .$$

$$(11.4)$$

The above probability generating function can also be derived by direct algebraic approach. In order to do so, let us first note that the rejection of an equipment can occur in one of the following three mutually exclusive ways:

**A:** 2 $F$'s (failures) within the first $r-1$ attempts,

**B:** 1 $F$ in the first $r-1$ attempt and the second $F$ on or after the $r$th attempt,

**C:** No $F$ in the first $r-1$ attempt, but on or after the $r$th attempt, 2 $F$'s occur (for the first time) within a window of $k$ attempts.

Clearly, the contribution of **A** to the probability generating function of $T$ is given by

$$\begin{aligned} G_A(z) &= q^2 z^2 + 2pq^2 z^3 + \cdots + (r-2)p^{r-3}q^2 z^{r-1} \\ &= \frac{q^2 z^2}{(1-pz)^2} \left\{ 1 - (r-1)(pz)^{r-2} + (r-2)(pz)^{r-1} \right\}. \end{aligned} \quad (11.5)$$

Next, for **B**, by noting that a typical sequence of outcomes is given by

$$\underbrace{1\ 2\ \cdots r-1}_{1\ F\ \&\ (r-2)S\text{'s}}\ \underbrace{r\ r+1\ \cdots}_{\geq 0\ S}\ F,$$

we obtain the contribution of **B** to the probability generating function of $T$ as

$$\begin{aligned} G_B(z) &= (r-1)qz(pz)^{r-2}\{1 + pz + (pz)^2 + \cdots\}qz \\ &= (r-1)\frac{(qz)^2(pz)^{r-2}}{1-pz} \; . \end{aligned} \quad (11.6)$$

Finally, for **C**, we note that a typical sequence of outcomes is given by

$$\underbrace{1\ 2\ \cdots\ r-1}_{\text{all } S\text{'s}}\ \underbrace{r\ r+1\ \cdots}_{\geq 0\ S}\ F$$

$$\underbrace{\underbrace{S\ \cdots\ S}_{\geq k-1}\ F\ \underbrace{S\ \cdots\ S}_{\geq k-1}\ F\ \cdots\cdots\ \underbrace{S\ \cdots\ S}_{\geq k-1}\ F}_{\ell \text{ times }, \ \ell \geq 0}\ \underbrace{S\ \cdots\ S}_{\leq k-2}\ F,$$

and its contribution to the probability generating function as

$$(pz)^{r-1}\{1 + pz + (pz)^2 + \cdots\}qz \left[ \left\{ (pz)^{k-1} + (pz)^k + \cdots \right\} qz \right]^{\ell}$$

$$\times \left\{ 1 + pz + (pz)^2 + \cdots + (pz)^{k-2} \right\} qz$$

$$= (pz)^{r-1}\frac{qz}{1-pz} \left\{ (pz)^{k-1}\frac{qz}{1-pz} \right\}^{\ell} \frac{1 - (pz)^{k-1}}{1-pz}\, qz.$$

Upon adding the above expression for all possible values of $\ell$ over 0 to $\infty$, we obtain the contribution of $\mathbf{C}$ to the probability generating function of $T$ as

$$
\begin{aligned}
G_C(z) &= (pz)^{r-1}(qz)^2 \frac{1-(pz)^{k-1}}{(1-pz)^2} \sum_{\ell=0}^{\infty} \left\{ (pz)^{k-1} \frac{qz}{1-pz} \right\}^{\ell} \\
&= (pz)^{r-1}(qz)^2 \frac{1-(pz)^{k-1}}{(1-pz)\{1-pz-qp^{k-1}z^k\}} \, .
\end{aligned}
\tag{11.7}
$$

By adding the expressions of $G_A(z)$, $G_B(z)$ and $G_C(z)$ in (11.5), (11.6) and (11.7), respectively, we simply obtain the probability generating function of the stopping time variable $T$ as presented earlier in Eq. (11.4).

Furthermore, by combining (11.1) and (11.2), we can show that

$$
\begin{aligned}
g(n) &= 0 \quad \text{for } n = 0,1 \\
&= (n-1)q^2 p^{n-2} \quad \text{for } 2 \leq n \leq k+r-1 \\
&= pg(n-1) + qp^{k-1}g(n-k) - (r-1)q^3 p^{n-3} \quad \text{for } n \geq k+r.
\end{aligned}
\tag{11.8}
$$

Thus, the probability mass function of $T$ can be determined easily in a recursive manner from (11.8).

Alternatively, by rewriting the probability generating function expression in Eq. (11.4) as

$$
\begin{aligned}
&(1-pz)^2(1-pz-qp^{k-1}z^k)G(z) \\
&= q^2 z^2 - q^2 pz^3 - q^3 p^{k-1}z^{k+2} - q^2 p^{r+k-2}z^{r+k} + q^2 p^{r+k-2}z^{r+k+1}
\end{aligned}
\tag{11.9}
$$

and comparing coefficients of $z^n$ on both sides of (11.9), we obtain the following relationships:

$$
g(2) = q^2,
$$

$$
g(3) - 3pg(2) = -q^2 p,
$$

$$
g(n) - 3pg(n-1) + 3p^2 g(n-2) - p^3 g(n-3) = 0 \text{ for } n = 4,\ldots,k+1,
$$

$$
g(k+2) - 3pg(k+1) + 3p^2 g(k) - p^3 g(k-1) - qp^{k-1}g(2) = -q^3 p^{k-1},
$$

$$
\begin{aligned}
&g(n) - 3pg(n-1) + 3p^2 g(n-2) - p^3 g(n-3) - qp^{k-1}g(n-k) \\
&\quad + 2qp^k g(n-k-1) - qp^{k+1}g(n-k-2) = 0 \\
&\qquad\qquad\qquad\qquad\qquad \text{for } n = k+3,\ldots,k+r-1,
\end{aligned}
$$

$$g(k+r) - 3pg(k+r-1) + 3p^2 g(k+r-2) - p^3 g(k+r-3)$$
$$- qp^{k-1}g(r) + 2qp^k g(r-1) - qp^{k+1}g(r-2) = -q^2 p^{k+r-2},$$

$$g(k+r+1) - 3pg(k+r) + 3p^2 g(k+r-1) - p^3 g(k+r-2)$$
$$- qp^{k-1}g(r+1) + 2qp^k g(r) - qp^{k+1}g(r-1) = q^2 p^{k+r-2},$$

$$g(n) - 3pg(n-1) + 3p^2 g(n-2) - p^3 g(n-3) - qp^{k-1}g(n-k)$$
$$+ 2qp^k g(n-k-1) - qp^{k+1}g(n-k-2) = 0$$
$$\text{for } n = k+r+2, k+r+3, \ldots \qquad (11.10)$$

It may be pointed out that, by repeated use of the relations in (11.10) and simplifying the resulting equations, (11.10) may be reduced to (11.8) which is certainly simpler for computational purposes.

---

## 11.4   Mean, Variance, and Moments of $T$

Upon differentiating the probability generating function of $T$ in (11.4) with respect to $z$ once and simplifying, we obtain

$$G'(z) = \frac{q^2\{2z - (r+1)p^{r-1}z^r + (r-1)p^r z^{r+1}\}}{(1-pz)^3}$$
$$+ \frac{q^2 p^{r-1}z^r\{(r+1) - rpz - (r+k)p^{k-1}z^{k-1} + (r+k-1)p^k z^k\}}{(1-pz)^2(1-pz-qp^{k-1}z^k)}$$
$$+ \frac{q^2 p^r z^{r+1}(1 - p^{k-1}z^{k-1})(1 + kqp^{k-2}z^{k-1})}{(1-pz)(1-pz-qp^{k-1}z^k)^2}. \qquad (11.11)$$

Setting now $z = 1$ in (11.11) and simplifying, we can derive the mean waiting time as

$$E(T) = G'(z)|_{z=1} = \frac{2}{q} + \frac{p^{k+r-2}}{q(1-p^{k-1})}. \qquad (11.12)$$

From (11.12), we observe that $E(T) \to 2$ as $q \to 1$ and $E(T) \to \infty$ as $q \to 0$. Further, it can also be readily verified that $E(T)$ is a monotonically decreasing function in $q$. In addition to revealing that the proposed start-up demonstration test procedure will, on an average, reject a bad equipment very early and take very long to reject a good equipment, the monotonicity of the mean in (11.12) will also enable us to develop a method-of-moment estimator for the parameter $p$ (assuming, of course, that $r$ and $k$ are fixed by the experimenter). This is discussed further in the next section.

Upon differentiating the expression of $G'(z)$ in (11.11) once more with respect to $z$ and setting $z = 1$, we derive the second factorial moment of $T$, viz., $E\{T(T-1)\}$. Then, upon adding to it the expression of $E(T) - \{E(T)\}^2$ obtained from (11.12) and simplifying the resulting expression, we obtain the variance of the waiting time to be

$$
\begin{aligned}
\mathrm{Var}(T) \;=\; & \frac{2p}{q^2} - \frac{p^{2k+2r-4}}{\{q(1-p^{k-1})\}^2} \\
& + \frac{p^{k+r-3}\{-3p + 5p^2 + 2(k+r)pq + (3-2r)p^k q\}}{\{q(1-p^{k-1})\}^2}\,. \quad (11.13)
\end{aligned}
$$

For the derivation of higher-order moments $\mu'_i = E(T^i)$ of $T$, we can establish a recurrence relation for the raw moments of $T$. For this purpose, with the aid of the recurrence relation for the probability mass function of $T$ presented in (11.8), we may write

$$
\begin{aligned}
\mu'_s \;=\; & E(T^s) = \sum_{n=2}^{\infty} n^s\, g(n) \\
=\; & \sum_{n=2}^{k+r-1} n^s\, g(n) + \sum_{n=k+r}^{\infty} n^s\, g(n) \\
=\; & q^2 \sum_{n=2}^{k+r-1} n^s(n-1)p^{n-2} + p \sum_{n=k+r}^{\infty} n^s\, g(n-1) \\
& + q\,p^{k-1} \sum_{n=k+r}^{\infty} n^s\, g(n-k) - (r-1)q^3 \sum_{n=k+r}^{\infty} n^s\, p^{n-3}. \quad (11.14)
\end{aligned}
$$

Noting now that

$$
\begin{aligned}
\sum_{n=k+r}^{\infty} n^s\, g(n-1) \;=\; & \sum_{n=k+r-1}^{\infty} (n+1)^s g(n) \\
=\; & \sum_{n=k+r-1}^{\infty} \sum_{i=0}^{s} \binom{s}{i} n^i g(n) \\
=\; & \sum_{i=0}^{s} \binom{s}{i} \left\{ \mu'_i - \sum_{n=0}^{k+r-2} n^i g(n) \right\}
\end{aligned}
$$

and

$$
\begin{aligned}
\sum_{n=k+r}^{\infty} n^s g(n-k) \;=\; & \sum_{n=r}^{\infty} (n+k)^s g(n) \\
=\; & \sum_{n=r}^{\infty} \sum_{i=0}^{s} \binom{s}{i} n^i k^{s-i} g(n) \\
=\; & \sum_{i=0}^{s} \binom{s}{i} k^{s-i} \left\{ \mu'_i - \sum_{n=0}^{r-1} n^i g(n) \right\},
\end{aligned}
$$

we may rewrite (11.14) as

$$\mu'_s = q^2 \sum_{n=2}^{k+r-1} n^s(n-1)p^{n-2} - (r-1)q^3 \sum_{n=k+r}^{\infty} n^s p^{n-3}$$

$$+ p \sum_{i=0}^{s} \binom{s}{i} \mu'_i - p \sum_{i=0}^{s} \binom{s}{i} \sum_{n=2}^{k+r-2} n^i g(n)$$

$$+ qp^{k-1} \sum_{i=0}^{s} \binom{s}{i} k^{s-i} \mu'_i - qp^{k-1} \sum_{i=0}^{s} \binom{s}{i} k^{s-i} \sum_{n=2}^{r-1} n^i g(n). \quad (11.15)$$

Now, upon using (11.8) and rearranging the terms, we derive a recurrence relation for the raw moments of $T$ as

$$\mu'_s = \frac{1}{q(1-p^{k-1})}\left\{ \sum_{i=0}^{s-1} \binom{s}{i} \left(p + qp^{k-1}k^{s-i}\right) \mu'_i \right.$$

$$+ q^2 \sum_{n=2}^{k+r-1} n^s(n-1)p^{n-2} - (r-1)q^3 \sum_{n=k+r}^{\infty} n^s p^{n-3}$$

$$- q^2 p \sum_{i=0}^{s} \binom{s}{i} \sum_{n=2}^{k+r-2} n^i(n-1)p^{n-2}$$

$$\left. - q^3 p^{k-1} \sum_{i=0}^{s} \binom{s}{i} k^{s-i} \sum_{n=2}^{r-1} n^i(n-1)p^{n-2} \right\}. \quad (11.16)$$

The recurrence relation in (11.16) can be used effectively to determine all the raw moments of $T$ in a recursive manner.

For example, by setting $s = 1$ in (11.16), we obtain

$$\mu'_1 = E(T) = \frac{1}{q(1-p^{k-1})} \left\{ p + kqp^{k-1} + q^2 \sum_{n=2}^{k+r-1} n(n-1)p^{n-2} \right.$$

$$- (r-1)q^3 \sum_{n=k+r}^{\infty} n\,p^{n-3} - q^2 p \sum_{n=2}^{k+r-2} (n-1)p^{n-2}$$

$$- q^2 p \sum_{n=2}^{k+r-2} n(n-1)p^{n-2} - kq^3 p^{k-1} \sum_{n=2}^{r-1}(n-1)p^{n-2}$$

$$\left. - q^3 p^{k-1} \sum_{n=2}^{r-1} n(n-1)p^{n-2} \right\}. \quad (11.17)$$

Upon using the identities

$$\sum_{n=2}^{m} n(n-1)p^{n-2} = -\frac{m(m+1)p^{m-1}}{q} - \frac{2(m+1)p^m}{q^2} + \frac{2(1-p^{m+1})}{q^3},$$

$$\sum_{n=m}^{\infty} n\,p^{n-1} = \frac{m\,p^{m-1}}{q} + \frac{p^m}{q^2}$$

and

$$\sum_{n=2}^{m}(n-1)p^{n-2} = -\,\frac{m\,p^{m-1}}{q} + \frac{1-p^m}{q^2}$$

in (11.17) and simplifying the resulting equation, we obtain

$$
\begin{aligned}
E(T) \;&=\; \frac{1}{q(1-p^{k-1})}\Big\{2(1-p^{k-1})+p^{k+r-2}\Big\}\\[2mm]
&=\; \frac{2}{q} + \frac{p^{k+r-2}}{q(1-p^{k-1})}
\end{aligned}
$$

as given earlier in (11.12).

Similarly, by setting $s = 2$ in (11.16), we obtain

$$
\begin{aligned}
\mu_2' = E(T^2) \;=\;& \frac{1}{q(1-p^{k-1})}\Big\{p + k^2qp^{k-1} + 2(p+kqp^{k-1})\mu_1'\\[2mm]
&+ q^2\sum_{n=2}^{k+r-1} n^2(n-1)p^{n-2} - (r-1)q^3\sum_{n=k+r}^{\infty} n^2p^{n-3}\\[2mm]
&- q^2p\sum_{n=2}^{k+r-2}(n-1)p^{n-2} - 2q^2p\sum_{n=2}^{k+r-2} n(n-1)p^{n-2}\\[2mm]
&- q^2p\sum_{n=2}^{k+r-2} n^2(n-1)p^{n-2} - k^2q^3p^{k-1}\sum_{n=2}^{r-1}(n-1)p^{n-2}\\[2mm]
&- 2kq^3p^{k-1}\sum_{n=2}^{r-1} n(n-1)p^{n-2} - q^3p^{k-1}\sum_{n=2}^{r-1} n^2(n-1)p^{n-2}\Big\}.
\end{aligned}
$$

$$\tag{11.18}$$

Then, from (11.18), upon using the identities

$$
\begin{aligned}
\sum_{n=2}^{m} n^2(n-1)p^{n-2} \;=\;& -\,\frac{m(m+1)^2p^{m-1}}{q} - \frac{(m+1)(3m+2)p^m}{q^2}\\[2mm]
&+ \frac{2-(6m+8)p^{m+1}}{q^3} + \frac{6(1-p^{m+1})}{q^4}
\end{aligned}
$$

and

$$\sum_{n=m}^{\infty} n^2\,p^{n-3} = \frac{m^2\,p^{m-3}}{q} + \frac{(m+1)p^{m-2}}{q^2} + \frac{m\,p^{m-1}}{q^2} + \frac{2p^m}{q^3}\,,$$

recalling the expression of the mean in (11.12), and doing algebraic simplifications, we obtain an expression for the second raw moment of $T$ from which the expression of the variance in (11.13) can be derived.

## 11.5 Inference on $p$

Suppose a random sample of $n$ equipments are placed on the proposed start-up demonstration test, and that the corresponding number of attempts until rejection are denoted by $T_1, T_2, \ldots, T_n$. In other words, they are the waiting times for rejection of the $n$ equipments. Then, based on these data, we develop here some inference procedures for the unknown probability of success $(p)$. Naturally, there are two types of data possible: one in which the entire sequence of binary outcomes leading to the rejection of each equipment is available, and the other in which only the waiting time for rejection of each equipment is available.

Let us now consider the first scenario in which it is assumed that the entire sequence of binary outcomes until the rejection of each equipment is available. Let $(S_1, F_1), (S_2, F_2), \ldots, (S_n, F_n)$ denote the number of successful and unsuccessful start-ups corresponding to the $n$ equipments under test; clearly, $S_i + F_i = T_i$ for $i = 1, 2, \ldots, n$. Then, the likelihood function is given by

$$L(p) = p^{S_1} q^{F_1} p^{S_2} q^{F_2} \cdots p^{S_n} q^{F_n} = p^S q^F, \tag{11.19}$$

where $S = \sum_{i=1}^{n} S_i$ is the total number of successful start-ups, $F = \sum_{i=1}^{n} F_i$ is the total number of unsuccessful start-ups, and $S + F = \sum_{i=1}^{n} T_i$. Note that $S$ is a sufficient statistic for $p$. From (11.19), we readily obtain the maximum likelihood estimator of $p$ as

$$\hat{p} = \frac{S}{S + F} = \frac{\sum_{i=1}^{n} S_i}{\sum_{i=1}^{n} T_i} . \tag{11.20}$$

The observed Fisher information is given by

$$I(\hat{p}) = \left( -\frac{d^2 \log L}{dp^2} \right)\bigg|_{p=\hat{p}} = \frac{S + F}{\hat{p}(1 - \hat{p})} = \frac{\sum_{i=1}^{n} T_i}{\hat{p}(1 - \hat{p})} . \tag{11.21}$$

From (11.20) and (11.21), upon invoking the asymptotic normality of the MLE $\hat{p}$, we have an approximate $100(1 - \alpha)\%$ confidence interval for $p$ as

$$\left( \hat{p} - \frac{z_{\alpha/2}}{\sqrt{I(\hat{p})}} , \hat{p} + \frac{z_{\alpha/2}}{\sqrt{I(\hat{p})}} \right), \tag{11.22}$$

where $z_{\alpha/2}$ is the upper $\alpha/2$ percentage point of the standard normal distribution.

Next, let us consider the second scenario in which it is assumed that only the waiting times $T_1, T_2, \ldots, T_n$ are available. Then, it is well known that $\bar{T} = \frac{1}{n} \sum_{i=1}^{n} T_i$ is an unbiased estimator of $E(T)$, say $h(p)$. Since $h(p)$ is a monotonically increasing function in $p$, the moment estimator of $p$ (or $q$) is unique and can be easily obtained as the solution of the equation

$$\bar{T} = h(p) = \frac{2}{1-p} + \frac{p^{k+r-2}}{(1-p)(1-p^{k-1})} \ . \qquad (11.23)$$

The moment estimate $\tilde{p}$ has to be determined numerically from (11.23). For the special case $k = r = 2$, however, (11.23) yields

$$(1-p)^2 = \frac{1}{\bar{T}-1}$$

from which an explicit moment estimator of $p$ can be derived as

$$\tilde{p} = 1 - (\bar{T}-1)^{-1/2}.$$

An approximate $100(1-\alpha)\%$ confidence interval can be developed based on the moment estimator as well. This may be done by using the central limit theorem and considering

$$\left(\bar{T} - z_{\alpha/2}\,\frac{\tilde{\sigma}}{\sqrt{n}} \ , \ \bar{T} + z_{\alpha/2}\,\frac{\tilde{\sigma}}{\sqrt{n}}\right), \qquad (11.24)$$

where $\tilde{\sigma}$ is the standard deviation of $T$ determined from (11.13) with $p$ replaced by $\tilde{p}$. Realizing that (11.24) is a confidence interval for $h(p)$, an approximate $100(1-\alpha)\%$ confidence interval for $p$ can be obtained as

$$\left(h^{-1}\left(\bar{T} - z_{\alpha/2}\,\frac{\tilde{\sigma}}{\sqrt{n}}\right) \ , \ h^{-1}\left(\bar{T} + z_{\alpha/2}\,\frac{\tilde{\sigma}}{\sqrt{n}}\right)\right) \qquad (11.25)$$

since $h(p)$ is a monotonically increasing function in $p$.

---

## 11.6    Results for Markov-Dependent Start-Ups

Until now, we have assumed that the attempted start-ups of an equipment are all independent trials with probability of a successful start-up $p$ and probability of an unsuccessful start-up $q$. However, when successive start-ups are attempted on an equipment, it is more realistic in many practical situations to assume some sort of dependence among them. In this section, we shall assume a two-state Markov dependence among the outcomes of successive start-ups and develop results analogous to those presented in Sections 11.2–11.4.

To this end, let us denote by $p_0$ ($q_0$) the probability of a successful (unsuccessful) start-up in the first attempt, by $p_1$ ($q_1$) the probability of a successful (unsuccessful) start-up in any attempt given that the earlier attempt was successful, and by $p_2$ ($q_2$) the probability of a successful (unsuccessful) start-up in any attempt given that the earlier attempt was unsuccessful. Once again, let us use $T$ for the waiting time until rejection of the equipment. Then, as in Section 11.3, we may find the contribution of $A$ to the probability generating function of $T$ as

$$
\begin{aligned}
G_A(z) \ &= \ q_0 q_2 z^2 + \sum_{j=3}^{r-1} q_0 p_1^{j-3} p_2 q_1 z^j \\
&\quad + \sum_{j=4}^{r-1} (j-3) p_0 p_1^{j-4} p_2 q_1^2 z^j + \sum_{j=3}^{r-1} p_0 p_1^{j-3} q_2 q_1 z^j \\
&= \ q_0 q_2 z^2 + (q_0 p_2 + p_0 q_2) q_1 z^3 \left\{ \frac{1 - (p_1 z)^{r-3}}{1 - p_1 z} \right\} \\
&\quad - \frac{(r-3) p_0 p_2 q_1^2 p_1^{r-4} z^r}{1 - p_1 z} + p_0 p_2 q_1^2 z^4 \left\{ \frac{1 - (p_1 z)^{r-3}}{(1 - p_1 z)^2} \right\} . \quad (11.26)
\end{aligned}
$$

Similarly, the contribution of $B$ to the probability generating function of $T$ is

$$
\begin{aligned}
G_B(z) \ &= \ \frac{q_0 p_2 q_1 p_1^{r-3} z^r}{1 - p_1 z} + \frac{(r-3) p_0 p_2 q_1^2 p_1^{r-4} z^r}{1 - p_1 z} \\
&\quad + p_0 p_1^{r-3} q_1 q_2 z^r + \frac{p_0 p_1^{r-3} q_1^2 p_2 z^{r+1}}{1 - p_1 z} . \quad (11.27)
\end{aligned}
$$

Finally, the contribution of $C$ to the probability generating function of $T$ is

$$
\begin{aligned}
G_C(z) \ &= \ p_0 p_1^{r-2} z^{r-1} \left\{ q_1 z + p_1 q_1 z^2 + p_1^2 q_1 z^3 + \cdots \right\} \\
&\quad \times \sum_{\ell=0}^{\infty} \left[ \left\{ p_2 p_1^{k-2} z^{k-1} + p_2 p_1^{k-1} z^k + \cdots \right\} q_1 z \right]^{\ell} \\
&\quad \times \left\{ q_2 z + p_2 q_1 z^2 + p_2 p_1 q_1 z^3 + p_2 p_1^2 q_1 z^4 + \cdots + p_2 p_1^{k-3} q_1 z^{k-1} \right\} \\
&= \ \frac{p_0 p_1^{r-2} q_1 z^r}{1 - p_1 z} \cdot \frac{q_2 z + (p_2 - p_1) z^2 - p_2 p_1^{k-2} q_1 z^k}{1 - p_1 z - p_2 p_1^{k-2} q_1 z^k} . \quad (11.28)
\end{aligned}
$$

Now, upon adding the expressions of $G_A(z)$, $G_B(z)$ and $G_C(z)$ presented in Eqs. (11.26)–(11.28), respectively, and simplifying the resulting expression, we derive the probability generating function of the waiting time $T$ as

$$
\begin{aligned}
G(z) \ &= \ \frac{q_0 q_2 z^2 + \{ q_0(p_2 - p_1) + q_2(p_0 - p_1) \} z^3 + (p_1 - p_0)(p_1 - p_2) z^4}{(1 - p_1 z)^2} \\
&\quad - \frac{p_0 p_1^{r-2} q_1 q_2 z^{r+1}}{1 - p_1 z} - \frac{p_0 p_1^{r-2} q_1^2 p_2 z^{r+2}}{(1 - p_1 z)^2} \\
&\quad + \frac{p_0 p_1^{r-2} q_1 z^r}{1 - p_1 z} \cdot \frac{q_2 z + (p_2 - p_1) z^2 - p_2 p_1^{k-2} q_1 z^k}{1 - p_1 z - p_2 p_1^{k-2} q_1 z^k} . \quad (11.29)
\end{aligned}
$$

It may be easily verified that in the case of independent start-ups (case $p_0 = p_1 = p_2 = p$ and $q_0 = q_1 = q_2 = q$), the above probability generating function of $T$ reduces to that presented earlier in (11.4). Also, as done earlier in Section 11.3, the generating function in (11.29) can be utilized to derive a set of recurrence relations that would facilitate the computation of the probability mass function of the waiting time $T$.

Furthermore, upon differentiating $G(z)$ in (11.29) once with respect to $z$, setting $z = 1$ and then simplifying the resulting expression, we obtain the mean waiting time to be

$$E(T) = \frac{2 + p_0 + p_2 - 2p_1}{q_1} + \frac{p_0 p_1^{r+k-4} p_2 (q_1 + p_2)}{q_1 (1 - p_2 p_1^{k-2})} \ . \tag{11.30}$$

It may be verified once again that in the case of independent start-ups (case $p_0 = p_1 = p_2 = p$ and $q_0 = q_1 = q_2 = q$), the above expression for the mean waiting time reduces to that presented earlier in (11.12).

Inference procedures for the parameters $p_0$, $p_1$ and $p_2$ can be developed along the lines followed in Section 11.5. However, we refrain from this discussion here.

---

## 11.7  Illustrative Example

In this section, we present a simulated start-up demonstration test data and use it to illustrate the inferential methods discussed in Section 11.5. We assume that $n = 20$ identical units with probability of successful start-up $p = 0.9$ are tested. If the equipment fails to start in any of the first 3 attempts ($r = 4$), an additional failure in any subsequent attempt will lead to its rejection. If the equipment starts successfully in the first 3 attempts, it will be rejected only if two attempted start-ups lying less than $k = 8$ places apart result in failures. The simulated data, along with the values of the statistics $S_i, F_i$ and $T_i$, are presented in Table 11.1. The ToF (type of failure) column in this table indicates which of the two termination conditions caused the rejection of the equipment: 10 indicates that a failure was observed in at least one of the first 3 attempts and the equipment was rejected immediately upon the occurrence of an additional start-up failure; 01 indicates that the equipment started successfully in the first 3 attempts, and was rejected upon the occurrence of two unsuccessful start-ups lying less than $k = 8$ places apart.

**Table 11.1:** Simulated start-up demonstration data for $n = 20$ identical units with $p = 0.9$, $r = 4$, and $k = 8$

| $i$ | ToF | $S_i$ | $F_i$ | $T_i$ | Individual start-up outcomes |
|---|---|---|---|---|---|
| 1 | 10 | 15 | 2 | 17 | 10111111111111110 |
| 2 | 01 | 32 | 4 | 36 | 111111111011111111111110111111101110 |
| 3 | 01 | 5 | 2 | 7 | 1111100 |
| 4 | 01 | 11 | 2 | 13 | 1111111101110 |
| 5 | 01 | 44 | 4 | 48 | 1111011111111110111111111111111111111111110110 |
| 6 | 10 | 5 | 2 | 7 | 0111110 |
| 7 | 01 | 20 | 2 | 22 | 11111111111111011110 |
| 8 | 10 | 6 | 2 | 8 | 10111110 |
| 9 | 01 | 123 | 7 | 130 | 11111111011111111111111111111111111101111111111011111111111111111111111111111111111111111111111111110111111111101111111111101111110 |
| 10 | 01 | 40 | 2 | 42 | 11111111111111111111111111111111111011111110 |
| 11 | 01 | 54 | 5 | 59 | 11110111111111111111111110111111111111111110111111111111100 |
| 12 | 01 | 5 | 2 | 7 | 1111100 |
| 13 | 01 | 11 | 2 | 13 | 1111111011110 |
| 14 | 10 | 2 | 2 | 4 | 1010 |
| 15 | 01 | 75 | 7 | 82 | 111101111111111111011111110111111101111111111111111111110111111111111010 |
| 16 | 01 | 12 | 2 | 14 | 11111111111010 |
| 17 | 01 | 28 | 3 | 31 | 111111111111111011111111111100 |
| 18 | 01 | 25 | 3 | 28 | 111111110111111111110111110 |
| 19 | 01 | 19 | 3 | 22 | 1111111111110111111100 |
| 20 | 01 | 66 | 5 | 71 | 11111111111111111101111111110111111110111111111111111111111111111101111110 |
| Total | | 598 | 63 | 661 | |

For these data, we have

$$S = \sum_{i=1}^{20} S_i = 598, \quad F = \sum_{i=1}^{20} F_i = 63, \quad T = \sum_{i=1}^{20} T_i = S + F = 661;$$

substituting these in (11.20) and (11.21), we obtain the MLE of $p$ and the observed Fisher information as

$$\hat{p} = \frac{S}{S+F} = \frac{\sum_{i=1}^{n} S_i}{\sum_{i=1}^{n} T_i} = 0.90469, \qquad I(\hat{p}) = \frac{\sum_{i=1}^{n} T_i}{\hat{p}(1-\hat{p})} = 7665.89.$$

An approximate 95% confidence interval for $p$ then follows from (11.22) as

$$(0.90469 - 1.96/\sqrt{7665.89}, 0.90469 + 1.96/\sqrt{7665.89}) = (0.8823,\ 0.9271).$$

The moment estimator of $p$ is computed by solving the equation $h(p) = \bar{T} = 661/20 = 33.05$ with $h(p)$ as given in (11.23). Thus,

$$\tilde{p} = h^{-1}(33.05) = 0.9134.$$

Moreover, by substituting this estimate in (11.13), we obtain an estimate of the standard deviation of $T$ as $\tilde{\sigma} = 24.5829$. Next, from (11.25), we obtain an approximate 95% confidence interval for $p$ (based on the moment estimate $\tilde{p}$) as

$$\left( h^{-1}\left( 33.05 - 1.96\,\frac{24.5829}{\sqrt{20}} \right), h^{-1}\left( 33.05 + 1.96\,\frac{24.5829}{\sqrt{20}} \right) \right) = (0.8866, 0.9279).$$

It is worth mentioning that, even for a sample as small as 20, both methods yield point estimates which are very close to the true value of $p$ ($p = 0.9$). Also, the confidence intervals are both quite narrow and close to each other; a significant improvement in the intervals' width is observed if the sample size is increased.

**Acknowledgments.** The second author wishes to thank the Natural Sciences and Engineering Research Council of Canada for funding this research. The authors also thank Mrs. Debbie Iscoe for the excellent typing of the manuscript and graduate student M. Boutsikas (University of Athens, Greece) for carrying out the numerical calculations of Section 11.7.

---

# References

1. Aki, S., Balakrishnan, N. and Mohanty, S. G. (1996). Sooner and later waiting time problems for success and failure runs in higher order Markov dependent trials, *Annals of the Institute of Statistical Mathematics*, **48**, 773–787.

2. Balakrishnan, N., Balasubramanian, K. and Viveros, R. (1995). Start-up demonstration tests under correlation and corrective action, *Naval Research Logistics*, **42**, 1271–1276.

3. Balakrishnan, N. and Koutras, M. V. (1999). *Runs and Patterns with Applications*, New York: John Wiley & Sons (to appear).

4. Balakrishnan, N., Mohanty, S. G. and Aki, S. (1997). Start-up demonstration tests under Markov dependence model with corrective actions, *Annals of the Institute of Statistical Mathematics*, **49**, 155–169.

5. Hahn, G. J. and Gage, J. B. (1986). Evaluation of a start-up demonstration test, *Journal of Quality Technology*, **15**, 103–105.

6. Koutras, M. V. (1996). On a waiting time distribution in a sequence of Bernoulli trials, *Annals of the Institute of Statistical Mathematics*, **48**, 789–806.

7. Viveros, R. and Balakrishnan, N. (1993). Statistical inference from start-up demonstration test data, *Journal of Quality Technology*, **22**, 119–130.

# Applications of the Scan Statistic in DNA Sequence Analysis

**Ming-Ying Leung and Traci E. Yamashita**

*University of Texas at San Antonio, San Antonio, TX*
*Johns Hopkins School of Hygiene and Public Health, Baltimore, MD*

**Abstract:** Advances of biochemical techniques have made available large databases of long DNA sequences. These sequences reflect conglomerates of random and nonrandom letter strings from the nucleotide alphabet { A, C, G, T }. As the databases expand, mathematical methods play an increasingly important role in analyzing and interpreting the rapidly accumulating DNA data. In this chapter, we discuss a specific example of identifying nonrandom clusters of palindromes in a family of herpesvirus genomes using the $r$-scan statistic. Palindrome positions on the genome are modeled by i.i.d. random variables uniformly distributed on the unit interval $(0,1)$. After a comparison of three Poisson-type approximations, the $r$-scan distribution is computed by a compound Poisson approximation proposed by Glaz (1994). Some of the significant palindrome clusters are located at genome regions containing origins of replication and regulatory signals of the herpesviruses.

**Keywords and phrases:** DNA sequence analysis, palindrome clusters, Poisson approximations

## 12.1 Introduction and Background

Since the elucidation of its double helical structure by Watson and Crick in 1953, DNA has repeatedly been confirmed to be the storage medium for genomic data. DNA stores all necessary information for controlling life processes, including its own replication. In the past two decades, molecular biologists have been able to delineate how genetic information in DNA is encoded, retrieved, and duplicated. This knowledge has led to a better understanding of the molecular mechanisms in many biological processes and genetic diseases.

DNA is deoxyribonucleic acid, which is made up of four different types of nucleotide bases: adenine (A), thymine (T), cytosine (C), and guanine (G). The bases A and T form a complementary pair, as do C and G. Two complementary bases are held together by hydrogen bonds to form a nucleotide base pair (bp). A large number of these base pairs are strung together to form a giant double stranded DNA molecule comprising two complementary polynucleotide sequences.

Recent advances in biochemical techniques have led to an exponential increase in the amount of sequence data. For example, in GenBank (the United States nucleic acid sequence database maintained by the National Institutes of Health), there are over a million sequences containing more than a billion nucleotide bases. As the genome databases expand, mathematical methods play an increasingly important role in obtaining, organizing, archiving, analyzing, and interpreting the rapidly accumulating DNA data. Excellent reviews of how various branches of mathematics contribute to the advancement of molecular biology can be found in Waterman (1989), Doolittle (1990), and Waterman (1995).

While probing for insights into the organization of a genome, one of the problems that arises is how to characterize anomalies in the spacings of markers in a long sequence of nucleotides. Here "markers" refer to any short sequence segments with a prescribed pattern. Spacing anomalies include properties of clumping (too many neighboring short spacings), overdispersion (too many long gaps between markers), and excessive regularity (too few short spacings and/or too few long gaps). The problems of identifying such anomalies in large DNA molecules as well as their biological significance are discussed by Karlin and Brendel (1992), who first suggested using $r$-scan lengths for evaluating their statistical significance.

The next section describes the $r$-scan lengths (or simply $r$-scans) and their close relationship to the traditional scan statistic. We shall illustrate the application of the $r$-scans by an example that identifies unusual palindrome clusters in a family of herpesvirus genomes. Since the exact probability distribution of the $r$-scans are not available, one must rely on an approximation when assessing statistical significance for the clusters. In Section 12.3, we shall compare the accuracy of three Poisson-type approximate distributions by contrasting the calculated approximate probabilities with simulation results.

## 12.2 $r$-Scans and DNA Sequence Analysis

### 12.2.1 Duality

For a set of points $X_1, \ldots, X_N$ distributed independently and uniformly over the unit interval $(0, 1)$, the $r$-scan is defined as the cumulative lengths of $r$ consecutive distances between the ordered statistics $X_{(1)}, \ldots, X_{(N)}$. Formally, let $D_i$ denote the distance between the ordered $i$th and $(i + 1)$th points, i.e., $D_i = X_{(i+1)} - X_{(i)}$, $i = 1, \ldots, N - 1$. For any fixed integer $r$ between 1 and $N - 1$, the $r$-scan at the point $X_{(i)}$ is $A_{r,i} = \sum_{j=i}^{i+r-1} D_j$, $i = 1, \ldots, N - r$ [Dembo and Karlin (1992)]. The order statistics of these $r$-scans are denoted by $A_{r,(i)}$. In particular, the minimal $r$-scan $A_{r,(1)} = \min\{A_{r,i}, i = 1, ..., N - r\}$ is most frequently used in DNA sequence analysis. For simplicity, we shall abbreviate $A_{r,(1)}$ as $A_r$ in this chapter.

$A_r$ is intimately related to the traditional scan statistic $S_w = \max_{0 < t < 1-w} Y_t(w)$, where $0 < w < 1$ is a prescribed window length and $Y_t(w)$ is the number of points in the interval $[t, t + w]$. Consider the event $\{Y_{X_{(i)}}(w) \geq r + 1\}$ for $i = 1, \ldots, N - r$, which says that there are at least $r + 1$ points contained in the window $[X_{(i)}, X_{(i)} + w]$. This is equivalent to the event $\{A_{r,i} \leq w\}$ which says that there are at least $r$ adjoining spacings, starting at $X_{(i)}$, whose cumulative length is no more than $w$. Since this equivalence holds for all $i$, certainly it will hold for the particular window holding the maximal number of points. Hence, we have the duality relation

$$\{S_w \geq r + 1\} = \{A_r \leq w\}$$

for fixed values of $w \in (0, 1)$ and $r = 1, \ldots, N - 1$.

By virtue of this duality relation, we will automatically know the distribution of $A_r$ if the distribution of $S_w$ is available. Thus, the traditional scan statistic and the minimal $r$-scan can be used interchangably. In DNA sequence analysis, the $r$-scan is usually preferred, principally due to convenience. We shall explain this further after we present an example in which palindrome clusters in herpesvirus genomes are analyzed using $A_r$.

### 12.2.2 DNA sequence analysis

Various applications of the $r$-scan theory to identify distributional anomalies of palindromes, close direct and inverted repeats, and over- and underrepresented DNA words in a yeast chromosome, bacterial sequences, as well as viral genomes have been discussed by Karlin *et al.* (1993), Karlin, Mrázek, and Campbell (1996, 1997), Karlin and Cardon (1994), Leung, Schachtel, and Yu (1994), and Leung, Marsh, and Speed (1996). Here, we present one example demonstrating how to use the distribution of $A_r$ to assess the statistical significance of unusual palindrome clusters in a family of completely sequenced herpesvirus genome DNA sequences.

The herpesvirus family includes several well-known viruses such as herpes simplex, chicken pox, Epstein-Barr, and cytomegalovirus which are associated with life threatening diseases such as AIDS and various cancers [Labrecque *et al.* (1995), and Vital *et al.* (1996)]. Each herpesvirus genome consists of a single DNA molecule which is wrapped inside an icosahedral capsid. A virus infects the host by introducing its genome into a suitable host cell. Inside the cell, herpesviruses may stay dormant most of the time and only become harmful after entering a lytic cycle in which they grow and replicate thousands of copies. There is biological evidence [Masse *et al.* (1992)] indicating that in some herpesvirus genomes, clusters of palindromes are harbored in the lytic origin of replication (i.e., the point in the genome at which DNA replication begins during the lytic cycle) and other gene regulatory regions.

A DNA molecule can be regarded as a long string of letters $a_1a_2a_3...a_n$, sampled from the four-letter nucleotide alphabet {A, C, G, T}. These four letters are grouped into pairs of complementary bases: A–T and C–G. Based on this complementary pairing, one can define a symmetric structure called a palindrome. A palindrome of length $2s$ is a sequence of $s$ bases followed immediately by its inverted complementary sequence. If we denote the complement of a base $b$ by $b'$, then a palindrome of length $2s$ has the form $b_1b_2...b_sb'_s...b'_2b'_1$. For example, GCGCATGCGC constitutes a length 10 palindrome.

Since short palindromes often occur by chance in a random letter string, one needs to focus only on reasonably long palindromes. Consider an i.i.d. random sequence $a_1a_2a_3...a_n$ where each $a_i$, $i = 1, 2, ..., n$, is drawn from the nucleotide alphabet with probabilities of getting A, C, G, T, being $p_A$, $p_C$, $p_G$, $p_T$, respectively. At each position, we expect to find a palindrome of length $2s$ ($<< n$) with probability $p = \lambda^s$, where $\lambda = 2(p_Ap_T + p_Cp_G)$. The probability $p$ attains its maximum value of $1/4^s$ when $p_A = p_T = p_C = p_G = 1/4$. Setting $s = 5$ yields a value of $p < 0.001$. So, taking only those palindromes of length $\geq$ 10 bp should be sufficient to screen out most of the random noise.

Our data set consists of seven herpesviruses whose complete genome DNA sequence is available. A hash-coding computer program [Leung *et al.* (1991)] is used to screen the entire genome of the herpesviruses for palindromes of length $\geq$ 10 bp. The number of palindromes observed in each genome, along with the genome lengths, are listed in Table 12.1.

**Table 12.1:** The number of palindromes of length $\geq 10$ bp in the viruses and their genome lengths. There are seven herpesviruses in our data set: Epstein-Barr (EBV), Equine herpes simplex (HSE), Ictalurid herpes (HSI), Herpes saimiri (HSS), Herpes simplex I (HSV1), Varicella-zoster (VZV), and Human cytomegalovirus (HCMV).

| Genome | Palindromes | Genome Length |
|--------|-------------|---------------|
| EBV | 113 | 172282 |
| HSE | 194 | 150223 |
| HSI | 111 | 134226 |
| HSS | 131 | 112930 |
| HSV1 | 220 | 152260 |
| VZV | 122 | 124885 |
| HCMV | 296 | 229354 |

The sliding window plot is a useful descriptive tool that helps identify clusters visually. The plot displays the frequency distribution of palindromes on the genome by reporting, at selected sequence positions (X), the counts (Y) of palindromes in a sliding window of size $w$. As an illustration, take the human cytomegalovirus (HCMV). We can use a window of size $w = 1000$ bp starting at the first base and sliding it forward in steps of 500 bp so that successive windows overlap by half. The number of palindromes contained in the window is counted at each step. A plot of the palindrome counts is shown in Figure 12.1.

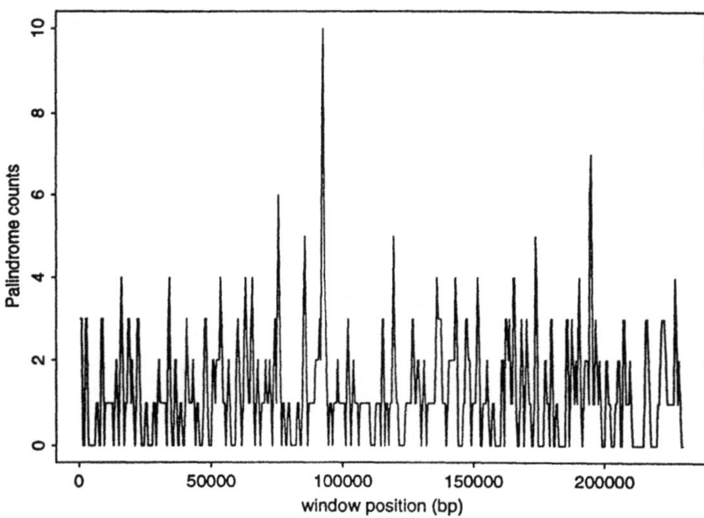



**Figure 12.1:** Sliding window plot of the palindrome counts of HCMV

Figure 12.1 shows the highest peak of 10 palindromes occurring at the window spanning the segment from 92001 bp to 93000 bp. Following this guide,

Masse *et al.* (1992) carried out detailed experimental assays around this part of the genome and characterized the segment between 92210 bp and 93715 bp as the lytic origin of replication for HCMV. Note also that the second highest peak is located at the window from 195000 to 196000 bp. This segment contains a well-known enhancer element which potentiates the transcriptional activity of certain genes [Weston (1988)].

If the identification of palindrome clusters is to serve as a computational tool for automated DNA sequence analysis by pinpointing biologically important regions of the genome, this underlying approach will require more rigorous statistical guidelines. Based on this premise, the following question arises: Does a peak observed in the sliding window plot actually represent a significant cluster, or could it solely be due to chance? Either the minimal $r$-scan $A_r$, or equivalently, the scan statistic $S_w$ will provide an inferential statistical means by which we can distinguish any significant clusters from those incurred by chance. We shall illustrate with the HCMV data how the analysis is performed using $A_r$.

### 12.2.3  Application of $A_r$

Palindromes, like many other markers of biological interest, are mostly very short compared to the length of the whole DNA sequence and are generally scattered quite evenly over the entire molecule. For illustration, Figure 12.2 displays the Q-Q plot of the location of the palindromes of HCMV versus the uniform quantiles. The palindromes are ordered according to their starting positions expressed in units of a thousand bases. The overall linearity displayed in the plot seems to justify modeling palindrome occurrence along a DNA sequence as points uniformly distributed over the unit interval. In this setting, one can apply $A_r$ to identify any nonrandom clusters.

There is a very simple asymptotic approximation for the distribution of $A_r$ [Cressie (1977), and Dembo and Karlin (1992)]: For any $x > 0$,

$$\lim_{N \to \infty} Pr\{A_r \leq \frac{x}{N^{1+1/r}}\} = e^{-x^r/r!}.$$

When $N$ is large, it yields the following approximation:

$$Pr\left\{A_r \leq w\right\} \approx 1 - \exp\left(\frac{-N^{r+1}w^r}{r!}\right). \tag{12.1}$$

Leung, Schachtel, and Yu (1994) made use of (12.1) to derive values of the critical $r$-spacings at various values of $r$ and levels of significance $\alpha$. Locations of palindrome clusters with $r$-spacings less than the critical value are identified.

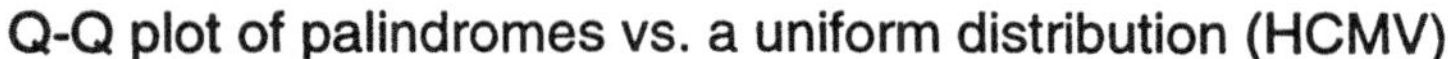

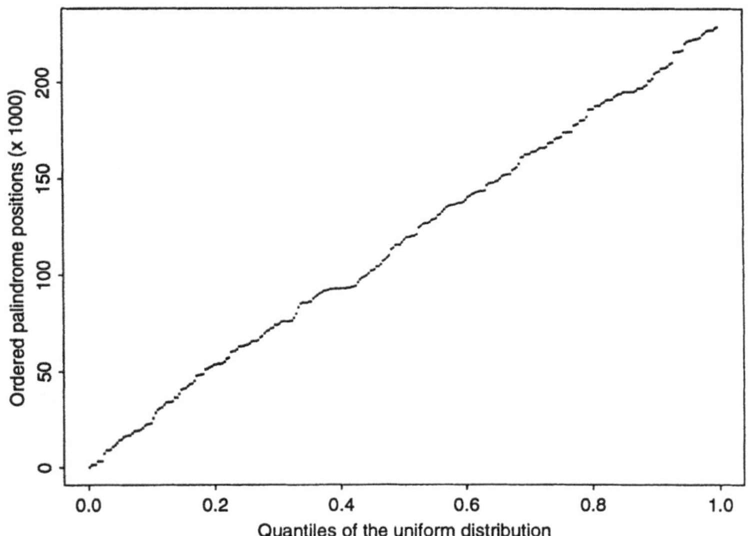



**Figure 12.2:** Q-Q plot of the ordered starting positions of 296 palindromes on the HCMV genome versus the uniform quantiles. The $i$th quantile is calculated by $(i - 0.5)/296$.

Table 12.2 is a listing of the positions of all statistically significant clusters at $\alpha = 0.05$ and $r = 1, ..., 15$ in the HCMV genome. These positions clearly indicate that the region between 91500 bp and 93000 bp is unusually rich in palindromes. This confirms that the highest peak of palindrome counts revealed by the sliding window plot is indeed a statistically significant cluster. Note also that the palindromes at the enhancer region are indicated by the 4- and 5-scans as significant clusters.

The identification of these nonrandom clusters can also be accomplished by using the traditional scan statistic $S_w$ instead of $A_r$. Note that, for $S_w$, we must choose an appropriate value of $w$ and, for $A_r$, we must also choose an appropriate value of $r$. In applying either statistic, it is difficult to know in advance which parameter choices are most appropriate. Hence, it is often necessary to perform the analysis using a variety of values. For $A_r$, our experience has shown that a range of $r$ from 1 to 15 is generally sufficient to capture all of the clusters of interest, regardless of the DNA sequence analyzed or the biological purpose of the investigation. In contrast, the continuous parameter $w$ must be obtained by dividing a chosen DNA window length (in bp) by the

**Table 12.2:** Starting positions of statistically significant ($\alpha = 0.05$) palindrome clusters on the HCMV genome identified by $A_r$ for $r = 1, ..., 15$

| $r$ | Positions of significant clusters |
|---|---|
| 1 | None |
| 2 | None |
| 3 | None |
| 4 | 92526 92570 92643 92701 195032 195112 |
| 5 | 92526 92570 92643 195032 |
| 6 | 92526 92570 92643 |
| 7 | 92526 92570 92643 |
| 8 | 91953 92526 92570 92643 92701 |
| 9 | 91637 91953 92526 92570 92643 |
| 10 | 91490 91637 91953 92526 92570 |
| 11 | 91490 91637 91953 92526 92570 |
| 12 | 91490 91637 91953 92526 |
| 13 | 91490 91637 91953 |
| 14 | 91490 |
| 15 | 91490 |

total genome length. Clearly, the choice of the range for $w$ would vary from application to application, depending on the data sequence and the biological objective of the analysis. Probably due to its convenience, $A_r$ seems to be used more often in these types of studies involving DNA sequence analysis.

When applying an asymptotic approximation in a practical situation where the number of points $N$ is finite, we must be concerned with the accuracy by which the asymptotic distribution approximates the true distribution of $A_r$. Leung, Schachtel, and Yu (1994) have compared some approximate probabilities computed using approximation (12.1) to simulated results of Glaz (1989). They observed that in many cases there are large discrepancies. It is, therefore, not advisable to routinely apply approximation (12.1) to evaluate the statistical significance of palindrome clusters. An alternative is provided by Dembo and Karlin (1992) using the Chen–Stein Poisson approximation technique. In a similar vein, Arratia, Goldstein, and Gordon (1990) and Glaz et al. (1994) put forth two other Poisson-type approximations to $A_r$. In the next section, we shall briefly review these approximations and present some simulation results specific to the family of herpesviruses, showing that Glaz's compound Poisson distribution produces the best approximation to $A_r$ in these cases. We have, therefore, adopted Glaz's result to evaluate the statistical significance of palindrome clusters in the seven herpesvirus genomes in our data set. These significant palindrome clusters found are listed in Table 12.3.

**Table 12.3:** Locations of palindrome clusters on the herpesviruses that were deemed statistically significant ($\alpha = 0.05$) by the compound Poisson approximate distribution of $A_r$. The corresponding features of biological interest found at those locations are listed in the third column for the three more extensively researched human herpesviruses.

| Genome | Cluster Location | Genome Feature |
|---|---|---|
| HCMV | 91490-92643 | origin of replication (oriLyt) |
| | 195032-195112 | transcriptional regulator |
| EBV | 52787-53311 | origin of replication (oriLyt) |
| | 85174 | |
| HSV1 | 129511 | transcriptional regulator |
| | 146228 | origin of replication (ori$_S$) |
| HSE | 115125-115893 | |
| | 144717-146485 | |
| HSS | 112418-112422 | |
| | 109081-109238 | |
| VZV | 1542 | |

### 12.2.4 Biological implication of the significant clusters

Once the locations of all the statistically significant palindrome clusters have been determined, one natural question arises: Do these locations correspond to important sites of genetic activity? This question cannot be addressed by mathematics but can only be answered through experimental results. We are able to gather some information on three of the herpesviruses: HCMV, EBV, and HSV1, for which extensive research has been done in the past.

It has been mentioned earlier that the two significant palindrome clusters on the HCMV genome contain an origin of replication [Masse *et al.* (1992)] and an enhancer element [Weston (1988)]. The genetic map of the EBV genome [Farrell (1993)] lists position 52787 as the start of the site of an origin of replication. For the HSV1 genome, position 129511 is the location of a transcriptional regulator, and position 146228 an origin of replication [McGeoch and Schaffer (1993)]. These regions correspond to the statistically significant locations determined for HCMV, EBV, and HSV1. Thus, it seems reasonable to expect that the regions of significant clusters in the other herpesviruses may be likely candidates for origins of replication or gene regulators.

## 12.3 Approximate Distributions of $A_r$

We shall look at three Poisson-type approximate distributions for $A_r$, all related in some way to the Chen–Stein Poisson approximation technique [Chen (1975)] for sums of dependent Bernoulli random variables. Arratia, Goldstein, and Gordon (1990) have provided an excellent introduction of this techique along with many interesting applications.

### 12.3.1   The finite Poisson approximation

Arratia, Goldstein, and Gordon (1989) have formulated a version of the Chen–Stein Poisson approximation theorem, which is applied by Dembo and Karlin (1992), to yield

$$Pr(A_r \leq w) \approx 1 - \exp\left\{-(N-r)\left[1 - e^{-(N+1)w}\sum_{i=0}^{r-1}\frac{w^i(N+1)^i}{i!}\right]\right\}. \quad (12.2)$$

Leung, Schachtel, and Yu (1994) have called this the finite Poisson approximation to distinguish it from the asymptotic approximation (12.1). Here, we briefly explain how it is obtained.

For every $\alpha$ in a finite or countable index set $I$, let $V_\alpha$ represent a Bernoulli random variable with success probability $p_\alpha$, and $B_\alpha \subset I$ a subset of indices containing $\alpha$. $B_\alpha$ may be thought of as a neighborhood about $\alpha$ such that, for each $\beta \in B_\alpha$, $V_\alpha$ and $V_\beta$ are dependent. Define

$$b_1 = \sum_{\alpha \in I}\sum_{\beta \in B_\alpha} p_\alpha p_\beta,$$

$$b_2 = \sum_{\alpha \in I}\sum_{\alpha \neq \beta \in B_\alpha} p_{\alpha\beta},$$

and

$$b_3 = \sum_{\alpha \in I} E|E\left\{V_\alpha - p_\alpha | \sigma(V_\beta : \beta \notin B_\alpha)\right\}|,$$

where $p_{\alpha\beta} = E[V_\alpha V_\beta]$ and $\sigma(V_\beta : \beta \notin B_\alpha)$ denotes the $\sigma$-algebra generated by the set of random variables $V_\beta$, $\beta \notin B_\alpha$.

The approximation errors in the Chen–Stein method are measured in terms of the total variational distance, which is defined for any two nonnegative integer valued random variables $W_1$ and $W_2$ to be

$$\begin{aligned}
\|\mathcal{L}(W_1) - \mathcal{L}(W_2)\| &= \sup_{\|h\|=1} |E[h(W_1)] - E[h(W_2)]| \\
&= 2\sup_A |P(W_1 \in A) - P(W_2 \in A)|,
\end{aligned}$$

where $\mathcal{L}(W)$ represents the distribution of $W$, $h$ is a real-valued function defined on the common set where the densities of both $W_1$ and $W_2$ are nonzero, $\|h\| = \sup_{k\geq 0}|h(k)|$, and $A$ is any subset of nonnegative integers.

**Theorem 12.3.1** *[Arratia, Goldstein, and Gordon (1989)] Let $W = \sum_{\alpha \in I} V_\alpha$ be the number of occurrences of dependent events, and let $Z$ be a Poisson random variable with $E[Z] = E[W] = \lambda < \infty$. Then,*

$$\|\mathcal{L}(W) - \mathcal{L}(Z)\| \leq 2\left[(b_1 + b_2)\frac{1 - \exp\{-\lambda\}}{\lambda} + b_3(1 \wedge 1.4\lambda^{-1/2})\right],$$

*where $(a \wedge b) = \min(a, b)$.*

This theorem implies that if the quantities $b_1, b_2$, and $b_3$ can be made small, $W$ will have a distribution close to that of the Poisson.

Now consider i.i.d. nonnegative random variables $X_1, X_2, ..., X_N$ with distribution function $F(x)$. For any $w > 0$, one can construct the (dependent) Bernoulli random variables

$$V_i = \begin{cases} 1 & \text{if } X_i + ... + X_{i+r-1} \leq w \\ 0 & \text{otherwise,} \end{cases}$$

$i = 1, 2, ..., N - r + 1$. Let $C_r(w) = \sum_{i=1}^{N-r+1} V_i$ represent the count of those $r$-sums $X_i + ... + X_{i+r-1}$ not exceeding $w$; and $Z_\lambda$ represent a Poisson random variable with mean $\lambda = (N - r + 1)F_r(w)$, where $F_r$ is the distribution of $\sum_{j=1}^{r} X_j$ which is the $r$-fold convolution of $F$. Dembo and Karlin (1992) have derived from the above theorem that

$$\|\mathcal{L}(C_r(w)) - \mathcal{L}(Z_\lambda)\| \leq (1 - e^{-\lambda})\left[(2r - 1)F_r(w) + 2\sum_{m=1}^{r-1} F_m(w)\right].$$

With $D_0, ..., D_N$ being the spacings between $N$ uniformly sampled points from the unit interval, the joint distribution of $(T_{N+1}D_0, ..., T_{N+1}D_N)$ is the same as that of $(E_1, ..., E_{N+1})$ where $E_i$'s are i.i.d. exponential random variables with parameter 1 and $T_{N+1}$ is a gamma$(N+1, 1)$ random variable independent of $D_0, ..., D_N$ [Karlin and Taylor (1981)]. By virtue of this distributional equivalence and the Berry–Essen estimates of the normal approximation to the gamma distribution, Dembo and Karlin (1992) obtained the following result: For $j = 1, ..., N - r$, let

$$V_j = \begin{cases} 1 & \text{if } A_{r,j} \leq w \\ 0 & \text{otherwise.} \end{cases}$$

Let $\tilde{C}_r(w) = \sum_{j=1}^{N-r} V_j$ be the count of $r$-spacings $A_{r,j}$, $j = 1, ..., N - r$, not exceeding $w$, and denote by $G_{(r,1)}$ the distribution function of the gamma$(r, 1)$ random variable. If we define

$$\lambda = (N - r)G_{(r,1)}((N + 1)w) = (N - r)\left[1 - e^{-(N+1)w}\sum_{j=0}^{r-1}\frac{w^j(N + 1)^j}{j!}\right],$$

then

$$\|\mathcal{L}(\tilde{C}_r(w)) - \mathcal{L}(Z_\lambda)\| \leq 4w(N + 1)(1 - e^{-\lambda}) + O\left(\sqrt{\frac{\log(N + 1)}{N + 1}}\right).$$

When $N$ is large and $w$ is suitably chosen that the right-hand side above is small, then $Z_\lambda$ will provide a good approximation for $\tilde{C}_r(w)$. So we have

$$Pr\{A_r > w\} = Pr\{\tilde{C}_r(w) = 0\} \approx Pr\{Z_\lambda = 0\},$$

yielding approximation (12.2).

## 12.3.2  Declumping approximation

If an $r$-spacing beginning at $X_{(j)}$ is exceedingly small, then there is a higher probability that the next $r$-spacing beginning at $X_{(j+1)}$ will also be small. And, depending upon the "smallness" of the first $r$-spacing, this dependence may well extend to the next $r - 1$ consecutive $r$-spacings, resulting in a clump of these small $r$-spacings. In terms of the previously quoted theorem of Arratia, Goldstein, and Gordon (1989), this would inflate the parameter $b_2$ and degrade the accuracy of the Poisson approximation. This difficulty also occurs in the investigation of the distribution of the longest run of 1's for i.i.d. Bernoulli trials by Arratia, Goldstein, and Gordon (1990) who propose a declumping technique to remedy the situation. Glaz *et al.* (1994) have adopted a similar declumping idea to derive an approximate distribution for $A_r$.

For $1 \le j \le N - r$, define new Bernoulli random variables

$$V_j^* = \begin{cases} 1 & \text{if } V_j = 1 \text{ and } V_i = 0 \text{ for } i = j + 1, ..., \min(j + r - 1, N - r) \\ 0 & \text{otherwise.} \end{cases}$$

Again, we will define the sum $\tilde{C}_r^*(w) = \sum_{j=1}^{N-r} V_j^*$. Note that $V_j^*$ will be 1 only if a "small" $r$-spacing is followed by $r - 1$ "large" $r$-spacings, "small" and "large" determined by the choice of $w$. Thus, every success will be separated by at least $r - 1$ failures which remove the clumping effect.

Glaz *et al.* (1994) have obtained the following:

$$Pr\{A_r \ge w\} = Pr\{\tilde{C}_r^*(w) = 0\} \approx \exp\{-\lambda\}, \tag{12.3}$$

where

$$\lambda = \sum_{j=1}^{r-1} Q_j + (N - 2r + 1)Q_r$$

for $2 \le r \le N/2$. Expressions for $Q_1$ and $Q_2$ have been calculated by Berman and Eagleson (1985) and can be expressed as

$$Q_1 = \sum_{j=r}^{N} b(j; N, w), \tag{12.4}$$

$$Q_2 = \sum_{j=r}^{N} (-1)^{r+j} b(j; N, w) \tag{12.5}$$

where $b(j; N, w)$ is the binomial probability $\binom{N}{j} w^j (1-w)^{N-j}$. For $3 \leq k \leq r \leq N/2$, Glaz (1992) has given

$$Q_k = b(r; N, w) - b(r+1; N, w) + \sum_{j=k}^{N-r} (-1)^j \prod_{i=1}^{k-2} \left[ 1 - \frac{j(j-1)}{i(i+1)} \right] b(r+j; N, w).$$

### 12.3.3  Compound Poisson approximation

Roos (1993) has shown that for $r \geq 2$ and in the presence of clumping, the rate of convergence between the distribution of $A_r$ and the Poisson approximation can be improved by using, instead, an appropriately chosen compound Poisson distribution. Define a compound Poisson random variable as $\sum_{i \geq 1} i N_i$, where each $N_i$ is an independent Poisson random variable with mean $\lambda_i$, $i = 1, 2, \ldots$, and $\sum_{i \geq 1} \lambda_i$ is finite. Roos (1993) has considered points uniformly distributed on the unit circle and has applied the general bounds on the compound Poisson approximation derived by Barbour, Holst, and Janson (1992). By using a coupling argument and explicitly choosing $\lambda_i$'s such that $i \lambda_i$ decreases to zero as $i$ increases to infinity, Roos has obtained bounds on the total variational distance between the count of small $r$-spacings and this compound Poisson distribution. The rate of convergence for the compound Poisson approximation is of the order $O(1/N)$ for all $r \geq 1$.

Glaz *et al.* (1994) have adapted Roos' results to consider $r$-spacings between points on the unit interval rather than the unit circle. With $1 \leq r \leq \frac{N+1}{2}$, $\frac{1}{N} < Nw < 1$, and an approximation for $\lambda_i$ suggested in Aldous (1989), they obtained the following compound Poisson approximation for the distribution of the minimal $r$-scan:

$$P\{A_r \geq w\} \approx \exp\{-(N-r)\pi(1 - p + p^r(r + p - rp))\}, \tag{12.6}$$

where $\pi = Q_1$, $p = 1 - \frac{Q_2}{Q_1}$, with $Q_1$ and $Q_2$ as defined in (12.4) and (12.5), respectively.

### 12.3.4  Comparison with simulated probabilities

It has already been mentioned by Glaz *et al.* (1994) and Leung, Schachtel, and Yu (1994) that the finite Poisson approximation (12.2) is not always sufficiently accurate. In order to evaluate the quality of the local declumping and compound Poisson approximations in terms of assessing the statistical significance of palindrome clusters, we compare the calculated approximation probabilities with those from a simulation with suitable values of $N$, $r$, and $w$ chosen to reflect the genome structure of the herpesviruses.

In HCMV, for example, there are 296 palindromes with length $\geq 10$ bp. We conducted a simulation consisting of $20,000$ trials which sampled $N = 296$ points from the uniform $(0,1)$ distribution using S-plus. The minimal $r$-spacing was determined for $r = 1, \ldots, 20$ and the probability, $P\{A_r \leq w\}$, obtained for window size $w = 1000/229354 \approx 0.004$. These values of $r$ and $w$ were chosen based on the observation that the functional sites of interest on these viral genomes usually span a sequence segment on the order of about 1000 bp and would extremely rarely, if at all, contain more than 20 palindromes of length $\geq 10$ bp. A comparison of the simulated probabilities with the Poisson and compound Poisson approximations given in (12.2), (12.3), and (12.6) are displayed with the corresponding errors in Table 12.4. Similar comparisons were also carried out for the HCMV genome by varying the value of $w$, as well as for the other herpesvirus genomes listed in Table 12.1 with appropriate adjustments for the number of palindromes $N$ and the genome length.

**Table 12.4:** Approximations and associated errors with respect to the corresponding simulated values $(SIM)$ of $Pr\{A_r \leq w\}$ for HCMV with $N = 296$, $w = 1000/229354$, using the compound Poisson $(CP)$, local declumping $(LD)$, and the finite $(F)$ Poisson approximations. The simulation consists of 20,000 trials.

| $w$ | $r$ | $SIM$ | $CP$ | $\lvert SIM - CP \rvert$ | $LD$ | $\lvert SIM - LD \rvert$ | $F$ | $\lvert SIM - F \rvert$ |
|---|---|---|---|---|---|---|---|---|
| $\approx 0.004$ | 1 | 1.0000 | 1.0000 | 0.0000 | — | — | 1.0000 | 0.0000 |
| | 2 | 1.0000 | 1.0000 | 0.0000 | 1.0000 | 0.0000 | 1.0000 | 0.0000 |
| | 3 | 1.0000 | 1.0000 | 0.0000 | 1.0000 | 0.0000 | 1.0000 | 0.0000 |
| | 4 | 0.9999 | 0.9997 | 0.0002 | 0.9990 | 0.0009 | 1.0000 | 0.0001 |
| | 5 | 0.8684 | 0.8627 | 0.0057 | 0.8442 | 0.0242 | 0.9529 | 0.0845 |
| | 6 | 0.3387 | 0.3445 | 0.0058 | 0.3354 | 0.0033 | 0.4698 | 0.1311 |
| | 7 | 0.0731 | 0.0747 | 0.0016 | 0.0731 | 0.0000 | 0.1077 | 0.0346 |
| | 8 | 0.0115 | 0.0124 | 0.0009 | 0.0121 | 0.0006 | 0.0179 | 0.0064 |
| | 9 | 0.0024 | 0.0018 | 0.0006 | 0.0017 | 0.0007 | 0.0025 | 0.0001 |
| | 10 | 0.0002 | 0.0002 | 0.0000 | 0.0002 | 0.0000 | 0.0003 | 0.0001 |
| | $\geq 11$ | 0.0000 | 0.0000 | 0.0000 | 0.0000 | 0.0000 | 0.0000 | 0.0000 |

It can be seen from Table 12.4 that the local declumping approximation and the compound Poisson approximation both perform better than the finite Poisson approximation. It is, however, not as obvious in Table 12.4 which of the local declumping or compound Poisson approximation works better than the other. In all of our 787 comparisons, over 87% of the cases indicate that the compound Poisson approximation error is no bigger than that of the local declumping approximation. We have, therefore, chosen to use the compound Poisson approximation in the previous section to determine the location of statistically significant clusters on the family of herpesvirus genomes.

Although not as frequently used in DNA sequence analysis as the minimal $r$-scan $A_r$, the higher order $r$-scans $A_{r,(k)}$, $k = 2, \ldots, N - r$, also have approximation distributions derived from the three methods above. Glaz *et al.* (1994) have remarked that the compound Poisson approximation provides an even more pronounced improvement over the local declumping approximation for the $k$th minimal $r$-scan length $A_{r,(k)}$ when $k > 1$.

## 12.4 Concluding Remarks

Presently, it is fairly common for computer DNA sequence analysis programs to include statistical modules for discerning significantly nonrandom sequence features from those occurring purely by chance. The statistical guidelines can then help design finely tuned laboratory experiments. However, one must keep in mind that statistical significance and biological relevance are two independent concepts. It is well possible that a palindrome cluster is highly significant from a statistical point of view but has little biological relevance. On the other hand, there are biologically active sequence sites that do not contain any statistically significant palindrome clusters. Since statistical assessment is always based on simplified model assumptions, significance levels should be regarded as mere benchmarks. While potentially important DNA features may be identified by statistics, the final conclusion has to come from biological experiments.

The comparison of the different Poisson-type approximations to the minimal $r$-scan distribution discussed in this chapter is an effort to incorporate the best available statistical criterion for assessing palindrome clusters. Recently, Huffer and Lin (1998a,b) have developed a recursive algorithm capable of calculating the $r$-scan probabilities to any desired degree of accuracy. Implementation of this algorithm will be more involved than the Poisson-type approximations, but it will offer a new dimension of flexibility that allows the user greater control over the accuracy of the statistical criterion employed.

This chapter has focused upon the scan statistic for i.i.d. uniformly distributed points over the unit interval because it seems to have a most immediate application in identifying origins of replication and gene regulatory sites. The works of Glaz and Naus (1991), Sheng and Naus (1994), and Naus and Sheng (1996) contain interesting discussions of the scan statistic for sequences of i.i.d. discrete random variables in relation to DNA sequence analysis. Their results are applied to the analysis of longest matching or almost matching words in nucleic acid and amino acid sequences, as well as in the identification of electric charge clusters on amino acid sequences. It is anticipated that the ability to identify significant charge clusters will help elucidate the three-dimensional structure and function of protein molecules. As the human and other genome projects evolve, one can envision many other applications of the scan statistic that will help extract useful information from the enormous, and still rapidly growing, amount of DNA sequence data.

**Acknowledgment.** This work was supported in part by the National Science Foundation (HRD-9628514).

# References


1. Aldous, D. (1989). *Probability Approximations via the Poisson Clumping Heuristic*, New York: Springer-Verlag.

2. Arratia, R., Goldstein, L. and Gordon, L. (1989). Two moments suffice for Poisson approximations: The Chen-Stein method, *Annals of Probability*, **17**, 9–25.

3. Arratia, R., Goldstein, L. and Gordon, L. (1990). Poisson approximation and the Chen-Stein method, *Statistical Science*, **5**, 403–434.

4. Barbour, A. D., Holst, L. and Janson, S. (1992). *Poisson Approximation*, Oxford: Clarendon Press.

5. Berman, M. and Eagleson, G. K. (1985). A useful upper bound for the tail probabilities of the scan statistic when the sample size is large, *Journal of the American Statistical Association*, **80**, 886–889.

6. Chen, L. H. Y. (1975). Poisson approximation for dependent trials, *Annals of Probability*, **3**, 534–545.

7. Cressie, N. (1977). The minimum of higher order gaps, *Austalian Journal of Statistics*, **19**, 132–143.

8. Dembo, A. and Karlin, S. (1992). Poisson approximations for $r$-scan processes, *Annals of Applied Probability*, **2**, 329–357.

9. Doolittle, R. F. (Ed.) (1990). Molecular Evolution: Computer Analysis of Protein and Nucleic Acid Sequences, *Methods of Enzymology*, **183**, San Diego: Academic Press.

10. Farrell, P. J. (1993). Epstein-Barr virus, In *Genetic Maps Sixth Edition, Book 1 Viruses* (Ed., S. J. Brien), pp. 1.120–1.133, Cold Spring Harbor, NY: Cold Spring Harbor Laboratory Press.

11. Glaz, J. (1989). Approximations and bounds for the distribution of the scan statistic, *Journal of the American Statistical Association*, **84**, 560–566.

12. Glaz, J. (1992). Approximations for tail probabilities and moments of the scan statistic, *Computational Statistics & Data Analysis*, **14**, 213–227.

13. Glaz, J. and Naus, J. (1991). Tight bounds and approximations for scan statistic probabilities for discrete data, *Annals of Applied Probability*, **1**, 306–318.

14. Glaz, J., Naus, J., Roos, M. and Wallenstein, S. (1994). Poisson approximations for the distribution and moments of ordered $m$-spacings, *Journal of Applied Probability*, **31**, 271–281.

15. Huffer, F. E. and Lin, C.-T. (1998a). Computing the exact distribution of the extremes of sums of consecutive spacings, *Computational Statistics & Data Analysis* (to appear).

16. Huffer, F. E. and Lin, C.-T. (1998b). Approximating the distribution of the scan statistic using moments of the number of clumps, *Journal of the American Statistical Association* (to appear).

17. Karlin, S., Blaisdell, B. E., Sapolsky, R. J., Cardon, L. and Burge, C. (1993). Assessments of DNA inhomogeneities in yeast chromosome III, *Nucleic Acids Research*, **21**, 703–711.

18. Karlin S. and Brendel, V. (1992). Chance and statistical significance in Protein and DNA sequence analysis, *Science*, **257**, 39–49.

19. Karlin S. and. Cardon, L. R. (1994). Computational DNA sequence analysis, *Annual Reviews of Microbiology*, **48**, 619–654.

20. Karlin, S., Mrázek, J. and Campbell, A. M. (1996). Frequent oligonucleotides and peptides of the Haemophilus influenzae genome, *Nucleic Acids Research*, **24**, 4263–4272.

21. Karlin, S., Mrázek, J. and Campbell, A. M. (1997). Compositional biases of bacterial genomes and evolutionary implications, *Journal of Bacteriology*, **179**, 3899–3913.

22. Karlin, S. and Taylor, H. M. (1981). *A Second Course in Stochastic Processes*, Second edition, New York: Academic Press.

23. Labrecque, L. G., Barnes, D. M., Fentiman, I. S. and Griffin, B. E. (1995). Epstein-Barr virus in epithelial cell tumors: a breast cancer study, *Cancer Research*, **55**, 39–45.

24. Leung, M. Y., Blaisdell, B. E., Burge, C. and Karlin, S. (1991). An efficient algorithm for identifying matches with errors in multiple long molecular sequences, *Journal of Molecular Biology*, **221**, 1367–1378.

25. Leung, M. Y., Schachtel, G. A. and Yu, H. S. (1994). Scan statistics and DNA sequence analysis: the search for an origin of replication in a virus, *Nonlinear World*, **1**, 445–471.

26. Leung, M. Y., Marsh, G. M. and Speed, T. P. (1996). Over- and under-representation of short DNA words in herpesvirus genomes, *Journal of Computational Biology*, **3**, 345–360.

27. Masse, M. J., Karlin, S., Schachtel, G. A. and Mocarski, E. S. (1992). Human cytomegalo virus origin of DNA replication (oriLyt) resides within a highly complex repetitive region, *Proceedings of the National Academy of Science USA*, **89**, 5246–5250.

28. McGeoch, D. J. and Schaffer, P. A. (1993). Herpes Simplex Virus, In *Genetic Maps Sixth Edition, Book 1 Viruses* (Ed., S. J. O'Brien), pp. 1.147-1.156, Cold Spring Harbor, NY: Cold Spring Harbor Laboratory Press.

29. Naus, J. I. and Sheng, K.-N. (1996). Screening for unusual matched segments in multiple protein sequences, *Communications in Statistics— Simulation and Computation*, **25**, 937–952.

30. Roos, M. (1993). Compound Poisson approximations for the numbers of extreme spacings, *Advances in Applied Probability*, **25**, 847–874.

31. Sheng, K. and Naus, J. (1994). Pattern matching between two non-aligned random sequences, *Bulletin of Mathematical Biology*, **56**, 1143–1162.

32. Vital, C., Monlun, E., Vital, A., Martin-Negrier, M. L., Cales, V., Leger, F., Longy-Boursier, M., Le Bras, M. and Bloch, B. (1995). Concurrent herpes simplex type 1 necrotizing encephalitis, cytomegalovirus ventriculoencephalitis and cerebral lymphoma in an AIDS patient, *Acta Pathologica*, **89**, 105–108.

33. Waterman, M. S. (Ed.) (1989). *Mathematical Methods for DNA Sequences*, Boca Raton: CRC Press.

34. Waterman, M. S. (1995). *Introduction to Computational Biology*, New York: Chapman and Hall.

35. Weston, K. (1988). An enhancer element in the short unique region of human cytomegalo virus regulates the production of a group of abundant immediate early transcripts, *Virology*, **162**, 406–416.

# 13

## On the Probability of Pattern Matching in Nonaligned DNA Sequences: A Finite Markov Chain Imbedding Approach


**James C. Fu, W. Y. Wendy Lou, and S. C. Chen**

*University of Manitoba, Winnipeg, Manitoba, Canada*
*Mount Sinai School of Medicine, New York, NY*
*National Donghwa University, Hualian, Taiwan, R.O.C.*



**Abstract:** Mathematically, a DNA segment can be viewed as a sequence of four-state $(A, C, G, T)$ trials, and a perfect match of size $M$ occurs when two DNA sequences have at least one identical subsequence (or pattern) of length $M$. Pattern matching probabilities are crucial for statistically rigorous comparisons of DNA (and other) sequences, and many bounds and approximations of such probabilities have recently been developed. There are few results on exact probabilities, especially for trials with unequal state probabilities, and no exact analytical formulae for the pattern matching probability involving arbitrarily long nonaligned sequences. In this chapter, a simple and efficient method based on the finite Markov chain imbedding technique is developed to obtain the exact probability of perfect matching for i.i.d. four-state trials with either equal or unequal state probabilities. A large deviation approximation is derived for very long sequences, and numerical examples are given to illustrate the results.

**Keywords and phrases:** Aligned and nonaligned DNA sequences, matching probability, finite Markov chain imbedding, large deviation approximation


## 13.1   Introduction

Deoxyribonucleic acid (DNA) molecules form the blueprint for life on earth, and each strand of the famous double helix is a linear combination of the polymerized nucleotides (bases) adenine $(A)$, guanine $(G)$, cytosine $(C)$, and thymine $(T)$. For our purposes, a DNA segment can be viewed as a sequence of trials over this four letter alphabet $\mathcal{B} = \{A, C, G, T\}$. Molecular biologists routinely compare newly discovered DNA sequences to existing databases such as Genbank (USA),

EMBL (Europe), and DDBJ (Japan), as genes with similar sequences often share similar function, resulting protein structure, and/or evolutionary origin. These databases presently contain hundreds of thousands of sequence entries corresponding to hundreds of millions of sequence nucleotides, and it is not unlikely that a new sequence will appear similar to a number of known ones. Such apparent similarities could, however, be due purely to chance, and hence there exists an urgent need to study matching probabilities for various chance models. To that end, this chapter formulates the exact probability of pattern matching between two nonaligned DNA sequences resulting from independent and identically distributed (i.i.d.) four-state trials. Here we discuss similarity based on primary structure or the linear sequence of nucleotides, as opposed to the secondary or tertiary spatial molecular structure, and focus on pattern matching instead of scoring matrices [e.g., Waterman (1995), and references therein].

The matching problem for simple patterns between two aligned sequences (i.e., without shifts) has been studied extensively in the literature. For instance, Erdös and Révész (1975), Gordon, Schilling, and Waterman (1986), and Karlin and Ost (1987, 1988) found asymptotic results. Naus (1974) derived an exact formula, and Fu and Curnow (1990) obtained recursive relations, but they are limited in use due to their computational complexity; Glaz and Naus (1991) recently gave accurate approximations and tight bounds. For nonaligned sequences (i.e., with shifts), the matching problem is more difficult since all possible alignments (shifts) need to be considered. Several bounds and approximations have been obtained. For example, Hunter (1976), Hoover (1990), and Glaz (1993) determined lower bounds; Mott, Kirwood, and Curnow (1990) and Sheng and Naus (1994) found good approximations. Asymptotic formulae were given by Arratia, Gordon, and Waterman (1986, 1990), and some results for matching among multiple sequences have been obtained by, for instance, Leung *et al.* (1991) and Naus and Sheng (1997). Most existing results for the probability of pattern matching between nonaligned sequences assume that the sequences are independent and that each sequence consists of i.i.d. four-state trials. The state probabilities are often assumed equal ($p_A = p_C = p_G = p_T = 1/4$) to reduce the computational complexity, but here we are able to incorporate unequal state probabilities with relative ease.

Computing exact pattern matching probabilities for nonaligned sequences has in the past required tremendous computational effort for anything but very short sequences, and large sample asymptotic approximations are only applicable to very long sequences. In this chapter, a simple, new and very efficient method, based on the finite Markov chain imbedding technique (FMCI) of Fu and Koutras (1994), is developed to derive the *exact* probability of pattern matching for arbitrarily long sequences. The FMCI technique has been used successfully to study the exact distributions of various types of runs and patterns in a sequence of two- or multistate Markov-dependent trials [Fu and

Koutras (1994), Koutras and Alexandrou (1995), Lou (1996) and Fu (1996), for example]. Our proposed approach to pattern matching is intuitive, naturally incorporates unequal state probabilities, and can easily be extended to Markov-dependent models. For very long sequences, a simple alternative approximation based on large deviations is suggested. Numerical results are given to illustrate the theoretical results.

This chapter is organized as follows: Definitions of patterns and finite Markov chain imbedding are given in Section 13.2, followed by the derivation of the exact probability for perfect matching in Section 13.3. In Section 13.4, the large deviation approximation is derived, and numerical results are presented in Section 13.5. A brief discussion is provided in the concluding section.

---

## 13.2 The Matching Problem and Notation

Consider two independent sequences, $W$ and $R$. Let $W = \{W_1, \cdots, W_d\}$ be a sequence of $d$ independent and identically distributed (i.i.d.) four-state trials having probability mass function

$$P(W_i = b) = p_{1b}, \quad i = 1, \cdots, d, \text{ and } b = \{A, C, G, T\}.$$

Similarly, let $R = \{R_1, \cdots, R_n\}$ be a sequence of $n$ i.i.d. four-state trials having probability mass function

$$P(R_j = b) = p_{2b}, \quad j = 1, \cdots, n, \text{ and } b = \{A, C, G, T\}.$$

For each pair of indices $(i, j)$, $1 \leq i \leq d$, $1 \leq j \leq n$, define a zero/one random variable

$$Z_{ij} = \begin{cases} 1 & \text{if } W_i = R_j \\ 0 & \text{otherwise.} \end{cases} \tag{13.1}$$

We say that there is a *match* between position $i$ of sequence $W$ and position $j$ of sequence $R$ when $Z_{ij} = 1$. Without loss of generality, we assume $d \leq n$ throughout this article.

For the two nonaligned sequences $W$ and $R$, the scan statistic $S_M$ [Naus (1965)] of window size $M$, $1 \leq M \leq d$, is given by

$$S_M(d, n) = \max_{1 \leq i \leq d-M+1, 1 \leq j \leq n-M+1} \sum_{t=0}^{M-1} Z_{i+t, j+t}, \tag{13.2}$$

and can be viewed as the maximum number of matching words within $M$ consecutive trials between two sequences where all possible shifts are considered.

When shifts are not allowed and the two sequences are of equal length ($d = n$), the scan statistic (13.2) for two aligned sequences simplifies to

$$S_M(d, d) = \max_{1 \leq i \leq d-M+1} \sum_{t=0}^{M-1} Z_{i+t,i+t}. \tag{13.3}$$

**Definition 13.2.1** For $1 \leq M \leq d$ and $1 \leq m \leq M$, an event $E_{d,n}(m, M)$ occurs when no $(m, M)$ matching exists between two nonaligned sequences of lengths $d$ and $n$, i.e.,

$$E_{d,n}(m, M) = \{S_M(d, n) < m\}. \tag{13.4}$$

For $m = M$, event $E_{d,n}^{c}(M, M)$ is called a "perfect" match, where the superscript "$c$" denotes the complement of the event, and for $m = M - 1$, event $E_{d,n}^{c}(M-1, M)$ is called an "almost perfect" match. In words, event $E_{d,n}(m, M)$ means that there are always less than $m$ matches between *any* subsequence of size $M$ from sequence $W$ and *any* subsequence of size $M$ from sequence $R$. Event $E_{d,n}^{c}(M, M)$ means that there exists at least one subsequence of size $M$ from each of the two sequences ($W$ and $R$) which are perfectly matched. Mathematically, the event $E_{d,n}^{c}(M, M)$ is equivalent to saying that there exists at least one pair $(i, j)$, $1 \leq i \leq d - M + 1$ and $1 \leq j \leq n - M + 1$, such that $\sum_{t=0}^{M-1} Z_{i+t,j+t} = M$. In what follows, the window size $M$ is also referred to as pattern size.

The probabilities of event $E_{d,n}(m, M)$, $m = 1, \cdots, M$, for various kinds of models are of great interest in molecular genetics applications. Here, we focus mainly on the exact probability for perfect matching $E_{d,n}^{c}(M, M)$, which is most commonly studied. In order to investigate the exact probability of perfect matching using the finite Markov chain imbedding technique, we first define two types of patterns.

**Definition 13.2.2** For a given positive integer $M$, if $\Lambda$ is composed of a specified sequence of $M$ symbols (or bases), i.e., $\Lambda = b_1 b_2 \cdots b_M$, then $\Lambda$ is a called a simple pattern of size $M$.

For example, $\Lambda = AGTA$ is a simple pattern of size 4, and $\Lambda = ACG$ is a simple pattern of size 3.

**Definition 13.2.3** Given $k$, if $\Lambda$ is a union of $k$ distinct simple patterns, i.e., $\Lambda = \Lambda_1 \cup \Lambda_2 \cup \cdots \cup \Lambda_k$, then $\Lambda$ is called a compound pattern.

For example, if $\Lambda_1 = ACA$ and $\Lambda_2 = AGTA$, then the compound pattern $\Lambda = \cup_{i=1}^{2} \Lambda_i$ means that either $ACA$ or $AGTA$ is considered. Note that the sizes of these simple patterns composing the compound pattern do not have to be the same. Hence, the compound pattern $\Lambda$ may not have a fixed length.

For a specified pattern $\Lambda$ (simple or compound), we now define an integer random variable $X_n(\Lambda)$ from a sequence of $n$ four-state trials $\{X_i\}_1^n$ as

$$X_n(\Lambda) = \text{the number of } \Lambda \text{ patterns that occurred in the sequence } \{X_i\}_1^n.$$

In general, the number of $\Lambda$ patterns that occurred in a sequence depends on the counting method, no-overlap-counting or overlap-counting. For example, consider a sequence of ten trials $\{AACACAGTAC\}$, and patterns $\Lambda_1 = ACA$ and $\Lambda_2 = AGTA$. Under no-overlap-counting, $X_{10}(\Lambda_1) = 1$ and $X_{10}(\Lambda) = X_{10}(\Lambda_1 \cup \Lambda_2) = 2$, while under overlap-counting, $X_{10}(\Lambda_1) = 2$ and $X_{10}(\Lambda) = 3$. Unless otherwise specified, throughout this chapter overlap-counting is used for $X_n(\Lambda)$.

**Definition 13.2.4** The random variable $X_n(\Lambda)$ is finite Markov chain imbeddable if

(i) there exists a finite Markov chain $\{Y_t : t = 1, \cdots, n\}$ defined on a finite state space $\Omega$ with transition probability matrices $M_t$, $t = 1, \cdots, n$, and initial probability $\pi_0$,

(ii) there exists a partition $\{C_x, x = 0, 1, \cdots, l\}$ on the state space $\Omega$ (where $C_x$ and $l$ may depend on $n$), and

(iii) for every $x = 0, 1, \cdots, l$,

$$P(X_n(\Lambda) = x) = P(Y_n \in C_x | \pi_0).$$

If $X_n(\Lambda)$ can be imbedded into a finite Markov chain $\{Y_t, t = 1, \cdots, n\}$, then it follows from the Chapman–Kolmogorov theorem [see Fu and Koutras (1994)] that the distribution of $X_n(\Lambda)$ can be obtained from

$$P(X_n(\Lambda) = x) = \pi_0 \left( \prod_{t=1}^{n} M_t \right) U'(C_x), \tag{13.5}$$

where $U'(C_x)$ is the transpose of $U(C_x)$, $U(C_x) = \sum_{a \in C_x} U(a)$, and $U(a) = (0, \cdots, 0, 1, 0, \cdots, 0)$ is a unit vector associated with state $a \in \Omega$.

Further, the moments (and hence the mean and variance) are given via

$$E(X_n^k(\Lambda)) = \pi_0 \left( \prod_{t=1}^{n} M_t \right) V'_x, \qquad k = 1, 2, \cdots, \tag{13.6}$$

where $V'_x$ is the transpose of $V_x$ and $V_x = \sum_{a \in C_x} x^k U(a)$.

In view of the above definition and (13.5), the exact probability for pattern matching can be obtained once one properly constructs the three essential components for imbedding:

- a proper finite state space $\Omega$ based on the structure of the pattern $\Lambda$,

- a proper partition $\{C_x\}$ on the state space $\Omega$ for each of the $x = 0, 1, \ldots, l$,

- a finite Markov chain and its transition probability matrices $M_t$, $t = 1, \ldots, n$.

---

## 13.3   Perfect Matching Probability

In the following, we assume for simplicity that the two nonaligned sequences $W$ and $R$ are i.i.d. and that they have common probability mass function

$$P(W_i = b) = P(R_j = b) = p_b, \qquad b = \{A, C, G, T\},$$

for all $i = 1, \cdots, d$, and $j = 1, \cdots, n$.

For a given pattern $\Lambda$ of size $M$, we now redefine, given the simplicity of this problem, $X_n(\Lambda)$ to be a zero/one random variable for the occurrence of pattern $\Lambda$ in sequence $R$. To illustrate how the FMCI technique described in the previous section can be used to study the exact probability for perfect matching of pattern size $M$, a simple example is given below.

**Example 13.3.1** Suppose that sequence $W$ is specified, say $W = \{AAGT\}$, and that sequence $R$ has length five, $R = \{R_j\}_1^5$. Given $M = 3$, there are two distinct simple patterns of size three associated with $W$, $\Lambda_1 = AAG$ and $\Lambda_2 = AGT$. Let $\Lambda$ be the compound pattern generated by $W$, that is $\Lambda = \Lambda_1 \cup \Lambda_2$. We first decompose $\Lambda_1$ and $\Lambda_2$ into four ending block states: $1 \overset{\triangle}{=} A$, $2 \overset{\triangle}{=} AA$, $3 \overset{\triangle}{=} AG$, $0 \overset{\triangle}{=}$ otherwise.

Define a Markov chain operating on $R$:

$$Y_t(R) = (X_t(\Lambda), E_t), \quad t = 1, 2, 3, 4, 5,$$

where $X_t(\Lambda)$ is the indicator of whether compound pattern $\Lambda$ has occurred in the first $t$ trials of sequence $R$ ($X_t(\Lambda)$ is zero if not, and one otherwise.) $E_t$ is the backward counting ending block [see Fu (1996)] of the first $t$ trials, and is defined according to the above decomposition of the pattern $\Lambda$ (hence $E_t$ takes on the values 0,1,2,3). The state space can be defined as

$$\Omega = \{(0, v) : v = 0, 1, 2, 3\} \cup \{(1, 0)\},$$

where $(1, 0)$ is an absorbing state for the destination of transitions when pattern matching has occurred. For instance, given the sequence $R = \{CAGTC\}$, the realization of the Markov chain $\{Y_t(R) : t = 1, 2, 3, 4, 5\}$ is $\{(0, 0), (0, 1), (0, 3),$

$(1,0)$, $(1,0)\}$. Note that once a perfect match has occurred, $Y_t$ stays at the absorbing state; i.e., $Y_t = (1,0)$ means that there is at least one pattern $\Lambda$ in the first $t$ trials. The partitions on the state space $\Omega$ are $C_0 = \{(0,v) : v = 0,1,2,3\}$ and $C_1 = \{(1,0)\}$.

Since all trials are considered to be homogeneous, the transition probability matrices of the imbedded Markov chain $Y_t$ are the same for all $t$, and can be expressed as

$$
M_t(W) = \begin{array}{c} \\ (0,0) \\ (0,1) \\ (0,2) \\ (0,3) \\ (1,0) \end{array}
\begin{array}{ccccc}
(0,0) & (0,1) & (0,2) & (0,3) & (1,0) \\
\end{array}
\left(
\begin{array}{ccccc}
1 - p_A & p_A & 0 & 0 & 0 \\
1 - p_A - p_G & 0 & p_A & p_G & 0 \\
1 - p_A - p_G & 0 & p_A & 0 & p_G \\
1 - p_A - p_T & p_A & 0 & 0 & p_T \\
0 & 0 & 0 & 0 & 1
\end{array}
\right), \; t = 1,2,3,4,5.
$$

For nonhomogeneous trials, we would simply replace the probabilities $p_b$ ($b = \{A,C,G,T\}$) with time-dependent probabilities $p_b(t)$ to form the transition probability matrices $M_t$.

Having completed the construction of the three essential components for imbedding ($\Omega$, $C_x$ and $M_t$), we can now obtain the exact probability for matching pattern $\Lambda$.

Given the initial probability $\pi_0 = (1,0,0,0,0)$, the conditional probability of perfect matching for a pattern size of 3 and specified $W$ can then be obtained via

$$
P(E_{4,5}^c(3,3)|W) = P(X_5(\Lambda) > 0|W) = \pi_0 \left( \prod_{t=1}^{5} M_t(W) \right) U'(C_1),
$$

where $U(C_1) = (0,0,0,0,1)$. For instance, given $p_A = 0.1$, $p_C = 0.2$, $p_G = 0.3$, and $p_T = 0.4$, then $P(E_{4,5}^c(3,3)|W = AAGT) = 0.0426$.

In the following, prior to giving a general expression for our FMCI approach, we denote

$$
\mathcal{G}_W = \{\Lambda_1, \cdots, \Lambda_k\}
$$

to be the collection of all distinct simple patterns of size $M$ in sequence $W$, where $M$ and $W$ are given, and further denote

$$
\Lambda_W = \cup_{\Lambda_i \in \mathcal{G}_W} \Lambda_i
$$

to be the compound pattern of size $M$ generated by $W$.

**Theorem 13.3.1** *Given sequence $W$, the random variable $X_n(\Lambda_W)$ is finite Markov chain imbeddable, and the conditional probability for perfect matching of pattern size $M$ is*

$$
P(E_{d,n}^c(M,M)|W) = P(X_n(\Lambda_W) > 0) = \pi_0 \left( \prod_{t=1}^{n} M_t(W) \right) U'_W, \qquad (13.7)
$$

*where $\pi_0 = (1, 0, \cdots, 0)$, $U'_W = (0, \cdots, 0, 1)$, and the transition probability matrices $M_t(W)$ are generated by the compound pattern $\Lambda_W$.*

PROOF. Given $W$, it follows from the definitions of $E_{d,n}(M, M)$ and $X_n(\Lambda_W)$ that $P(E^c_{d,n}(M, M)|W) = P(X_n(\Lambda_W) > 0)$. It is easy to see that the random variable $X_n(\Lambda_W)$ is finite Markov chain imbeddable, and the result (13.7) follows immediately from the Chapman–Kolmogorov equation. This completes the proof.                                                                                ∎

If sequence $W$ is not specified, then all possible sequences of length $d$ have to be considered. Denote

$$\mathcal{F}_d = \{W = (W_1, \cdots, W_d) : W_i = A, C, G, T\}$$
$$= \text{the collection of all possible sequences of length } d.$$

It follows from Theorem 13.3.1 that the following corollary holds.

**Corollary 13.3.1** *If sequence $W$ is not specified, then the probability (unconditional) of perfect matching for pattern size $M$ between two nonaligned sequences is given by*

$$P(E^c_{d,n}(M, M)) = \sum_{W \in \mathcal{F}_d} P(X_n(\Lambda_W) > 0)P(W)$$
$$= \sum_{W \in \mathcal{F}_d} \pi_0 \left( \prod_{t=1}^{n} M_t(W) \right) U'_W \times P(W), \qquad (13.8)$$

*where $P(W)$ is the probability associated with $W$, $W \in \mathcal{F}_d$.*

Pattern sizes of interest in the matching of DNA sequences often range from 2 to 9 ($2 \leq M \leq 9$). Theorem 13.3.1 and Corollary 13.3.1 provide a simple and efficient procedure for computing the exact conditional and unconditional perfect matching probabilities for $2 \leq M \leq 9$ and moderate $d$, as will be illustrated via numerical examples in Section 13.5.

---

## 13.4   A Large Deviation Approximation

Various upper and lower bounds for the matching probability have been proposed [e.g., Hunter (1976), Hoover (1990), Glaz (1993), and Sheng and Naus (1994)], and several approximations have been derived [e.g., Arratia, Gordon, and Waterman (1990), Mott, Kirwood, and Curnow (1990), and Sheng and Naus (1994)]. Numerical comparisons for some selected bounds and approximations have been performed by Sheng and Naus (1994), using simulations to obtain the matching probability.

With the FMCI approach, we are able to efficiently obtain the exact probability of perfect matching for sequences of moderate length. For example, for $d = n = 9$ and $M = 3$, we obtain the exact probability $P(E_{9,9}^c(3,3)) = 0.4716$, which differs by about one percent (relative error) from the estimate derived via simulations by Sheng and Naus (1994), as shown in Table 13.1 along with various other bounds.

**Table 13.1:** $P\{E_{9,9}^c(3,3)\}$ with $p_A = p_C = p_G = p_T = 0.25$

| Bonferroni-Sheng-Naus bound | Sheng*-Naus bound | Simulat. probab. | Exact probab. | Glaz-Hoover bound | Hunter-Glaz-Hoover bound | Bonferroni upper bound |
|---|---|---|---|---|---|---|
| 0.3771 | 0.4524 | 0.4747 | 0.4716 | 0.5289 | 0.5575 | 0.7329 |

* For details, see Sheng and Naus (1994).

In general, for $M \approx d$ and large $n$, the bounds are reasonable, but they become less accurate when $M$ is much smaller than $d$ and $n$ is moderate. The exact matching probabilities, on the other hand, can be easily obtained via Theorem 13.3.1 and Corollary 13.3.1, even for moderate or large $n$. When $n$ is very large, bounds and approximations are very useful, and for such cases, we suggest a simple alternative based on the large deviation method to approximate the exact probability.

The idea of the large deviation approximation is based on the fact that, for any given $d$ and $1 \leq m \leq M$, the exact probability $P(E_{d,n}(m,M))$ converges to zero exponentially (or obeys the first large deviation principle) in the following sense: there exists a positive constant $\beta > 0$, for which

$$\lim_{n \to \infty} \frac{1}{n} \log P(E_{d,n}(m,M)) = -\beta, \qquad (13.9)$$

where $\beta$ is called the exponential rate for the sequence of probabilities of events $E_{d,n}(m,M)$. The exponential rate $\beta$ usually depends on $m$, $d$, and the state probabilities $p_b$ in a very complex way.

Before deriving the approximation, we first study the exponential rate of $P(E_{d,n}(M,M)|W)$ where the sequence $W$ is specified.

**Theorem 13.4.1** *Given* $W = \{W_1, \cdots, W_d\}$,

$$\lim_{n \to \infty} \frac{1}{n} \log P(E_{d,n}(M,M)|W) = -\beta_W, \qquad (13.10)$$

*where* $\beta_W = min(\beta_{\Lambda_1}, \cdots, \beta_{\Lambda_k})$ *and* $\beta_{\Lambda_i} = -log(1-p_{\Lambda_i})$, $\Lambda_i \in \mathcal{G}_W$, $i = 1, \cdots, k$, *k is the number of distinct patterns in* $\mathcal{G}_W$, *and* $p_{\Lambda_i}$ *is the probability of simple pattern* $\Lambda_i$ *(e.g., if* $\Lambda_i = GTA$, *then* $p_{\Lambda_i} = p_G p_T p_A$).

PROOF. Consider a simple pattern $\Lambda$ with probability $p_\Lambda$ of occurring. Using the method developed by Fu (1986), Chao and Fu (1989), and Papastavridis and Koutras (1992), it can be shown that for arbitrarily small $\varepsilon > 0$, there exists a positive integer $n_0$ such that the following inequality holds:

$$(1 - p_\Lambda)^n \leq P(X_n(\Lambda) = 0) \leq (1 - p_\Lambda + \varepsilon)^{n-n_0}.$$

It follows that

$$\lim_{n \to \infty} \frac{1}{n} \log P(X_n(\Lambda) = 0) = -\beta_\Lambda, \tag{13.11}$$

where $\beta_\Lambda = -log(1 - p_\Lambda)$. Since $S_M(d,n) < M$ if and only if $X_n(\Lambda_i) = 0$ for all $\Lambda_i \in \mathcal{G}_W$, (13.10) follows directly from (13.11) and the following inequality,

$$\max_{\Lambda_i \in \mathcal{G}_W} P(X_n(\Lambda_i) = 0) \leq P(S_M(d,n) < M|W)$$
$$\leq k \times \max_{\Lambda_i \in \mathcal{G}_W} P(X_n(\Lambda_i) = 0).$$

This completes the proof. $\blacksquare$

Theorem 13.4.1 implies that $P(E_{d,n}(M,M)|W) \approx exp\{-n\beta_W\}$ for very large $n$. This yields a large deviation approximation for the perfect matching probability of pattern size $M$ between two nonaligned sequences.

**Theorem 13.4.2** *For fixed $d$ and very large $n$,*

$$P(E_{d,n}^c(M,M)) \approx 1 - \sum_{W \in \mathcal{F}_d} P(W) \exp\{-n\beta_W\}, \tag{13.12}$$

*where $\beta_W$ is as defined in Theorem 13.4.1.*

PROOF. Since $P(E_{d,n}^c(M,M)) = 1 - P(E_{d,n}(M,M))$ and $P(E_{d,n}(M,M)) = \sum_{W \in \mathcal{F}_d} P(W)P(E_{d,n}(M,M)|W)$, (13.12) is a direct consequence of Theorem 13.4.1. This completes the proof. $\blacksquare$

---

## 13.5 Numerical Results

The computational algorithm associated with the finite Markov chain imbedding technique is very simple. In addition to construction of the state space and the transition probability matrices, it basically only requires matrix multiplications. To illustrate our theoretical developments of the previous sections, several numerical results are presented here.

The exact perfect matching probabilities $P(E_{d,n}^c(M,M))$ for various $n$, $d$, and $M$ are given in Table 13.2 for equal state probabilities (i.e., $p_A = p_C = p_G = p_T = 0.25$) and in Table 13.3 for a set of unequal state probabilities ($p_A = 0.15$, $p_C = 0.25$, and $p_G = p_T = 0.3$).

**Table 13.2:** $P\{E^c_{d,n}(M, M)\}$ with $p_A = p_C = p_G = p_T = 0.25$

| $n$ | Matching prob. for $M = 3$ | | | Matching prob. for $M = 4$ | |
| --- | --- | --- | --- | --- | --- |
| | $d = 3$ | $d = 4$ | $d = 5$ | $d = 4$ | $d = 5$ |
| 10 | 0.1188 | 0.2027 | 0.2793 | 0.0270 | 0.0479 |
| 50 | 0.5316 | 0.7368 | 0.8521 | 0.1682 | 0.2766 |
| 100 | 0.7869 | 0.9326 | 0.9780 | 0.3162 | 0.4868 |
| 500 | 0.9996 | 1 | 1 | 0.8571 | 0.9665 |
| 1500 | 1 | 1 | 1 | 0.9971 | 0.9999 |

$M$: pattern size, $n$: length of sequence $R$, and $d$: length of sequence $W$.

**Table 13.3:** $P\{E^c_{d,n}(M, M)\}$ with $p_A = 0.15, p_C = 0.25, p_G = p_T = 0.3$

| $M$ | $n$ | Matching Probability | | |
| --- | --- | --- | --- | --- |
| | | $d = 3$ | $d = 4$ | $d = 5$ |
| 3 | 10 | 0.1384 | 0.2318 | 0.3155 |
| | 50 | 0.5733 | 0.7653 | 0.8702 |
| | 100 | 0.8043 | 0.9319 | 0.9754 |
| | 1000 | 0.9997 | 1 | 1 |
| 4 | 10 | | 0.0339 | 0.0592 |
| | 100 | | 0.3688 | 0.5440 |
| | 1000 | | 0.9654 | 0.9939 |
| 5 | 100 | | | 0.1164 |
| | 1000 | | | 0.6785 |
| | 4000 | | | 0.9631 |

$M$: pattern size, $n$: length of sequence $R$,
and $d$: length of sequence $W$.

Note that $M \leq d \leq n$, and hence the lower-left corner of Table 13.3 has no values. It may be seen from Tables 13.2 and 13.3 that (a) for fixed $M$ and $n$, the matching probability increases as $d$ increases, (b) for fixed $d$ and $n$, the matching probability decreases rather quickly with increasing $M$, and (c) for fixed $M$ and $d$, as $n$ increases, the matching probability increases exponentially fast to one.

To show the last point (c), the exponential rate of convergence, and Theorem 13.4.1, a graph of the conditional probability $P(E^c_{d,n}(M, M)|W)$ is given in Figure 13.1, where sequence $W$ is specified as $W = ACG$.

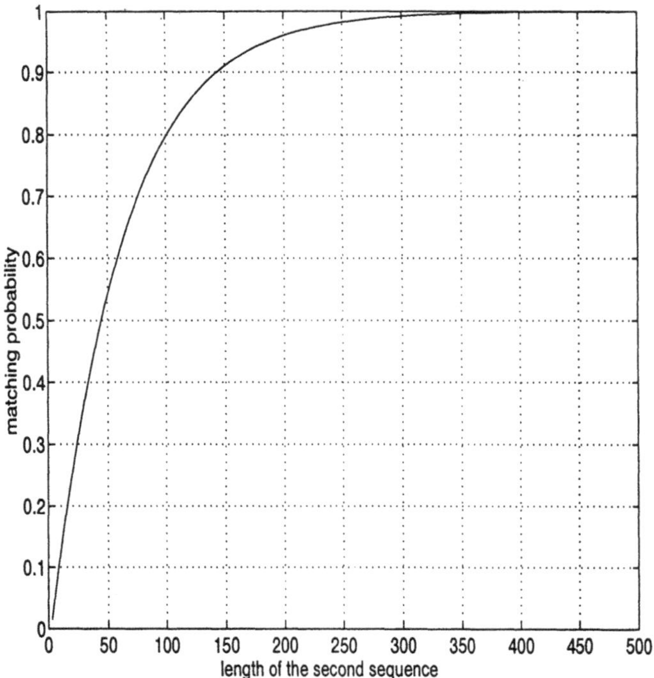


**Figure 13.1:** Rate of convergence for $P(E^c_{d,n}(M,M)|W)$ with $W = AGC$, and $p_A = p_C = p_G = p_T = 0.25$

Since $P(E^c_{d,n}(M,M))$ is a weighted linear combination of $P(E^c_{d,n}(M,M)|W)$, the unconditional probability $P(E^c_{d,n}(M,M))$ also tends to one exponentially fast. This phenomenon can also be seen in Tables 13.2 and 13.3.

When $n$ is very large, the conditional probability for perfect matching of pattern size $M$ can be approximated via Theorem 13.4.1 for both equal and unequal state probabilities. For instance, for $d = M = 3$ and specified $W = AGC$, given $p_A = 0.1$, $p_C = 0.2$, $p_G = 0.3$, and $p_T = 0.4$, the conditional perfect matching probability can be approximated by the quantity

$$P(E^c_{3,n}(3,3)|W = AGC) \approx 1 - \exp\{-n\beta_W\},$$

where $\beta_W = -\log(1 - p_A p_G p_C) = 0.0060180723$. Comparisons between exact and approximated probabilities are listed in Table 13.4, and show that the large deviation approximations are highly accurate.

For equal state probabilities ($p_A = p_C = p_G = p_T = 0.25$), if also $M = d$, a much simpler approximation for this special case can be used. It follows from Theorem 13.4.2 that

$$P(E^c_{M,n}(M,M)) \approx 1 - \left(1 - \frac{1}{4^M}\right)^n. \tag{13.13}$$

**Table 13.4:** Comparisons between exact probabilities and large deviation approximations for $P(E^c_{3,n}(3,3)|W = AGC)$ with $p_A = 0.1, p_C = 0.2, p_G = 0.3, p_T = 0.4$

| Length of $R$-seqs. $n$ | Matching probability | |
|---|---|---|
| | Exact | Large Deviation |
| 500 | 0.95186467 | 0.95066079 |
| 1000 | 0.99771131 | 0.99756564 |
| 1500 | 0.99989118 | 0.99987989 |
| 2000 | 0.99999999 | 0.99999941 |

Comparisons between the exact probabilities and approximation (13.13) are given in Table 13.5.

**Table 13.5:** Comparisons between exact probabilities and large deviation approximations for $P\{E^c_{d,n}(M,M)\}$ with $p_A = p_C = p_G = p_T = 0.25$

| $n$ | Matching prob. for $M = d = 3$ | | Matching prob. for $M = d = 4$ | |
|---|---|---|---|---|
| | Exact | Large Devi. Approx. | Exact | Large Devi. Approx. |
| 10 | 0.1188 | 0.1457 | 0.0270 | 0.0384 |
| 50 | 0.5316 | 0.5450 | 0.1682 | 0.1777 |
| 100 | 0.7869 | 0.7930 | 0.3162 | 0.3239 |
| 500 | 0.9995 | 0.9996 | 0.8571 | 0.8587 |

$M$: pattern size, $n$: length of sequence $R$, and $d$: length of sequence $W$.

The numerical results presented in Table 13.5 show that the simplified large deviation approximation performs extremely well even when $n$ is only moderately large. Based on our computational experience, if $n \geq 50$, the large deviation formula (13.13) provides a good approximation.

## 13.6   Discussion

The theorems and numerical results presented herein are based on the assumption that the sequences of four-state trials are independent and identically distributed with either equal or unequal state probabilities ($p_A, p_C, p_G$, and $p_T$). In real biological applications, the trials within sequences ($W$ and $R$) could be dependent. In view of our formulation for the matching problem using the finite Markov chain imbedding technique, with some modifications our results could be extended to study the matching probability when the trials within sequences are Markov-dependent.

For simplicity, we have assumed that the two nonaligned sequences ($W$ and $R$) have a common probability mass function, i.e., $p_{1b} = p_{2b} = p_b$, $b =$

$\{A, C, G, T\}$. Relaxing this assumption, i.e., $p_{1b} \neq p_{2b}$ for some $b$, the formula (13.8) could still be used to obtain the exact probability of perfect matching, where $P(W)$ is computed under $p_{1b}$ and $P(E_{d,n}^c(M, M)|W)$ is computed under $p_{2b}$.

Here, we view DNA segments as sequences of four-state trials. In principle, our FMCI approach is applicable to other multi-state trials, as in the matching of protein sequences where $\mathcal{B}$ is an alphabet of twenty amino acids. It is obvious that our FMCI approach for nonaligned sequences could also be used for aligned sequences. We feel that the derivation is rather simple, and hence leave it to the reader.

Our numerical results were computed on a UNIX system using the programming language MATLAB, and for the examples we have considered, the CPU time on a SPARC 1000 machine is negligible when sequence $W$ is specified. For unspecified sequence $W$, $3 \leq M \leq 5$, and moderate $d$ ($3 \leq d \leq 9$), computing the exact matching probabilities requires a CPU time of only about a few minutes even when $n$ is very large. For unspecified $W$, and moderate $M$ ($6 \leq M \leq 12$), the computational time increases rapidly with $d$ to hours, since all possible sequences $W$ have to be considered. We believe that the computational time could be significantly reduced via faster computer hardware and more efficient numerical implementation. For very large $n$, the proposed large deviation approximation provides a very simple and efficient alternative.

**Acknowledgments.** This work was supported in part by the NATO Collaborative Research Grants Programme CRG-970278 and the National Science and Engineering Research Council of Canada Grant A-9216.

# References

1. Arratia, R., Gordon, L. and Waterman, M. S. (1986). An extreme value theory for sequence matching, *Annals of Statistics*, **14**, 971–993.

2. Arratia, R., Gordon, L. and Waterman, M. S. (1990). The Erdös-Rényi law in distribution, for coin tossing and sequence matching, *Annals of Statistics*, **18**, 539–570.

3. Chao, M. T. and Fu, J. C. (1989). A limit theorem of certain repairable systems, *Annals of the Institute of Statistical Mathematics*, **4**, 809–818.

4. Erdös, P. and Révész, P. (1975). On the length of the longest head-run, *Topics in Information Theory, Colloquia of Mathematical Society János Bolyai*, **16**, 219–228, Keszthely, Hungary.

5. Fu, J. C. (1986). Bounds for reliability of large consecutive-$k$-out-of-$n$:F systems with unequal component reliability, *IEEE Transactions on Reliability*, **35**, 316–319.

6. Fu, J. C. (1996). Distribution theory of urns and patterns associated with a sequence of multi-state trials, *Statistica Sinica*, **6**, 957–974.

7. Fu, J. C. and Koutras, M. V. (1994). Distribution theory of runs: A Markov chain approach, *Journal of the American Statistical Association*, **89**, 1050–1058.

8. Fu, Y. X. and Curnow, R. N. (1990). Locating a changed segment in a sequence of Bernoulli variables, *Biometrika*, **77**, 295–304.

9. Glaz, J. (1993). Approximations for the tail probabilities and moments of the scan statistic, *Statistics in Medicine*, **12**, 1845–1852.

10. Glaz, J. and Naus, J. I. (1991). Tight bounds and approximations for scan statistic probabilities for discrete data, *Annals of Applied Probability*, **1**, 306–318.

11. Gordon, L., Schilling M. F. and Waterman, M. S. (1986). An extreme value theory for long head runs, *Probability Theory and Related Fields*, **72**, 279–287.

12. Hoover, D. R. (1990). Subset complement addition upper bounds – an improved inclusion-exclusion method, *Journal of Statistical Planning and Inference*, **24**, 195–202.

13. Hunter, D. (1976). An upper bound for the probability of a union, *Journal of Applied Probability*, **13**, 597–603.

14. Karlin, S. and Ost, F. (1987). Counts of long aligned word matches among random letter sequences, *Advances in Applied Probability*, **19**, 293–351.

15. Karlin, S. and Ost, F. (1988). Maximal length of common words among random sequences, *Annals of Probability*, **16**, 535–563.

16. Koutras, M. V. and Alexandrou, V. A. (1995). Runs, scans, and runs models: a unified Markov chain approach, *Annals of the Institute of Statistical Mathematics*, **47**, 743–766.

17. Leung, M. Y., Blaisdell, B. E., Burge, C. and Karlin, S. (1991). An efficient algorithm for identifying matches with errors in multiple long molecular sequences, *Journal of Molecular Biology*, **221**, 1367–1378.

18. Lou, W. Y. W. (1996). On runs and longest run tests: a method of finite Markov chain imbedding, *Journal of the American Statistical Association*, **91**, 1595–1601.

19. Mott, R. F., Kirwood, T. B. L. and Curnow, R. N. (1990). An accurate approximation to the distribution of the length of the longest matching word between two random DNA sequences, *Bulletin of Mathematical Biology*, **52**, 773–784.

20. Naus, J. I. (1965). The distribution of the size of the maximum cluster of points on a line, *Journal of the American Statistical Association*, **60**, 532–538.

21. Naus, J. I. (1974). Probabilities for a generalized birthday problem, *Journal of the American Statistical Association*, **69**, 810–815.

22. Naus, J. I. and Sheng, K. N. (1997). Matching among multiple random sequences, *Bulletin of Mathematical Biology*, **59**, 483–496.

23. Papastavridis, S. G. and Koutras, M. V. (1992). Consecutive-$k$-out-$n$ systems with maintenance, *Annals of the Institute of Statistical Mathematics*, **44**, 605–612.

24. Sheng, K. N. and Naus, J. I. (1994). Pattern matching between two non-aligned random sequences, *Bulletin of Mathematical Biology*, **56**, 1143–1162.

25. Waterman, M. S. (1995). *Introduction to Computational Biology: Maps, Sequences, and Genomes*, London: Chapman and Hall.

# 14

# Spatial Scan Statistics: Models, Calculations, and Applications

**Martin Kulldorff**

*National Cancer Institute, Bethesda, MD*

**Abstract:** A common problem in spatial statistics is whether a set of points are randomly distributed or if they show signs of clusters or clustering. When the locations of clusters are of interest, it is natural to use a spatial scan statistic.

Different spatial scan statistics have been proposed. These are discussed and presented in a general framework that incorporates two-dimensional scan statistics on the plane or on a sphere, as well as three-dimensional scan statistics in space or in space–time. Computational issues are then looked at, presenting efficient algorithms that can be used for different scan statistics in connection with Monte Carlo-based hypothesis testing. It is shown that the computational requirements are reasonable even for very large data sets. Which scan statistic to use will depend on the application at hand, which is discussed in terms of past as well as possible future practical applications in areas such as epidemiology, medical imaging, astronomy, archaeology, urban and regional planning, and reconnaissance.

**Keywords and phrases:** Spatial statistics, geography, spatial clusters, space–time clusters, maximum likelihood, likelihood ratio test

## 14.1    Introduction

The scan statistic is a statistical method with many potential applications, designed to detect a local excess of events and to test if such an excess can reasonably have occurred by chance. The scan statistic was first studied in detail by Naus (1965a,b), who looked at the problem in both one and two dimensions.

In two or more dimensions, which is the topic of this chapter, the events may
be cases of leukemia, with an interest to see if there are geographical clusters
of the disease; they may be antipersonnel mines, with an interest to detect
large mine fields for removal; they could be Geiger counts, with an interest to
detect large uranium deposits; they could be stars or galaxies; they could be
breast calcifications showing up in a mammography, possibly indicating a breast
tumor; or they could be a particular type of archæological pottery. Later on
we will discuss each of these and several other applications and the type of scan
statistic that is suitable in each situation.

Three basic properties of the scan statistic are the geometry of the area
being scanned, the probability distribution generating events under the null
hypothesis, and the shapes and sizes of the scanning window. We present a
general framework in which most multidimensional scan statistics fit. Depending on the application, different models will be chosen, and depending on the
model, the test statistic may be evaluated either through explicit mathematical derivations and approximations or through Monte Carlo sampling. In the
latter case, random data sets are generated under the null hypothesis, and the
scan statistic is calculated in each case, comparing the values from the real and
random data sets to obtain a hypothesis test.

While computer intensive, the Monte Carlo approach need not be overly so.
In this chapter, we present a set of efficient algorithms which can be used to
calculate the spatial scan statistic for a set of different models with a circular
window. One of these with a continuously variable radius, required 163 minutes
of computing time on a 100 MHz Pentium PC, when applied to 65,040 cases of
melanoma in the 3,053 counties of the continental United States.

Section 14.2 is essentially a review of the existing literature, while Section 14.3 presents mostly new material. Section 14.4 describes how the spatial
scan statistic can be utilized in practice in an attempt to inspire its use in
current as well as new areas of application.

## 14.2  Models

### 14.2.1  A general model

As mentioned above, the three basic properties of the scan statistic are the
geometry of the area being scanned, the probability distribution generating
events under the null hypothesis, and the shapes and sizes of the scanning
window.

Kulldorff (1997) defined a general model for the multidimensional scan
statistic. Let $A$ be the area in which events may occur, a subset of Euclidean
space where different dimensions may represent either physical space or time.

For example, $A$ could be particular geographical area during a ten-year period, where events are recorded both geographically and temporally.

On $A$ define a measure $\mu$, representing a known underlying intensity that generates events under the null hypothesis. For a homogeneous Poisson process on a rectangle $A$, we have $\mu(x) = \lambda$ for all $x \in A$ and some constant $\lambda$. The measure could also be discrete, so that it is only positive on a finite number of *population points*, where $\mu(B)$ is the combined measure of the population points located in area $B \subset A$. We require that $\mu(B) > 0$ for all areas $B$.

Let $X$ denote a spatial point process where $X(B)$ is the random number of events in the set $B \subset A$. Two different probability models are considered, based on Bernoulli counts and the Poisson process, respectively.

For the Bernoulli model, we consider only discrete measures $\mu$ such that $\mu(B)$ is an integer for all subsets $B \subset A$. Each unit of measure corresponds to an "entity" or "individual" who could be in either one of two states, for example with or without some disease, or being of a certain species or not. Individuals in one of these states are defined as events, and the location of those individuals constitute the point process. Under the null hypothesis, the number of events in any given area is binomially distributed, so that $X(B) \sim Bin(\mu(B), p)$ for some value $p$ and for all sets $B \subset A$.

For the Poisson model, events are generated by a homogeneous or nonhomogeneous Poisson process. Under the null hypothesis, $X(B) \sim \text{Poisson}(p\mu(B))$, for some value $p$ and for all sets $B \subset A$. The measure $\mu$ may either be defined continuously so that events may occur anywhere, or discretely so that events may occur only at prespecified locations, or as a combination of the two. The discrete case is useful when we are dealing with individual counts or with aggregated data.

The window of a scan statistic is often thought of as an interval, area, or volume of fixed size and shape, which then moves across the study area. As it moves, it defines a collection $\mathcal{W}$ of zones $W \subset A$. To be more general, we allow for windows of variable size and shape, by defining the window as a collection $\mathcal{W}$ of zones $W \subset A$ of any size and shape. What defines it as a scan statistic is that the different zones overlap each other and jointly cover the whole area $A$.

Conditioning on the observed total number of events, $X(A)$, the definition of the scan statistic is the maximum likelihood ratio over all possible zones

$$S_{\mathcal{W}} = \frac{\max_{W \in \mathcal{W}} L(W)}{L_0} = \max_{W \in \mathcal{W}} \frac{L(W)}{L_0} , \qquad (14.1)$$

where $L(W)$ is the likelihood function for zone $W$, expressing how likely the observed data are given a differential rate of events within and outside the zone, and where $L_0$ is the likelihood function under the null hypothesis.

Let $X(A \setminus W) = X(A) - X(W)$ and $\mu(A \setminus W) = \mu(A) - \mu(W)$. For the Bernoulli model,

$$\frac{L(W)}{L_0} =$$

$$\frac{\left(\frac{X(W)}{\mu(W)}\right)^{X(W)} \left(1-\frac{X(W)}{\mu(W)}\right)^{\mu(W)-X(W)} \left(\frac{X(A\setminus W)}{\mu(A\setminus W)}\right)^{X(A\setminus W)} \left(1-\frac{X(A\setminus W)}{\mu(A\setminus W)}\right)^{\mu(A\setminus W)-X(A\setminus W)}}{\left(\frac{X(A)}{\mu(A)}\right)^{X(A)} \left(1-\frac{X(A)}{\mu(A)}\right)^{\mu(A)-X(A)}}$$

$$(14.2)$$

if $X(W)/\mu(W) > X(A \setminus W)/\mu(A \setminus W)$, and $L(W) = 1$ otherwise. For the Poisson model,

$$\frac{L(W)}{L_0} = \frac{\left(\frac{X(W)}{\mu(W)}\right)^{X(W)} \left(\frac{X(A\setminus W)}{\mu(A\setminus W)}\right)^{X(A\setminus W)}}{\left(\frac{X(A)}{\mu(A)}\right)^{X(A)}} \qquad (14.3)$$

if $X(W)/\mu(W) > X(A\setminus W)/\mu(A\setminus W)$, and $L(W) = 1$ otherwise. The expression $X(W)/\mu(W) > X(A \setminus W)/\mu(A \setminus W)$ simply states that there are more than the expected number of events within the window as compared to outside the window. If we were scanning for areas with a low number of events, then ">" would change to "<." For details and derivations as a likelihood ratio test, see Kulldorff (1997), who has also proved some optimal properties for these test statistics.

When the window size is fixed in terms of the expected number of events, that is, if $\mu(W) = \mu(W')$ for all $W, W' \in \mathcal{W}$, then the scan statistic is

$$S'_{\mathcal{W}} = \max_{W \in \mathcal{W}} X(W),$$

the maximum number of events in the window over all possible locations. Note that $S'_{\mathcal{W}} \neq S_{\mathcal{W}}$, but for any two realizations of the point process, say $\omega_1$ and $\omega_2$, $S'_{\mathcal{W}}(\omega_1) > S'_{\mathcal{W}}(\omega_2)$ if and only if $S_{\mathcal{W}}(\omega_1) > S_{\mathcal{W}}(\omega_2)$. This means that, when the window size is fixed, then a hypothesis test based on $S'_{\mathcal{W}}$ is identical to one based on $S_{\mathcal{W}}$.

For a Poisson model with continuous measure, a lower bound on the window size is needed. If not, then a window containing a sequence of increasingly smaller zones all containing the same event will in the limit give an infinite valued test statistic. It is also natural to put an upper bound on the window size. A window $W$ that contains almost all of $A$ makes little sense, and should be interpreted as a lack of events outside of $W$ rather than as an excess inside.

### 14.2.2   Special cases

Both one and multidimensional scan statistics are special cases of the above model. Many features of it originated in connection with one-dimensional scan statistics; see, for example, Saperstein (1972), Naus (1974), Weinstock (1981), Wallenstein, Weinberg, and Gould (1989b), and Glaz and Naus (1991). Here, we review the multi-dimensional literature.

In terms of the area $A$ being scanned, Naus (1965b), Loader (1991), Alm (1997, 1998) and Anderson and Titterington (1997) all considered a rectangle. Alm (1998) also looked at a three-dimensional rectangular volume. Chen and Glaz (1996) looked at a regular grid of discrete points within a rectangular area. Turnbull *et al.* (1990) used an irregular grid, where points may be anywhere within an arbitrarily shaped area.

Under the null hypothesis, Naus (1965b), Loader (1991), and Alm (1997, 1998) looked at a homogeneous Poisson process, Turnbull *et al.* (1990) considered a nonhomogeneous Poisson process, while Anderson and Titterington (1997) considered both types. Chen and Glaz (1996) considered a Bernoulli model.

As for the scanning window, Naus (1965b), Loader (1991), Chen and Glaz (1996), Alm (1997, 1998) and Anderson and Titterington (1997) all considered rectangles. In addition, Alm (1997, 1998) also looked at circles, triangles, and other convex shapes. Turnbull *et al.* (1990) considered a circular window centered at any of the grid points making up the data. The window is, in all cases, of fixed shape as well as of fixed size in terms of the expected number of events, with the exception of Loader (1991), who also considered a variable size window.

In terms of applications, the general model has been applied in a number of different settings, the first of which was presented at the SPRUCE conference in 1992 and later published by Kulldorff and Nagarwalla (1995). For all of these, the data are located on an irregular grid within an arbitrarily shaped area. Kulldorff and Nagarwalla (1995) and Section 6.1 of Kulldorff (1997) used the Bernoulli model, while Section 6.2 of Kulldorff (1997), Hjalmars *et al.* (1996), Kulldorff *et al.* (1997, 1998), and Walsh and Fenster (1997) used a nonhomogeneous Poisson process. In terms of the scanning window, all used a variable size circle centered on the grid points, except for Kulldorff *et al.* (1998), who used a three-dimensional cylinder where the size of both the base and the height is variable independently of each other.

The choice of scan statistic will depend on the particular application at hand, a topic we will turn to in Section 14.4.

### 14.2.3  Related methods

As part of a "geographical analysis machine," Openshaw *et al.* (1987) used a number of overlapping circular zones of different radii. The purpose is the same as with a spatial scan statistic, to detect clusters of events, but a separate test is performed for each of the many zones. This leads to multiple testing, and even under the null hypothesis we would expect a large number of "significant" clusters, but as a descriptive geographical analysis tool the method is useful. Turnbull *et al.* (1990) solved the problem of the multiple testing for circles with fixed expected number of cases, while Kulldorff and Nagarwalla (1995) and Kulldorff (1997) solved it for variable size circles.

Priebe (1998) proposed a spatial scan statistic for stochastic scan partitions. In a two-step procedure, one set of data is first used to create a set of non-overlapping zones, called scan partitions, while another set of data containing the events is used to see if any of these partitions have a statistically significant excess of events. Because the zones are nonoverlapping, the calculations for the second part are more simple than for a standard scan statistic. It is necessary to have the additional data set though, used in the first step, and under the null hypothesis the two data sets need to be independent of each other for the test to be valid.

In other related problems, Eggleton and Kermack (1944), Besag and Newell (1991), Månsson (1996), and many others have studied the number of clusters of some prespecified magnitude. Lawson (1997) applied a Bayesian framework to investigate the number of clusters and their locations. Adler (1984), Worsley *et al.* (1992), and some others have investigated the supremum of a Gaussian random field.

Wallenstein, Gould, and Kleinman (1989a) used a scan statistic in the time dimension to improve on a previously proposed space–time clustering test, but the test itself is not a scan statistic. Rather than taking a maximum over the geographical zones, the degree of clustering in each zone is summed over all zones, making it a global clustering test. Such tests are useful for quite different purposes, when the locations of clusters are not of interest.

---

## 14.3   Calculations

### 14.3.1   Probabilistic approximations

The mathematics for obtaining the distribution of the scan statistic is quite complex, and exact derivations have proved elusive for all but the simplest scenarios. There are some very interesting and impressive probabilistic approximations though. Starting with Naus (1965b), later results have been obtained by Loader (1991), Chen and Glaz (1996), and Alm (1997,98). Månsson (1996) has derived some limit results. Details of these developments can be found in Chapter 5 of this volume by Sven-Erick Alm, and in Chapter 10 by Marianne Månsson.

### 14.3.2   Monte Carlo-based hypothesis testing

When probabilistic approximations are not available, Monte Carlo-based hypothesis testing is. In principle, this can be applied to any special case of the general model presented in Section 14.2. Generating random cases is typically not a problem, but calculating the value of the test statistic can be a complex undertaking, depending on the model chosen. For the descriptive cluster detection method described earlier, Openshaw *et al.* (1987) used a Cray supercomputer even though their approach is conceptually simpler than a scan statistic. By using efficient statistical algorithms, the calculation times can be substantially reduced.

Monte Carlo-based hypothesis testing was proposed by Dwass (1957), who pointed out that the probability of falsely rejecting the null hypothesis is exactly according to the significance level, in spite of the simulation involved. Mantel (1967) proposed its use in terms of spatial point processes, while Turnbull *et al.* (1990) was the first to use it in the context of a multidimensional scan statistic. Monte Carlo hypothesis testing for a scan statistic is a four-step procedure:

1. Calculate the value of the test statistic for the real data.

2. Create a large number of random data sets generated under the null hypothesis.

3. Calculate the value of the test statistic for each of the random replications.

4. Sort the values of the test statistic, from the real and random data sets, and note the rank of the one calculated from the real data set. If it is ranked in the highest $\alpha$ percent, then reject the null hypothesis at $\alpha$ percent significance level.

The key in terms of minimizing computing time is Step 3, as it can be complex in nature, and most of all, because it must be repeated once for each random replication of the data set. Anderson and Titterington (1997) presented the following algorithm for a circular window of fixed diameter $d$ on a homogeneous Poisson process:

**Algorithm 14.3.1** (Anderson-Titterington: Circular window. Fixed size. Homogeneous Poisson process.)

1. *Identify the locations $(x, y)$ of two events no more than distance $d$ apart.*

2. *Construct the two circles of diameter $d$ for which $x$ and $y$ lie on the circumference.*

3. *Identify the number of events that lie on or inside each of the two circles and let $n$ be the larger of those two numbers.*

4. *Repeat Steps 1 to 3 for all relevant pairs of locations and report the largest of the resulting $n$-values as being the scan statistic.*

The complexity of one visit to Step 3 is of the order $O(N)$, where $N = X(A)$, the total number of events. Steps 1–3 must be repeated $O(N^2)$ times for each of $R$ Monte Carlo replications, so the total complexity is $O(RN^3)$. When $N$ is large, a more efficient algorithm is:

**Algorithm 14.3.2** (Circular window. Fixed size. Homogeneous Poisson process.)

1. *Identify the location $x$ of an event and construct a large circle with radius $d$ centered at $x$. Pick an arbitrary location on the large circle, $x_0$, and denote the angle from $x$ to $x_0$ as $0°$.*

2. *Create a smaller circle of radius $d/2$ within the larger one. Imagine the smaller circle moving clockwise completely within the larger circle in such a way that $x$ is always on its circumference. Denote by $x_a$, $0° < a < 360°$, the single point that is on the circumference of both circles, where $a$ is the angle from $x$ to $x_a$.*

3. *For each event on or inside the larger circle, note the two angles of the line from $x$ to $x_a$ when the event enters and departs the smaller moving circle. Sort the angles in increasing order, keeping track of whether the angle corresponds to an entrance or a departure.*

4. *For the smaller circle which has both $x$ and $x_0$ on its circumference, count the number of events inside it. Then go through the array of sorted angles from $0°$ to $360°$, adding one to the count for each entrance, subtracting one for each departure. Denote the maximum count by $n$.*

5. *Repeat Steps 1 to 4 for all events, and report the largest of the resulting $n$-values as the scan statistic.*

6. *Repeat Steps 1 to 5 for each Monte Carlo replication.*

Each visit to Steps 2 and 4 is $O(N)$ while the sorting in Step 3 is $O(N \log N)$. There are $N$ iterations of Steps 1 to 5 for each of $R$ replications, and hence the total complexity is $O(RN)[O(N) + O(N \log N) + O(N)] = O(RN^2 \log N)$.

In most practical applications, the cluster size is unknown a priori. For a homogeneous Poisson process, the simplest algorithm to program would be to pick all triplets of events, in turn, and for each triplet construct the circle for which all three events lie on the circumference, then counting the number of events within that circle. Based on the number of events and the circle size, it is then possible to calculate the likelihood according to (14.3), and the largest likelihood over all possible triplets is the scan statistic. Such an algorithm is $O(RN^4)$. A more efficient algorithm, with complexity $O(RN^3 \log N)$, is as follows.

**Algorithm 14.3.3** (Circular window. Variable size. Homogeneous Poisson process.)

1. *Identify the locations $(x, y)$ of two events, and construct the straight line $L$ between the two where each point on the line is equal distance from $x$ and $y$. Denote one end of the line as the left end.*

2. *For each remaining event $z$, construct the circle such that all of $(x, y, z)$ lie on the circumference. Note where on $L$ lies the circle centroid corresponding to $z$, and whether event $z$ enters or departs the circle as the centroid moves toward the left.*

3. *Sort the circle centroids on $L$ from right to left, keeping track of whether that centroid corresponds to an entrance or a departure.*

4. *Calculate the number of events in the circle with its centroid farthest to the right, as well as the circle size. Then move down the sorted array of circles centroids adding or subtracting events as they enter or depart the circle. For each circular area $W$, register the number of events $n$ as well as the circle measure $\mu(W) = \int_W \mu \, dy = \mu \pi r^2$, where $r$ is the radius.*

5. *Repeat Steps 1 to 4 for all pairs of events, and report the largest likelihood based on all $(n, \mu(W))$-pairs as the scan statistic, where the likelihood is calculated according to (14.3).*

6. *Repeat Steps 1 to 5 for each Monte Carlo replication.*

So far, we have presented algorithms for homogeneous Poisson processes. A simple case of a nonhomogeneous Poisson process is a gradual linear shift in intensity so that $\mu(x) = a + bx$ for some $a$ and $b$. Algorithm 14.3.3 can be easily modified to account for this by calculating the measure of the circular area $W$ centered at $x$ as $\mu(W) = \int_W \mu(y)dy = \mu(x)\pi r^2 dx$, where $r$ is the circle radius.

Another form of nonhomogeneity is the discrete case in which the measure is concentrated on a finite set of population points. The following algorithm is similar to Algorithm 14.3.3 but based on the location of the population points containing positive measure, rather than on the location of events. We can no longer calculate the measure simply from the circle size, and hence, we need to keep track of the amount of measure in the window simultaneously with the number of events.

**Algorithm 14.3.4** (Circular window. Variable size. Discrete nonhomogeneous Bernoulli or Poisson process.)

1. *Identify the locations $(x, y)$ of two population points, and construct the straight line $L$ between the two where each point on the line is equal distance from $x$ and $y$. Denote one end of the line as the left end.*

2. *For each remaining population point $z$, construct the circle such that all of $(x, y, z)$ lie on the circumference. Note where on $L$ lies the circle centroid and whether the population point enters or departs the circle as the centroid moves toward the left.*

3. *Sort the circle centroids located on $L$ from right to left, keeping track of whether that centroid corresponds to an entrance or departure.*

4. *Calculate the number of events $n$ in the circle with its centroid farthest to the right on the line, as well as the measure $\mu(W)$ for that circle. Then move down the sorted array of circles centroids adding and subtracting events and measure as population points enter and depart the circle. For each circular area $W$, register the number of events $n$ as well as the population measure $\mu(W)$.*

5. *Repeat Steps 1 to 4 for all pairs of population points, and report the largest likelihood based on all $(n, \mu(W))$-pairs as the scan statistic, where the likelihood is calculated according to (14.2) in the case of a Bernoulli model, and according to (14.3) for the Poisson model.*

6. *Repeat Steps 1 to 5 for each Monte Carlo replication.*

The complexity of this algorithm is $O(RM^3 logM))$, where $M$ is the number of population points.

For most applications, it is not crucial to include all possible circles in the set of zones constituting the window, and an alternative is to use only a subset of closely overlapping circles. This reduces the computing time. In the following two algorithms, the window contains only those circles that are centered at any of a number of prespecified irregular grid points. The radius of the circles still vary continuously.

**Algorithm 14.3.5** (Circular window. Variable size. Circle centroids on grid. Homogeneous Poisson process.)

1. *Pick a grid point. Calculate the distance to the different events and sort in increasing order.*

2. *Create a circle centered at the grid point and continuously increase the radius. For each event entering the circle, note the number of events $n$ and the measure $\mu(W) = \mu\pi r^2$ inside the circle.*

3. *Repeat Steps 1 and 2 for each grid point. Report the largest likelihood based on all $(n, \mu(W))$-pairs as the scan statistic, where the likelihood is calculated according to (14.3).*

4. *Repeat Steps 1 to 3 for each Monte Carlo replication.*

The complexity of this algorithm is $O(RGNlogN)$, where $G$ is the number of grid points. For a discrete nonhomogeneous process, we have the following:

**Algorithm 14.3.6** (Circular window. Variable size. Circle centroids on grid. Discrete nonhomogeneous Bernoulli or Poisson process.)

1. *Pick a grid point. Calculate the distance to the different population points and sort those in increasing order. Memorize the sorted population points in an array.*

2. *Repeat Step 1 for each grid point.*

3. *Pick a grid point.*

4. *Create a circle centered at the grid point and continuously increase the radius. For each population point entering the circle, update the number of events $n$ and the measure $\mu(W)$ inside the circular area $W$.*

5. *Repeat Steps 3 and 4 for each grid point. Report the largest likelihood based on all $(n, \mu(W))$-pairs as the scan statistic, where the likelihood is calculated according to (14.2) or (14.3).*

6. *Repeat Steps 3 to 5 for each Monte Carlo replication.*

The complexity of Steps 1 and 2 is $O(GMlogM)$, as this does not have to be repeated for each Monte Carlo replication. The complexity of Steps 3 to 6 is $O(RGM)$.

Algorithms 14.3.5 and 14.3.6 also work for three-dimensional spherical windows by simply defining the population points in three-dimensional space. The complexity remains the same but for a complete coverage, the number of grid points $G$ may have to be larger.

In space–time applications, one option is simply to define time as a third dimension and use a spherical window on that three-dimensional space. One problem with this is that the result will depend on the relative units of spatial and temporal distances. Another problem is that a sphere would represent a cluster starting with zero spatial size, then growing steadily over time until a maximum spatial size is reached, after which it gradually shrinks back to zero size again. It is more natural to scan for clusters using the intersection of a spatial circle and a temporal interval, leading to a cylindrical window. Algorithm 14.3.6 can be adjusted for this purpose, if for each geographical circle, we also scan the time-dimension using a variable size temporal interval. It also means that the geographical and temporal size can vary independently of each other. The complexity of Steps 3 to 6 then becomes $O(RGMN^2)$ if exact times are known, and $O(RGMI^2)$ if times are aggregated into $I$ time intervals.

Algorithms 14.3.1 to 14.3.6 extend to circular windows on the surface of a sphere, by simply defining the events and population points in three dimensions on the spherical surface, and by adjusting the calculations of circle sizes and distances accordingly. This is very useful for geographical applications, avoiding the need for two-dimensional map projections.

Scanning for low rates can also be handled by any of the mentioned algorithms. For Algorithms 14.3.4 and 14.3.6, it is just a question of changing the sign of the inequality when calculating the likelihood $L(W)$. For the other algorithms, it is also necessary to subtract the number of events on the border of the circle from the circle total.

A circular window has the advantage of being invariant under a rotation of the space. There are applications though where other shapes are of interest. Anderson and Titterington (1997) gave an $O(RN^2)$ algorithm for a square window of fixed size with sides parallel to the axes of the coordinate system. A scan statistic with a fixed shape variable size ellipsoidic window can be calculated using any of Algorithms 14.3.3 to 14.3.6, by rescaling one of the axes in the underlying coordinate system. We leave it for future research to present algorithms for other models.

## 14.3.3   Software

For certain multidimensional scan statistics, Kulldorff and Williams (1997) have developed *SaTScan*. This software is available free of charge from the authors, or from the World Wide Web at `http://dcp.nci.nih.gov/BB/SaTScan.html`.

SaTScan uses Algorithm 14.3.6, and is based on a nonhomogeneous Poisson process defined on an irregular grid; it can be used to analyze the following types of multidimensional scan statistics: (i) a scan statistic on the plane with a circular window of variable size with centroids on an arbitrarily defined regular or irregular grid, (ii) same on the surface of a sphere such as the earth, (iii) a three-dimensional scan statistic with variable size spheric windows centered on an arbitrary irregular grid, (iv) a space–time scan statistic with a variable size cylinder, where the base of the circle corresponds to a geographical area, and the height to a time interval, and where the sizes of the circle and interval are variable independently of each other.

The software will, in all cases, adjust for any number of covariates specified by the user, and it is possible to scan for areas with a large number of events as well as for areas with a low number of events. Certain one-dimensional scan statistics can also be analyzed by putting all data on a single line. A future version will also include the Bernoulli model.

Using SaTScan, the calculations for the New Mexico example below took 8 seconds on a 100 MHz Pentium PC. With 1175 events in only 32 census areas, it is a rather small data set though. For the same type of analysis but with 65,040 cases of melanoma in the United States, aggregated to 3053 counties, SaTScan used 163 minutes of computer time when the maximum window size was set to 50% of the total, and 80 minutes when it was set to 10%. For 1592 cases of leukemia in 2507 Swedish parishes, it used 62 and 15 minutes, respectively. The cylinder based space–time scan statistic used 21 hours for the Swedish data set with years as the temporal unit and 10% as the maximum geographic window size. The number of Monte Carlo replications were in all cases 999.

This shows that the computational requirements for the spatial scan statistic is quite reasonable in practical applications with very large data sets.

---

## 14.4   Applications

### 14.4.1   Epidemiology

There is a long history of geographical surveillance of disease by publishing disease atlases. If there are areas with exceptionally high rates, they may give us clues to the etiology of the disease, it may indicate areas where health care needs improvement, or it may indicate areas to be targeted for preventive measures. In those atlases that are not purely descriptive, analysis is often done by dividing the study region into nonoverlapping districts, making a separate test of hypothesis for each district to see if it has an excess incidence or mortality [Choynowski (1959)]. With a spatial or space–time scan statistic, we can do the surveillance adjusting for the multiplicity of possible cluster locations, without being limited by the boundaries of prespecified districts, and without defining the size of potential cluster a priori.

Events may be cases diagnosed of some disease or deaths due to that disease. The measure is by nature nonhomogeneous, reflecting the geographical distribution of the population at risk. In most situations, we want to adjust for covariates that are known risk factors such as age or sex. We might have individual locations for cases and all non-cases, or of cases and a random set of controls, but more often the data are aggregated at some small geographical level such as census tracts, parishes or postal code areas. In either case, we can use Algorithm 14.3.4 or 14.3.6.

If the population at risk is all births and the events are occurrences of sudden infant death syndrome [Kulldorff (1997)] or birth defects, then we should use the Bernoulli model. If on the other hand, we are looking at fatal cardiac arrest in a population, we choose the Poisson model since such individuals are no longer part of the population numbers after the event occurs. Most applications fall somewhere in between the two, but whenever the number of events is small compared to the population at risk, the two models approximate each other so that either could be chosen.

In terms of practical epidemiological applications, the spatial scan statistic has been used to study leukemia in Upstate New York by Turnbull *et al.* (1990) using a fixed size window, and by Kulldorff and Nagarwalla (1997) using a variable size window. Hjalmars *et al.* (1996) have looked at childhood leukemia incidence in Sweden, Kulldorff (1997) studied sudden infant deaths in North Carolina, Kulldorff *et al.* (1997) have looked at breast cancer mortality in the northeastern United States, while Walsh and Fenster (1997) have studied mortality from systemic sclerosis in the southeastern United States. All these use a variable size circular window. Using a fixed size square window, Anderson and Titterington (1997) looked at laryngeal cancer in South Lancashire, England. The space–time scan statistic, using a variable size cylindrical window, has been applied to brain cancer incidence in New Mexico by Kulldorff *et al.* (1998).

## 14.4.2  Example: Brain cancer in New Mexico

To give an example, we look at the geographical distribution of brain cancer incidence in New Mexico. In 1989, a local resident detected an excess of brain cancer in Los Alamos during the previous year. This cluster alarm was evaluated statistically by Kulldorff *et al.* (1998) using a space–time scan statistic, without finding a significant space–time cluster in Los Alamos. Here, we will use a purely spatial scan statistic in more of a surveillance setting.

Broken down by age and sex, brain cancer and population data are available from 1973 to 1992 at the aggregated level of 32 counties. A circular variable size window was used. The circle centroids are limited to the county centroids, while the radius varies continuously from zero and up until it includes 50% of the total population at risk. Using a Poisson model, the analysis is adjusted for age and sex. One analysis was done scanning for areas with high rates (clusters) and another scanning for areas with low rates.

When scanning for areas with high rates, a cluster was found in and around Albuquerque, containing Bernadillo, Cibola-Valencia, Los Alamos, Sandoval, San Miguel, Santa Fe, Socorro, and Torrance counties (Figure 1.1), almost half the total state population. With 642 cases when 583.2 were expected, this area had a rate 10 percent higher than the New Mexico average, and it is significant with $p = 0.030$. As the New Mexico mortality rate was 16 percent lower than the United States average during 1986–90 [Miller *et al.* (1993)], this cluster may indicate that the Albuquerque area is more similar to the rest of the United States in terms of brain cancer than other parts of New Mexico.

When scanning for areas with low rates, the likelihood took on its maximum value for Lea and Eddy counties combined (Figure 14.1). With 72 cases when 97.4 were expected, these counties had an incidence rate 26 percent lower than the state average, with $p = 0.221$, a nonsignificant result.

When interested in areas with either high or low rates, then we can either do two one-sided tests as we have done above, or we can do a single two-sided test, which is recommended. The clusters found will be the same, but not the $p$-value. For the two-sided test, $p = 0.067$.

Note from Figure 14.1 that the detected clusters are not perfect circles even though we used a circular window. This is because the data are aggregated to the county level, so that all of a county is considered to be within the window when the centroid is, and vice versa. The only way to obtain perfect circles is to have non-aggregated data.

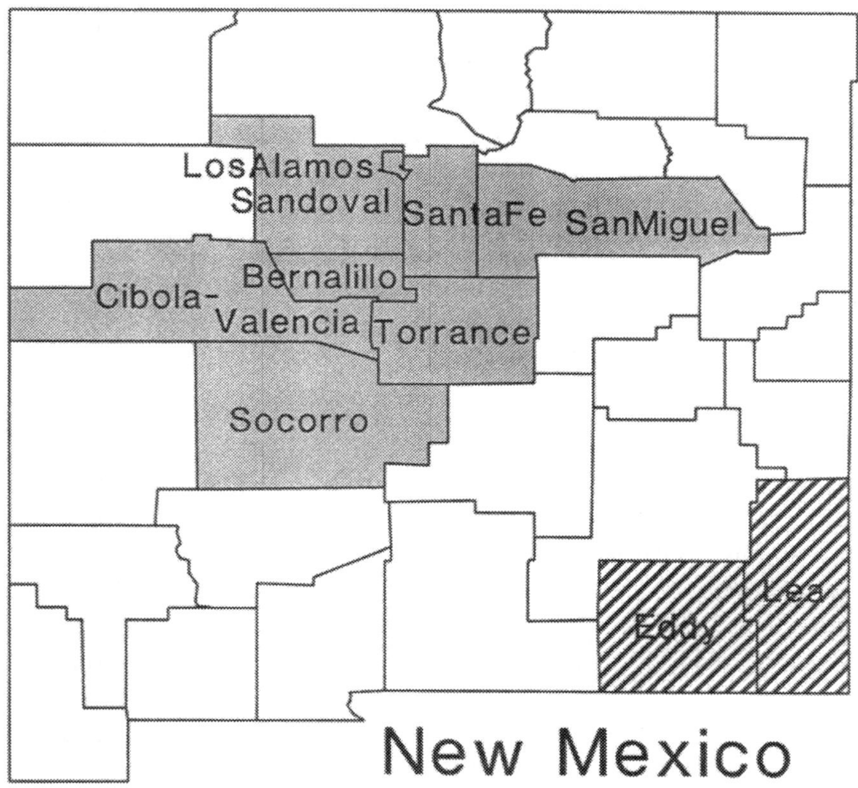


**Figure 14.1:** Brain cancer incidence in New Mexico 1973–1991: The most likely cluster around Albuquerque in Bernadillo county ($p = 0.030$) and the most likely area with exceptionally low rate in Lea and Eddy counties ($p = 0.221$)

### 14.4.3 Medical imaging

In medical imaging, the aim may be to detect tumors using mammography, or areas of activation in a brain scan related to certain physical or mental activities. There are applications in both two and three dimensions. Priebe (1998) applied his scan statistic based on random scan partitions on mammography images, looking for clusters of breast calcifications, and using the texture of the breast to define the scan partitions. Worsley *et al.* (1992) and others have looked at the supremum of a Gaussian random field to determine centers of activity in the brain. The multidimensional scan statistic is a complementary approach to these problems, where each specific application will determine the best method to use.

### 14.4.4    Astronomy

The three-dimensional scan statistic can be used for two different types of astronomy problems. We could be interested to see if stars, galaxies, or some other type of heavenly object are randomly distributed, or whether there are significant local clusters. This leads to a homogeneous Poisson model and Algorithm 14.3.5. It can also be of interest to know whether a particular type of star or galaxy is randomly distributed after adjusting for the locations of all stars/galaxies. Then we should use the nonhomogeneous Bernoulli model and Algorithm 14.3.6.

### 14.4.5    Archaeology and history

Alt and Vach (1991) studied the location of graves containing individual, with a certain genetically determined odontological feature, comparing them to the locations of all graves within a prehistoric burial site. The purpose was to see if biologically related persons, who are more likely to share the same odontological feature, were buried close to each other. Using a test for global clustering, their main purpose was to test for spatial correlation without any interest in cluster locations. If we are interested in the latter, we would instead use a spatial scan statistic based on a discrete Bernoulli model with calculations based on Algorithms 14.3.4 or 14.3.6.

Other potential archaeological and historical applications include the geographical distribution of a certain type of pottery as compared to the distribution of all discovered pottery, to locate areas where that type is significantly abundant, the geographical location of cities or castles in relation to the population distribution, or the geographical distribution of villages with a certain name ending as compared to the distribution of all villages.

### 14.4.6    Urban and regional planning

Post offices, elementary schools, voting locations and many other establishments need to be fairly spread out so they can be conveniently reached by most people. By applying the spatial scan statistic to look for areas with an exceptionally low number of them, adjusting for the underlying population distribution, we may find underserved populations where additional localizations are warranted. Businesses could also use such an approach to help determine appropriate locations for restaurants, grocery stores, health clubs, hairdressers, etc.

### 14.4.7  Reconnaissance

Antipersonnel mines injure thousands of people each year long after the war for which they were intended has ended. It is of great importance to detect mines so they can be deactivated and removed. It is possible to scan a large area for possible mines from the air, but of the point locations obtained, only some will reflect true mines while others will be false detections. By using a scan statistic, areas most likely to contain mines can be detected.

For such an application, we have a homogeneous Poisson process under the null hypothesis. As the size of possible minefields are hard to know a priori, we should use a variable size window, leading to Algorithms 14.3.3 or 14.3.5.

If there are boundary features in the landscape in such a way that it is unlikely that a minefield would cut across such borders, then it is advantageous to use those to create scan partitions as suggested by Priebe (1998). That will increase the power of the test.

Another type of reconnaissance for which a spatial scan statistic can be useful is when searching for mineral, oil, or uranium deposits.

### 14.4.8  Power

Wallenstein, Naus, and Glaz (1993, 1994a,b) have provided simple approximations for the power of the one-dimensional scan statistic against a rectangular pulse alternative, and Sahu, Bendel, and Sison (1993) have shown that it has good power against other pulse alternatives such as triangles. This may indicate that multidimensional scan statistics also have good power against pulse alternatives, but that has never been thoroughly investigated. For one special case, it has been confirmed by Kulldorff and Nagarwalla (1995) who compared their model using a variable window size with the fixed window size model used by Turnbull *et al.* (1990). The variable size model had good power irrespective of the true cluster size. The fixed size model had higher power if the specified size was within about 20 percent of the true cluster size. Neither model had a problem detecting a square shaped cluster even though both used a circular window.

---

# References


1. Adler, R. J. (1984). The supremum of a particular Gaussian field, *Annals of Probability*, **12**, 436–444.

2. Alm, S. E. (1997). On the distribution of the scan statistic of a two dimensional Poisson process, *Advances in Applied Probability*, **29**, 1–16.

3. Alm, S. E. (1998). On the distribution of scan statistics for Poisson processes in two and three dimensions, *Extremes* (to appear).

4. Alt, K. W. and Vach, W. (1991). The reconstruction of 'genetic kinship' in prehistoric burial complexes—problems and statistics, In *Classification, Data Analysis, and Knowledge Organization* (Eds., H. H. Bock and P. Ihm), Berlin: Springer-Verlag.

5. Anderson, N. H. and Titterington, D. M. (1997). Some methods for investigating spatial clustering with epidemiological applications, *Journal of the Royal Statistical Society, Series A*, **160**, 87–105.

6. Besag, J. and Newell, J. (1991). The detection of clusters in rare diseases, *Journal of the Royal Statistical Society, Series A*, **154**, 143–155.

7. Chen, J. and Glaz, J. (1996). Two dimensional discrete scan statistics, *Statistics & Probability Letters*, **31**, 59–68.

8. Choynowski, M. (1959). Maps based on probabilities, *Journal of the American Statistical Association*, **54**, 385–388.

9. Dwass, M. (1957). Modified randomization tests for nonparametric hypotheses, *Annals of Mathematical Statistics*, **28**, 181–187.

10. Eggleton, P. and Kermack, W. O. (1944). A problem in the random distribution of particles, *Proceedings of the Royal Society, Edinburgh Section*, **62**, 103–115.

11. Glaz, J. and Naus, J. (1991). Tight bounds and approximations for scan statistic probabilities for discrete data, *Annals of Applied Probability*, **1**, 306–318.

12. Hjalmars, U., Kulldorff, M., Gustafsson, G. and Nagarwalla, N. (1996). Childhood leukemia in Sweden: Using GIS and a spatial scan statistic for cluster detection, *Statistics in Medicine*, **15**, 707–715.

13. Kulldorff, M. (1997). A spatial scan statistic, *Communications in Statistics—Theory and Methods*, **26**, 1481–1496.

14. Kulldorff, M., Athas, W. F., Feuer, E. J., Miller, B. A. and Key, C. R. (1998). Evaluating cluster alarms: A space-time scan statistic and brain cancer in Los Alamos, *American Journal of Public Health* (submitted).

15. Kulldorff, M., Feuer, E. J., Miller, B. A. and Freedman, L. S. (1997). Breast cancer clusters in Northeast United States: A geographic analysis, *American Journal of Epidemiology*, **146**, 161–170.

16. Kulldorff, M. and Nagarwalla, N. (1995). Spatial disease clusters: Detection and inference, *Statistics in Medicine*, **14**, 799–810.

17. Kulldorff, M. and Williams, G. (1997). *SaTScan v 1.0, Software for the Space and Space-Time Scan Statistics*, Bethesda, MD: National Cancer Institute.

18. Lawson, A. (1997). Cluster modeling of disease incidence via MCMC methods, *Journal of Statistical Planning and Inference* (submitted).

19. Loader, C. R. (1991). Large-deviation approximations to the distribution of scan statistics, *Advances in Applied Probability*, **23**, 751–771.

20. Månsson, M. (1996). On Clustering of Random Points in the Plain and in Space, *Ph.D. Thesis*, Department of Mathematics, Chalmers University of Technology and Gothenburg University, Gothenburg.

21. Mantel, N. (1967). The detection of disease clustering and a generalized regression approach, *Cancer Research*, **27**, 209–220.

22. Miller, B. A., Gloeckler Ries, L. Y., Hankey, B. F., Kosary, C. L., Harras, A., Devesa, S. S. and Edwards, B. K. (1993). *SEER Cancer Statistics Review 1973–1990*, Bethesda, MD: National Cancer Institute.

23. Naus, J. (1965a). The distribution of the size of maximum cluster of points on the line, *Journal of the American Statistical Association*, **60**, 532–538.

24. Naus, J. (1965b). Clustering of random points in two dimensions, *Biometrika*, **52**, 263–267.

25. Naus, J. (1974). Probabilities for a generalized birthday problem, *Journal of the American Statistical Association*, **69**, 810–815.

26. Openshaw, S., Charlton, M., Wymer, C. and Craft, A. (1987). A Mark 1 Geographical Analysis Machine for the automated analysis of point data sets, *International Journal of Geographical Information Systems*, **1**, 335–358.

27. Priebe, C. (1998). A spatial scan statistic for stochastic scan partitions, *Journal of the American Statistical Association* (to appear).

28. Sahu, S. K., Bendel, R. B. and Sison, C. P. (1993). Effect of relative risk and cluster configuration on the power of the one-dimensional scan statistic, *Statistics in Medicine*, **12**, 1853–1865.

29. Saperstein, B. (1972). The generalized birthday problem, *Journal of the American Statistical Association*, **67**, 425–428.

30. Turnbull, B., Iwano, E. J., Burnett, W. S., Howe, H. L. and Clark, L. C. (1990). Monitoring for clusters of disease: Application to leukemia incidence in Upstate New York, *American Journal of Epidemiology*, **132**, S136–S143.

31. Wallenstein, S., Gould, M. S. and Kleinman, M. (1989a). Use of the scan statistic to detect time-space clustering, *American Journal of Epidemiology*, **130**, 1057–1064.

32. Wallenstein, S., Weinberg, C. R. and Gould, M. (1989b). Testing for a pulse in seasonal event data, *Biometrics*, **45**, 817–830.

33. Wallenstein, S., Naus, J. and Glaz, J. (1993). Power of the scan statistic for detection of clustering, *Statistics in Medicine*, **12**, 1819–1843.

34. Wallenstein, S., Naus, J. and Glaz, J. (1994a). Power of the scan statistic in detecting a changed segment in a Bernoulli sequence, *Biometrika*, **81**, 595–601.

35. Wallenstein, S., Naus, J. and Glaz, J. (1994b). Power of the scan statistics, *ASA Proceedings of the Section of Epidemiology*, **81**, 70–75.

36. Walsh, S. J. and Fenster, J. R. (1997). Geographical clustering of mortality from systemic sclerosis in the Southeastern United States, 1981-90, *Journal of Rheumatology* (to appear).

37. Weinstock, M. A. (1981). A generalized scan statistic test for the detection of clusters, *International Journal of Epidemiology*, **10**, 289–293.

38. Worsley, K. J., Evans, A. C., Marrett, S. and Neelin, P. (1992). A three-dimensional statistical analysis for CBF activation studies in human brain, *Journal of Cerebral Blood Flow and Metabolism*, **12**, 900–918.

# *Index*

GPSR Compliance
The European Union's (EU) General Product Safety Regulation (GPSR) is a set
of rules that requires consumer products to be safe and our obligations to
ensure this.

If you have any concerns about our products, you can contact us on

ProductSafety@springernature.com

In case Publisher is established outside the EU, the EU authorized
representative is:

Springer Nature Customer Service Center GmbH
Europaplatz 3
69115 Heidelberg, Germany

www.ingramcontent.com/pod-product-compliance
Lightning Source LLC
Chambersburg PA
CBHW080231020625
27560CB00006B/446